Sources and Applications of Ultraviolet Radiation

Sources and Applications of Ultraviolet Radiation

Roger Phillips

Research and Development Division
Metal Box plc
Wantage
Oxfordshire
United Kingdom

1983

Academic Press

A Subsidiary of Harcourt Brace Jovanovich, Publishers

London New York
Paris San Diego San Francisco São Paulo
Sydney Tokyo Toronto

ACADEMIC PRESS INC. (LONDON) LTD.
24/28 Oval Road
London NW1

United States Edition published by
ACADEMIC PRESS INC.
111 Fifth Avenue
New York, New York 10003

British Library Cataloguing in Publication Data

Phillips, Roger
Sources and applications of ultraviolet radiation.
1. Photochemistry
I. Title
541.3′5 QD708.2

ISBN 0–12–553880–4
LCCCN 83–70338

Text set in 10/12 pt Linotron 202 Times, printed and bound in Great Britain at The Pitman Press, Bath

Preface

This book has been written primarily for the benefit of those who have a responsibility for, or an interest in, the design, construction, purchase or operation of the "hardware" associated with applied photochemistry. The emphasis is on technologically viable sources of actinic radiation, and the way in which they are used in the processes and procedures based on applied photochemistry. Existing texts on pure and applied photochemistry are concerned primarily with the materials involved, the nature, lifetime and reactivity of excited intermediates, and the chemical reactions which result directly or indirectly from irradiation. The methods by which photons of the required wavelength and flux density are generated are either ignored completely or are treated in a manner relevant only to laboratory experimentation. However, as photochemical processes developed in the laboratory are translated into routine and often large scale processes in the working environment, the satisfactory design of the "hardware" involved assumes a greater importance, and introduces the concept of "photochemical engineering".

Perhaps the most successful application of photochemistry in industry which involves the use of a radiation source to bring about chemical change is the modification by chemical reaction of the physical and chemical properties of thin sections of materials. The use of photoinitiated polymerization to provide a rapid, low energy and pollution-free method of drying inks and other organic surface coatings was one of the more significant developments in printing and surface coating technology during the 1970s. Developments in photoresist technology during the same period allowed spectacular achievements in the electronics industry as well as in the graphic arts. The new UV curing techniques, when they became technologically viable at the beginning of the 1970s, were of particular relevance to the large scale converter of packaging materials who routinely prints or coats vast quantities of often impermeable substrates. Metal Box has maintained an active research and development programme in the area of applied photochemistry and "photochemical engineering", and this has allowed not only the widespread application of UV curing in the printing and coating processes used in the company's plants, but also the design and construction of irradiator systems for a wide range of applications. I have been involved with this programme since joining the company in 1974, and have had the opportunity to study in some depth both the materials and the hardware involved in the photopolymerization process. This book began, in concept, as a monograph on the photopolymerization of organic surface coatings, with the emphasis being on the chemistry involved. However, it soon became clear that this was an area which was already well documented by numerous excellent reviews and by books which had either already appeared or were known to be in the process of preparation. I therefore felt that a more valuable and unique contribution would be possible if the emphasis was placed on the sources of radiation suitable for photopolymerization. This concept however created problems associated with the limited scope. My imagination had now been captured by the possibilities of a text on the radiation sources used in applied photochemistry. My professional interests were expanding into areas such as sterilization and chemical synthesis, and it seemed logical and appropriate to widen the range of applications covered by the book. This, it was hoped, would not only allow a wider readership but provide a rationale for a more comprehensive and academically better balanced treatment of actinic photon sources.

Chapters 1–4 of the book provide the essential background material for the subsequent detailed discussion (Chapters 5–13) of the various types of actinic radiation sources. Chapter 1 is a general introductory review of the technological applications of photochemistry. A brief and elementary treatment of the mechanisms of photon generation is given in Chapter 2 for readers who are not familiar with the concepts of quantum theory. Chapter 3 begins with the basic concepts of photochemistry and goes on to discuss the scientific aspects of photopolymerization, with the emphasis being on the mechanisms of photoinitiation. Chapter 4 describes the methods of radiation measurement which are relevant to sources of actinic radiation. Chapter 14 explains how light sources are used in photochemical reactors, in other words how the radiation from a source is directed to the surface of a material to be processed. The final chapter (Chapter 15) is concerned with the physiological hazards of UV radiation and the recommended methods of protecting personnel from any harmful effects of exposure.

It is a pleasure to acknowledge the help and encouragement I received from many friends and colleagues during the preparation of the manuscript. I am particularly grateful to my colleagues Neil Gerrard and Tony Lott, who made major contributions to the gathering of material and with whom I had many valuable discussions. I am indebted to Don Major of Primarc Jigs and Lamps Ltd, who supplied many experimental lamps and much helpful advice, and to E. J. G. Beeson and Dr R. J. Hall (Thorn), Professor R. S. Davidson (City University), F. Harlen and Dr A. McKinley (National Radiological Protection Board), Dr R. A. J. Keijser (Philips), R. G. King and R. W. Thomas (British Aluminium), Professor A. Ledwith and Dr C. F. H. Vickers (Liverpool University) and Dr M. A. West (Royal Institution), who kindly read parts of the initial draft, pointed out mistakes and made many suggestions for improvement. Finally I would like to thank Metal Box plc for providing me with experimental and library facilities, and permission to publish.

May 1983

Roger Phillips
Wantage, Oxfordshire

Contents

1 Applications of Ultraviolet Radiation

1.1 Making Use of UV

This book is concerned with the use of electromagnetic radiation as "actinic" radiation to bring about chemical or biological change, and with the methods of generating this actinic radiation at the wavelengths and intensities required. The essential property of actinic radiation is its ability to excite molecules into electronically excited and chemically reactive states. The photon energies required are such that for many purposes the appropriate wavelengths are in the UV region between 200 and 400 nm. The scope for using radiation below 200 nm is restricted by attenuation in the atmosphere. The word "ultraviolet" is used in the title of this book because it is the wavelength region of greatest interest. However, the upper wavelength limit at 400 nm is determined by the physiology of the human eye, a subject of limited interest in this context. Some photochemical reactions of practical value are initiated by visible radiation and these will be discussed when appropriate.

It is instructive to classify the essential properties of UV and the fundamental processes which are involved in the practical uses of UV. Any particular application may of course involve more than one property and more than one process.

1.1.1 Physical optics

Being electromagnetic radiation, UV obeys the laws of geometrical optics, and can be used simply as an optical ray, where for one reason or another UV has an advantage over visible radiation. UV can be used to produce a magnified image in optical microscopy. The shorter wavelength allows better ultimate resolution. Other applications benefit from the invisibility of UV, or the possibility of using it without the necessity to exclude ambient visible light.

Some uses of UV as actinic radiation are totally dependent on its optical properties, as in the preparation of photoresists by imaging through masks.

1.1.2 Photoelectric effects

Photons in the UV and visible regions have sufficient energy to cause various types of photoelectric effect. An example of practical importance is the ability of UV or visible radiation to cause some insulators to become conductors by the phenomenon of photoconductivity. This finds use in photocopying machines where electrostatic charge on a surface is selectively removed by irradiation. Another example is EPROM (erasable programmable read only memory) erasure in microelectronics by UV (Ultra-Violet Products Inc., 1977).

1.1.3 Excitation/physical decay

One of the most significant aspects of UV is that the photons have energies sufficient to raise atoms and molecules into electronically excited states, and thereby to alter their properties in a very profound way. An atom or molecule in an electronically excited state is unstable and has a strong tendency to lose its energy, either by returning to the ground state, or by chemical reaction. It may return to the ground state by a variety of *physical* decay processes, and the third category involves situations where electronic excitation followed by physical decay takes place. The simplest example is UV absorption spectroscopy, used for qualitative and quantitative analysis. Sometimes the physical decay process involves the emission of light, a phenomenon known as luminescence. Luminescence occurs at wavelengths longer than the incident radiation (Stokes' law). Luminescence in this context is of two types, *fluorescence* and *phosphorescence.*

Sometimes the physical decay process involves another molecule. An example is *photosensitization* to produce an excited state in the other molecule.

1.1.4 Excitation/chemical change

The excited state resulting from the absorption of a UV photon may undergo a *chemical* change by either intramolecular or intermolecular reaction, if these processes can compete effectively with the various possible *physical* decay processes. The chemically reactive state may not be the one resulting immediately from photon absorption, but may be produced by a subsequent rearrangement, internal conversion and/or intersystem crossing. It is possible by irradiation to produce very energetic molecules at ambient temperature. To produce molecules of such energy by purely thermal means would in principle require temperatures of many thousands of kelvins, well above the temperature at which vibrational energy would destroy the molecule (Wayne, 1970, p. 6). Irradiation can produce very energetic electronic states without simultaneously producing very energetic vibrational and rotational states, because thermodynamic equilibrium does not apply. The excess energy is located in one region of the molecule. Electronically excited states have electronic structures which are quite different from that of the ground state and therefore are not only *more* reactive, but are capable of taking part in *different* reactions. Reactions which otherwise would require much higher temperatures are able to take place at or near ambient temperature, with more selectivity and fewer undesirable by-products. Also, reactions which cannot be obtained by any other means are able to occur.

The applications in this fourth category involve the degradation, or photolysis, of molecules, and the synthesis of new materials. Some applications emphasize the destruction of a material rather than the formation of a new material, and photodegradation finds practical applications where the starting material has undesirable properties. Perhaps the most familiar example is the sterilization of materials by short wavelength UV (see Section 1.3). Photodegradation is a widespread and often unwelcome phenomenon, particularly in the deterioration of materials in sunlight, and as Pape (1975) has said, "it is hardly surprising that more chemists are working at stopping and alleviating the damaging effects of light than are looking for new ways of exploiting light". However, photodegradation can be turned to advantage in the formulation of photodegradable plastics for ecologically acceptable packaging materials.

Some applications demonstrate a unique type of selectivity which the photochemical process provides. Reaction occurs only where a material is irradiated, and therefore a *spatial* selectivity is possible. An excellent example is provided by the use of diazonium salts for photo-imaging (Jacobson and Jacobson, 1976, p. 159). When diazo copying paper is irradiated according to the pattern of an image, the diazonium salt decomposes in the areas irradiated. The paper is then developed with ammonia or an alkali to allow the undecomposed diazonium salt to react with a phenol to form a stable coloured azo compound, e.g.

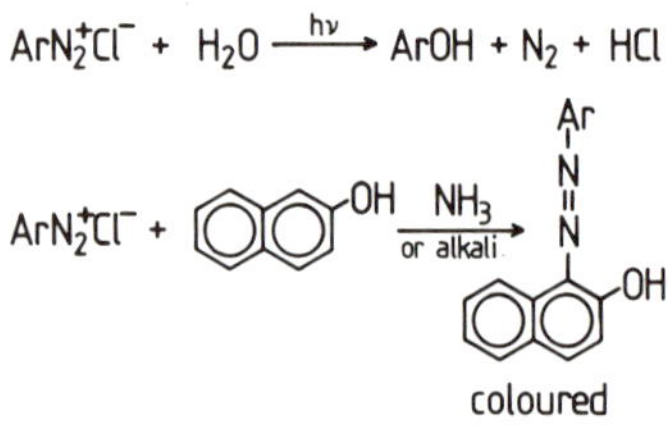

Spatial selectivity is of great value in the preparation of photoresists (Section 1.7).

Photochemical synthesis can be divided into two main areas. First, there is the synthesis of individual molecules, and secondly the production of macromolecular systems by photopolymerization.

The applications of lasers will be discussed in Section 12.5.

1.2 Fluorescence and Phosphorescence

Fluorescence and phosphorescence are phenomena which have many familiar applications which make use of the wavelength conversion which takes place. Most important is the fluorescent lighting tube, the inside wall of which is coated with an inorganic fluorescent powder which converts the short wavelength UV radiation emitted by the low pressure mercury discharge into a wide band of visible radiation. Another similar application is in optical brighteners (Pape, 1975) which are added to paper, detergents, fabrics, etc. Optical brighteners are compounds which absorb ambient light in the 300–400 nm region and fluoresce in the 400–450 nm region to counteract poor reflectance in the blue region of the spectrum. Fluorescent dyes are used in inks and coatings to produce luminous colours for advertising, warning signs, marker pens and so on. The commonest colours are orange, yellow and green, colours to which the eye is very sensitive, and the brightness is obtained by wavelength conversion of UV and blue light.

Fluorescence and phosphorescence are widely used in the laboratory where they provide very sensitive techniques for qualitative and quantitative analysis of both organic and inorganic materials (Mellor *et al.*, 1974). Fluorescence analysis can typically detect the presence of a material down to one part in 10^8, which provides a sensitivity two or three orders of magnitude better than is possible with absorption methods. The sensitivity of absorption methods is limited by the need to detect a very small fractional decrease in the light transmitted by a material, whereas with fluorescence analysis advantage can be taken of the high sensitivity of photomultipliers for the detection of faint emissions of light, and the availability of intense sources for excitation.

The ability to detect very small concentrations of fluorescent dyes in solution allows techniques for examining the flow patterns of liquids for purposes as diverse as detecting fuel leaks, detecting cracks in castings and monitoring ocean currents.

Fluorescence can be used to reveal invisible identification marks, such as signatures, product codes and laundry marks. Invisible fluorescent pattern markings are used in the textile industry for the cutting and stitching of fabrics. Invisible fluorescent dyes are sometimes included in the formulations of clear protective coatings, so that after application to the surface the coating can be examined under UV light to check for completeness and uniformity of coverage.

Examination under UV light can often reveal differences between genuine and counterfeit documents, or indicate that a document has been fraudulently altered. It can reveal oil stains on textiles and residual food stains on cooking utensils and tableware after washing. It can be used to detect spoilage of foodstuffs. The urine trails of rodents fluoresce, and this can help in their detection and eradication.

1.3 Medical, Biological and Environmental Applications

1.3.1 Medical uses

The medical uses of actinic radiation date back to the 1890s, when Niels Finsen discovered that it is the ultraviolet component in sunlight, rather than the radiant heat, which is responsible for sunburn, and also that ultraviolet radiation has a bactericidal effect useful in the treatment of certain skin disorders, for instance tuberculosis of the skin (lupus vulgaris). Finsen was awarded the Nobel prize for medicine in 1903. Actinotherapy was used to treat a wide variety of ailments during the early years of this century, but its significance in clinical practice faded somewhat in the 1940s and 1950s following the introduction of antibiotic drugs, and the realization that some of the successes claimed for the technique were of dubious validity (Diffey, 1982, p. 6). Modern techniques of phototherapy and photochemotherapy, based on research in photochemistry and photobiology, are of proven value in the treatment of a limited number of diseases, some of which are very common.

The penetration of light into the human body is only slight, so that the *primary* photochemical processes take place near the surface. However, the *ultimate* effects may be more widespread if the products of the photochemical reactions can diffuse to other parts of the body, and possibly undergo dark reactions.

1.3.1.1 Tanning

Exposure to ultraviolet radiation may cause erythema and tanning of the skin, and the latter has a cosmetic appeal in modern Western culture. This subject is treated in more detail in Chapter 15.

1.3.1.2 Diagnostic techniques

The phenomenon of fluorescence can be used in diagnosis. When irradiated with UVA (long wavelength UV) from a low intensity source, normal skin is seen to fluoresce blue white. Certain skin diseases cause the skin to fluoresce a different colour (Parrish *et al.*, 1978, p. 234; Diffey, 1982, pp. 126–128). Fluorescence can also be used in dental diagnosis for revealing caries, plaque and calculus (Diffey, 1982, p. 128) and the presence of gum disease.

1.3.1.3 Phototherapy and photochemotherapy

Ultraviolet radiation by itself often has a beneficial effect in the treatment of common skin ailments such as acne and eczema, and certain types of pruritis. More serious conditions, and in particular psoriasis, are treated by photochemotherapy, which involves the use of a photosensitizing medication in conjunction with irradiation. Psoriasis was for many years treated using the Goeckerman regime, which involved the topical application of a solution of coal tar to the affected region followed by irradiation. The application of coal tar was particularly messy and unpleasant. A more modern version of this technique uses the topical application of dithranol. A major development in photochemotherapy during the 1970s has involved the oral administration of photosensitizing drugs. The most important of these are the psoralens, and in particular 8-methoxypsoralen. The drug is taken by the patient, and after about 2 h, when the photosensitizing effect is at its maximum, the body is exposed to UVA radiation (Parrish *et al.*, 1974). The most effective wavelengths are between 340 and 350 nm, and the treatment is known as PUVA (Psoralen Ultra Violet A). PUVA has been mainly used for the treatment of psoriasis. It does not cure the disease but does provide medium term alleviation of the symptoms. The effectiveness of PUVA for psoriasis treatment has been compared with that of dithranol by Rogers *et al.* (1979). The photochemistry of the psoralens has been extensively studied (Song and Tapley, 1979; Parsons, 1980). PUVA is sometimes effective in the treatment of vitiligo (Levin and Parrish, 1975; Epstein, 1977), and some other disorders (Anderson and Voorhees, 1980). There are worries that PUVA may lead to skin cancer (Stern *et al.*, 1979). The radiation sources used for PUVA treatment are almost invariably low pressure, low intensity fluorescent tubes of the type used in sunbeds, although there has been some use of high intensity, medium pressure sources (Fry, 1977). Diffey (1978) attempted to determine the dose required in PUVA treatment from published data but found this difficult because of the inadequate dosimetry employed by the workers who had experimented with PUVA.

Ultraviolet radiation converts 7-dehydrocholesterol in the skin to pre-vitamin D_3, which in turn isomerizes in a dark reaction to vitamin D_3 (cholecalciferol), a hormone necessary for healthy bone formation and the prevention of rickets (Havinga, 1973; Holick *et al.*, 1980; Holick, 1981). The most effective wavelength range is 270–300 nm (Knudson and Benford, 1938). Normal exposure to sunlight is sufficient to prevent vitamin D_3 deficiency (Lawson, 1978), but artificial sources of UV may be helpful for the housebound or institutionalized (Boyle, 1980). Unfortunately, the action spectrum for pre-vitamin D_3 formation is so similar to that for kerato-conjunctivitis and adverse skin reactions that there are dangers involved in the use of commercial sunlamps for promoting pre-vitamin D_3 synthesis in the skin (Devgun *et al.*, 1981).

Pre-vitamin D_3 formation is an example of a photochemical process causing an ultimate effect well away from the irradiation site. Another example is the treatment of hyperbilirubinemia (neonatal jaundice) in infants, particularly those born prematurely (Sisson, 1976; Jewess, 1978). The degradation of red blood cells releases haemoglobin, which itself degrades to a yellow substance known as bilirubin. Normally the bilirubin would be processed by the liver and excreted from the body. However, in some infants the liver may not be fully functional during the first few days of life. If the bilirubin is allowed to accumulate in the blood stream, brain damage (kernicterus) may result. It has been found that if the child is irradiated for several hours a day with blue light from fluorescent lamps, the bilirubin is photochemically degraded to harmless products which can be excreted (Cremer *et al.*, 1958; Tabb *et al.*, 1972; Myshkin and Sakharov, 1982). The treatment may be required for several years in the case of congenital hyperbilirubinemia (Crigler Najjar disease) (Lucey, 1972). The most effective wavelength is about 450 nm (Thorington *et al.*, 1971;

Lightner *et al.*, 1980), which corresponds to the absorption maximum of bilirubin (McDonagh, 1976). Considerable effort has been expended in designing lamps specifically for the treatment of hyperbilirubinemia (Sausville *et al.*, 1972; Wanmaker *et al.*, 1975).

A quite different use of UV has been for the hardening of self-supporting lightweight orthopaedic casts formed from a glass fibre fabric impregnated with an unsaturated polyester/styrene mix and incorporating a photoinitiator (Beightol, 1971). The rapid and easily controllable hardening of resin systems by UV finds applications in dentistry (Patel *et al.*, 1976; Kilian, 1981). The resins used normally contain methacrylate monomers and oligomers, and are used for pit and fissure sealing (Buonocore, 1970), for restoring and repair of anterior teeth, and for bonding agents for brackets. They offer the advantages of long working times and shorter setting times than conventional materials. The radiation sources used are compact source mercury or xenon arcs with filters to cut out short wavelength UV and much of the visible and infra-red, and with light pipes or fibre optics to carry the light to the area of tooth being treated, or alternatively tungsten halogen lamps or low pressure mercury "superactinic" fluorescent tubes with peak output in the UVA. It appears that some of the early radiation units exposed patients and dentists to hazardous levels of short wavelength UV, and this was a cause of official concern (American Dental Association Council on Dental Materials and Devices, 1976).

1.3.2 Disinfection and sterilization

Short wavelength UV (mainly 254 nm radiation from low pressure mercury lamps) has long been used for the disinfection of air and water (Paulus, 1942; Johansen and Myhrstad, 1978) and the sterilization of the surfaces of materials (Phillips and Novak, 1956; Bachmann, 1975). UV may for instance be used to disinfect sewage effluents (Ho and Bohm, 1981), drinking water (Braendli *et al.*, 1981), water for the cosmetics industry (Schenck, 1979), and for bathing pools (Gemne *et al.*, 1981). The lethal effects of sunlight on bacteria were first observed by Downes and Blunt (1877). It was many years later when it was shown that the effect on the microorganism was a result of direct photochemical action by the UV component in sunlight. The "action spectrum" for the bactericidal effect follows the absorption curve for DNA. Active wavelengths are below 300 nm, with a peak at about 260 nm (Koller, 1965, p. 236). The effect follows the Bunsen–Roscoe reciprocity law (Luckiesh *et al.*, 1949).

UV radiation damages the nucleic acids and prevents replication. Bacteria vary greatly in their susceptibility to UV radiation (Russell, 1982). They may be killed permanently by a lethal dose, or may be merely deactivated by a sub-lethal dose, with the possibility of subsequent reactivation, by, for instance, a photorepair process brought about by exposure to visible light (Jagger, 1958).

The use of UV has a number of disadvantages when compared with the use of chemical disinfectants such as chlorine. The bactericidal action of UV lasts only as long as the period of irradiance, and therefore recontamination can occur after treatment. UV is only effective if it can be absorbed by the microorganism and therefore it is difficult for instance to sterilize water with a high concentration of suspended solids. On the other hand, in some instances the presence of a persistent chemical disinfectant may be undesirable, and indeed there is growing concern about the contamination of drinking water by trihalomethanes, which result from chlorination (Cotruvo, 1981; Fiessinger *et al.*, 1981; Lahl *et al.*, 1981). Also, in some biotechnology processes, e.g. fermentation, it may be desirable to sterilize a water supply and then inoculate the process water with desirable bacteria. This would not be possible with a persistent disinfectant (Legan, 1982). In a recirculating system regular use of chlorine may cause corrosion problems and the build-up of oxidant residuals (Garey and DeHart, 1979).

While UV is effective in killing bacteria, for example coliforms in sewage effluent, it will not by itself necessarily change the total organic content (TOC) of water (Lee *et al.*, 1982), unless an organic constituent can be photolysed to volatile products (Matsuura and Smith, 1970). However, if the water is aerated so that there is the possibility of ozone or hydrogen peroxide being formed on irradiation (Cannon, 1965), these oxidizing agents and active species formed from them will tend to reduce the TOC. Ozone and hydrogen peroxide and other powerful oxidizing agents are of course powerful bactericides, and the effectiveness of ultraviolet radiation may be much enhanced by the simultaneous use of oxidants. Bayliss and Waites (1979) have suggested the combination of hydrogen peroxide and UV for killing bacterial spores on packaging materials. The active sterilizing agent appears to be hydroxyl radicals produced by photolysis of the peroxide. It might therefore be more realistic to think of the UV enhancing the effect of the peroxide rather than vice versa. A combination of UV and ozone has been suggested for disinfecting shellfish quarantine effluent (Blogoslawski and Alleman, 1979), and for treating both drinking water and wastewater in army field hospitals (Lee *et al.*, 1978), and spacecraft (Zeff *et al.*, 1976). Schenck (1981) has recommended a combination of UV and iodine for disinfection of drinking water.

1.3.3 Water purification

The combination of UV and ozone has a very powerful oxidizing action, particularly in the presence of water vapour or liquid water. Oxygen atoms and hydroxyl radicals are formed, both of which are very effective oxidants. Hydroxyl radicals have an oxidizing potential relative to hydrogen of 2·87 V, compared with 2·07 V for ozone (Legan, 1982). The UV/O_3 combination has been found to be very effective in removing contaminants from water, both supplies and wastewater effluent. Under the right conditions it is possible to convert virtually all organic carbon to carbon dioxide (Prengle and Mauk, 1978), even with refractory species (Prengle *et al.*, 1975). Examples of contaminants which may be removed are EDTA from plating plants (Macur *et al.*, 1980), polynuclear aromatic hydrocarbons (Leitis *et al.*, 1981), nitro-organics from ammunition plants (Roth and Murphy, 1979), dyes (Horikawa *et al.*, 1976), pesticides (Mauk *et al.*, 1976; Prengle and Mauk, 1976), and polychlorinated biphenyls (Prengle *et al.*, 1976; Arisman *et al.*, 1980).

The UV/O_3 combination is capable of reducing the organic content of water to extremely low levels, and indeed can produce water of lower TOC than is possible by any other purification technique, with values of less than one part in 10^9 achievable (Barker, 1981). There is an increasing need for very pure water for use in the fabrication of microcircuits (Taubenest and Ubersax, 1980), for pharmaceuticals and medical preparations, and even for use in boilers (Jonas, 1982), and this technique may prove to be an economical method of converting moderately pure water as supplied by water authorities into ultrapure water. The UV/O_3 combination is suitable not only for treating water, but also for deodorizing air streams (Ohtsuka and Seki, 1977; Barker, 1979).

The UV/H_2O_2 combination also yields hydroxyl radicals and is capable of removing organics from water (Malaiyandi *et al.*, 1980). It has been recommended for the treatment of boiler feed makeup water (Clarke and Knowles, 1982), and sugar-containing effluent from bottling plants (Lipowicz, 1981).

The combination of UV and an oxidant can be used as the basis of an analytical technique for estimating TOC. Mar (1981) used a combination of UV irradiation and sodium persulphate and measured carbon dioxide sparged from the solution.

1.3.4 Surface cleaning

The UV/O_3 combination has been found to be effective in removing traces of organic contaminants on quartz and metal surfaces left behind after conventional solvent cleaning processes (Sowell *et al.*, 1974; Vig and LeBus, 1976), and has found an application in the cleaning of quartz crystal resonators prior to the deposition of metal film electrodes.

1.4 Industrial Synthesis

Many organic syntheses which involve a photochemical step have been devised in the laboratory, but few have so far been successfully translated into profitable industrial production (Clements, 1980). This is disappointing in view of the potential benefits of photochemical pathways, and the very large effort which has been expended in recent years in academic research, both in the area of organic synthesis (Carless, 1980) and in the analysis of reactor design (see Chapter 14). The potential benefits include greater selectivity of products, lower thermal load and better control (Mullin, 1968; Pape, 1975). Also, there are some organic compounds which can be synthesized only by photochemical means. The reasons for the modest impact of photochemistry on industrial synthesis are in part associated with the difficulty of design of photochemical reactors (Arnold and de Mayo, 1971). Scaling up from laboratory and pilot plant scale presents problems (Clements, 1980), particularly if the reactants must compete for radiation with other components in the system (Pape, 1975). Industrial secrecy has been an inhibiting factor. Fremed (1955), in introducing the first major review of industrial photochemical synthesis (Doede and Walker, 1955), said that "the techniques and know-how of the photochemical engineer are obscured by trade secrecy". The same appears to be true today and it is not at all easy to establish the extent to which photochemical engineering is employed in commercial chemical synthesis.

Commercial syntheses using photochemical reactors may be divided into *large scale* processes producing low priced bulk chemicals, and *small scale* processes producing high priced fine chemicals. The large scale processes are normally carried out in *continuous* reactors, and the chemical reactions which take place usually, but not always, allow a very high overall quantum yield. In other words "the effect of the primary photochemical process is amplified by a large number of successive chemical reactions" (Fischer, 1978), and this allows an economical use of light. Small scale processes are normally carried out in *batch* reactors which are easier to design. The high value of the product makes the cost of light insignificant.

1.4.1 Large scale processes

1.4.1.1 Photochlorination

The chlorination of hydrocarbons is probably the most important application of industrial photochemical synthesis, and has been in use for many years (Anderson, 1947; Shearon *et al.*, 1949; Doede and Walker, 1955). Chlorine molecules absorb long wavelength UV and blue radiation (Gibson and Bayliss, 1933) and are photolysed to chlorine atoms, which can initiate a chain reaction:

$$Cl_2 \xrightarrow{h\nu} 2Cl\cdot$$

$$Cl\cdot + RH \longrightarrow R\cdot + HCl$$

$$R\cdot + Cl_2 \longrightarrow RCl + Cl\cdot$$

Each chlorine atom photolysed may lead to the formation of 10^2–10^3 molecules of RCl. The reaction is highly exothermic and rapid and a small continuous reactor with a radiation source of modest output is capable of high output (Hirschkind, 1949; Hutson and Logan, 1972). Photochlorination of toluene allows *substitution* of chlorine atoms in the methyl group, to the exclusion of chlorination of the aromatic ring (Shearon *et al.*, 1949). Benzyl chloride, benzylidene dichloride and benzotrichloride are possible products, and these are valuable intermediates (Fischer, 1978; Bloomfield and Owsley, 1982). Photochlorination can be used to increase the chlorine content in alkyl chlorides, for instance to convert 1,1-dichloroethane to the valuable solvent 1,1,1-trichloroethane (Richtzenhain and Stephan, 1971; Christiaens and Martens, 1973).

The photochlorination of benzene involves an *addition* reaction:

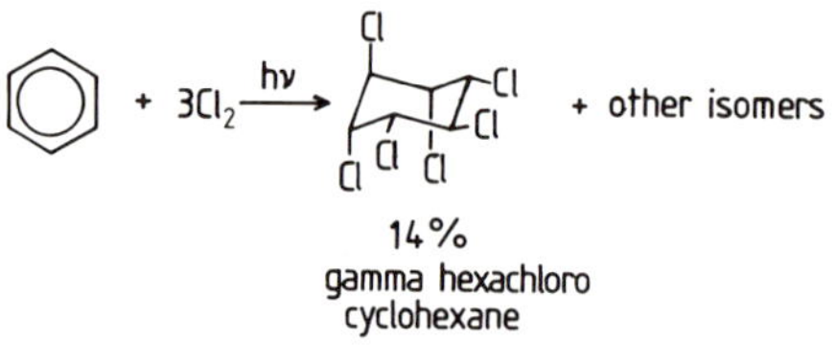

The photochemical route allows lower reaction temperatures and higher yields of the gamma isomer (Governale and Clarke, 1956). This isomer is commonly known as gammexane, or lindane, and is a valuable, biologically degradable insecticide (Brooks, 1974; Colson, 1979).

1.4.1.2 Photohydrosulphonation

Hydrosulphonation is used to convert terminal olefins to commercially useful thiols (Vaughan and Rust, 1942; Elad, 1969):

$$H_2S \xrightarrow{h\nu} HS\cdot + H\cdot$$

$$HS\cdot + RCH{=}CH_2 \longrightarrow R\dot{C}H{-}CH_2SH$$

$$R\dot{C}H{-}CH_2SH + H_2S \longrightarrow RCH_2{-}CH_2SH + HS\cdot$$

This chain reaction has a high quantum yield. Thiols (or mercaptans) are useful intermediates, and the strong smell of the more volatile primary mercaptans makes them suitable trace additives to natural gas supplies (Arnold and de Mayo, 1971).

1.4.1.3 Photosulphochlorination

Sulphochlorination is another high quantum yield chain reaction which involves the photolysis of chlorine molecules:

$$Cl_2 \xrightarrow{h\nu} 2Cl\cdot$$

$$Cl\cdot + RH \longrightarrow R\cdot + HCl$$

$$R\cdot + SO_2 \longrightarrow RSO_2\cdot$$

$$RSO_2\cdot + Cl_2 \longrightarrow RSO_2Cl + Cl\cdot$$

Photosulphochlorination was first described by Reed (1936). The sulphonyl chlorides produced are intermediates in the synthesis of alkane sulphonates, used as emulsifiers and surfactants. Sulphochlorination competes with straight chlorination, and it is important to choose reaction conditions which favour the sulphochlorinated product (Boynton *et al.*, 1959).

1.4.1.4 Photonitrosation

Although most large scale photochemical syntheses involve chain reactions with high overall quantum yield, there is one often-quoted example of a high tonnage process which has proved to be commercially successful, despite a quantum yield which does not exceed unity. This is the manufacture of ε-caprolactam by a process which involves the photochemical conversion of cyclohexane to its oxime by reaction with nitrosyl chloride:

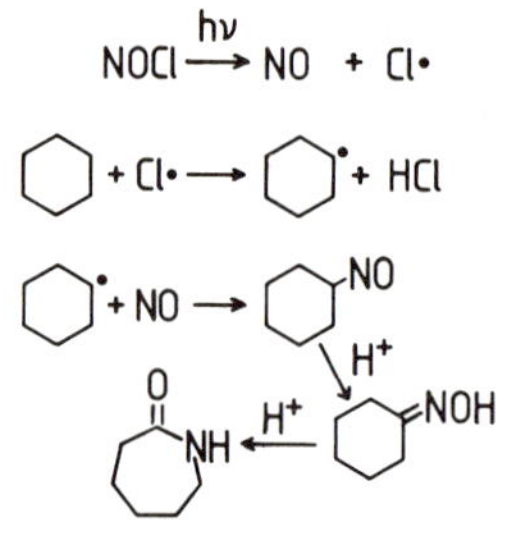

Caprolactam is the cyclic amide used in the production of Nylon 6. The justification for the large scale photochemical synthesis of caprolactam, developed by Toyo Rayon in Japan, is that the cyclohexane can be converted to the oxime under mild conditions in one step (Marcus *et al.*, 1962; Ito and Matsuda, 1969; Jubb, 1971). Because of the low quantum yield (Miyama *et al.*, 1968, 1969; Mosher and Bunce, 1971) high intensity light sources are required. A plant producing 10 000 t a^{-1} uses fifty 60 kW lamps (Fischer, 1978). Nitrosyl chloride has a broad absorption in the UV and visible parts of the spectrum (Goodeve and Katz, 1939). Light up to 600 nm is effective, and to maximize the efficiency of the process metal halide lamps doped with thallium iodide, with maximum output at 535 nm, are used (Pape, 1975; Fischer, 1978). Thallium iodide lamps will be discussed in Chapter 9. It is necessary to exclude light at wavelengths below 365 nm to avoid tar formation (Hulme and Turner, 1968), and this can be done by adding filters to the lamp cooling water. The scale-up of a plant from laboratory to production scale has been described by Fischer (1978).

1.4.2 Small scale processes

1.4.2.1 Vitamin D synthesis

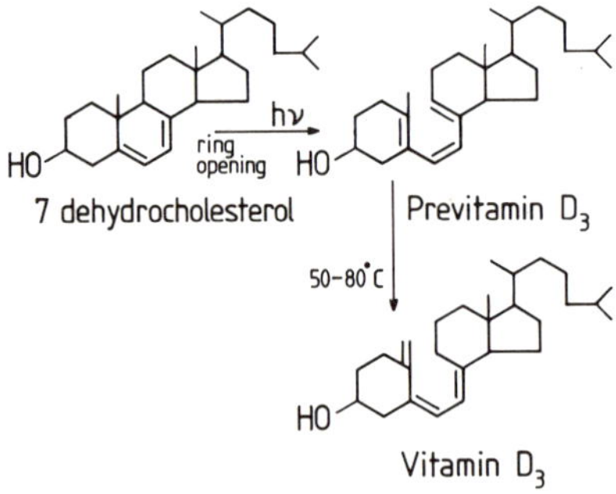

The natural conversion of sterols in the skin to vitamin D has its counterpart in industrial synthesis. Vitamin D_2 is obtained from ergosterol, and vitamin D_3 from 7-dehydrocholesterol. Vitamin D_3 is of greater importance than vitamin D_2 because it can be added to animal feedstuffs (Fisher, 1978).

1.4.2.2 Singlet oxygen reactions

The ground state oxygen molecule is unusual in being a triplet, i.e. it has two unpaired electrons. Oxygen has a singlet excited state (i.e. no unpaired electrons) which can be formed when oxygen is irradiated in the presence of a dye sensitizer, such as rose bengal. Singlet oxygen is a valuable and selective oxidizing agent, and is used for the synthesis of rose oxide (present in natural rose and geranium oils, and used in perfumes) from citronellol (Ohloff *et al.*, 1961; Clements, 1980).

1.5 Polymer Systems

1.5.1 Linear and crosslinked polymers

A *polymer* may be defined as a large molecule (a macromolecule), the structure of which consists of the regular repetition of a simple unit. Polymers are therefore synthesized by the linking together, by covalent bonds, of simple molecules known as *monomers* by the process of *polymerization*. The above definition of the word polymer describes a particular type of molecule. The word is also used to describe a *material* which is composed of these molecules.

Authors of texts in polymer science and engineering are fond of classifying both the polymer materials and the mechanisms by which they are synthesized. A common distinction is made between *thermoplastic* polymers and *thermosetting* polymers. A thermoplastic material is one which will flow under the influence of temperature and/or pressure, and therefore can be moulded into any desired shape by *physical* means, *after* the polymerization process is complete. Thermoplastic polymer materials consist of predominantly linear chains, these being held together by the van der Waals forces. Each polymer chain can be thought of as a separate and *separable* molecule. The mobility of the individual molecules increases with temperature, and so the material will soften on heating, although there may be no distinct melting point, and there may also be a degree of solubility in solvents of appropriate solubility parameter. Thermoplastics include most of the high tonnage materials—PVC, polyethylene, poly-

general enhance properties such as adhesion. However, the chemical reactions which take place are essentially within the coating, rather than between coating and substrate.

Grafting, in contrast, involves a chemical reaction between monomer and substrate. The substrate may be a plastic film or moulding, or a textile fibre. The purpose of grafting is to modify the properties of the substrate. In the case of a plastic film the modification required may be increased wettability or improved antistatic properties; in the case of a textile fibre, improved wrinkle resistance, flame resistance, acceptance of dyes etc. Grafting has been an important industrial process for many years, with chemical reactions brought about thermally or by the use of ionizing radiation. In recent years there has been a growth of interest in the use of actinic radiation to graft vinyl-type monomers to plastic films, for example acrylic acid to polypropylene, polyethylene and polyethylene terephthalate (Ogiwara *et al.*, 1982), and methyl methacrylate, styrene, methacrylic acid, acrylamide and trimethylolpropane triacrylate to cellulose acetate (Herold and Fouassier, 1981a). Most interest however has been focused on textile fibres, with synthetic fibres, e.g. polyester (Siddiqui *et al.*, 1981) and Nylon 6 (Lenka and Nayak, 1982), receiving less attention than cellulose-based fibres (Arthur, 1980), such as wool (Garnett and Leeder, 1977; Barker *et al.*, 1981) and cotton (Abdel-Hay *et al.*, 1980; Herold and Fouassier, 1980, 1981b). Photografting is normally carried with the monomer in solution, and with the use of a photoinitiator.

An intriguing use of photografting is the immobilization of enzymes on graft copolymers (Barker *et al.*, 1982; D'Anguiro and Cremonesi, 1982). This may provide an economical way of using enzymes to catalyse biochemical processes.

1.6 Surface Coatings

The application of a relatively thin film of one material to the surface of another may be done for a wide variety of reasons. Some of the more important are listed in Table 1.1.

The categories listed in Table 1.1 are by no means mutually exclusive, and obviously a particular coating application often fulfils several functions. Organic coatings have been used to fulfil the requirements of all the categories shown in the table, and in most cases the coating is applied in a more or less fluid form which must be converted into a more or less solid form to provide the functional performance required. Various methods are available to give the required increase in viscosity, the main ones being (a)

Table 1.1 Reasons for Surface Coating

1. Protection against chemical attack
2. Provision of barrier properties
3. Reduction of porosity
4. Modification of optical properties
5. Improvement of surface mechanical properties
6. Modification of surface energy, i.e. "wettability"
7. Modification of surface electrical properties
8. Information, i.e. printing
9. Decoration

solvent evaporation, (b) solvent absorption into an absorbent substrate, (c) chemical reaction leading to an increase in molecular weight, (d) decrease in temperature (e.g. sealing wax).

Many conventional coating materials contain large quantities of non-reactive solvents, the principal purpose of which is simply to reduce the viscosity to a value suitable for the coating application. The solvent plays no part in the final cured coating and it is seldom feasible to recycle it. The solvent therefore represents a wasteful use of raw materials and a pollution and fire hazard. There are strong economic and environmental pressures to reduce or eliminate the use of non-reactive solvents in coatings, and the attractions of a formulation in which all the components end up as part of the cured film are very powerful. The elimination of solvents implies in most practical applications a system which cures by an increase in molecular weight—by crosslinking or polymerization.

Photopolymerization provides the mechanism for the "*UV curing*" *of coatings* (Roffey, 1982). Many potential advantages have been claimed for UV curing, but almost all are based on (a) the lack of non-reactive, volatile solvents in the coating formulations and/or (b) the possibility of more rapid cure and/or a lower energy/space/temperature requirement. Polymerizing systems are chosen for which the propagation reaction proceeds rapidly at ambient temperature, and energy input is required simply to activate an initiator.

Electromagnetic radiation which is of the right wavelength to be selectively absorbed by the photoinitiator is a much more elegant and efficient source of energy than convected or conducted heat. This is especially so when the polymerizable material is in the form of a very thin section on a much thicker substrate, such as an ink film on a sheet of tinplate, the latter having a much higher heat capacity than the former. An analysis of the energy savings possible with UV curing has been made by Parrish (1977). Energy accountancy should take into account not only the energy to cure a coating, but also the energy required to manufacture the resins, monomers and other components which make up the formulations. Happily these are no more energy intensive than the components of conventional coatings (North *et al.* 1976).

Heat, as an energy source, apart from being unselective and wasteful, has other problems associated with it. Ovens suffer from thermal inertia; they require time to warm up to operating temperature and time to cool down. Also a hot coating on a hot substrate may need to cool down before it can be handled.

UV is not the only type of radiation which can be used for curing. Infra-red is used extensively, but only as a method of heating the coating. Microwave energy can be used to heat coatings if they contain highly polar components such as water or alcohols. In recent years there has been a growth of interest in electron beam (EB) curing, which from a chemical point of view has much in common with UV curing. The formulations used are similar to those used for UV curing, except that no photoinitiator is required. A beam of high energy electrons (energy in excess of 150 keV) is directed at the coating, and by a rather complex mechanism (Vrancken, 1977) monomer radicals are produced and a chain polymerization takes place which leads to a cured coating. EB curing has many attractions. The penetration, i.e. the thickness of the section that can be cured, is a function of the electron energy and the density of the coating, and is not affected by pigmentation except inasmuch that some inorganic pigments such as titanium dioxide have a high density. Even the lowest energy commercial EB units are able to cure pigmented coatings up to 200 μm in thickness. As photoinitiators are not required, formulations are cheaper, and the problems which photoinitiators may cause, such as yellowing, are avoided. Energy requirements are lower than with UV curing. However, the capital expenditure is very much higher, and it is necessary to inert the surface of the coating, usually with nitrogen, to exclude air, otherwise oxygen inhibition will cause poor surface cure. Also, high energy electrons cause some X-ray emission and adequate shielding is required to eliminate any possible hazard.

1.6.1 UV curing of inks and coatings

1.6.1.1 Antecedents

The concept of using UV to dry surface coatings is not new. Indeed, it is a technique which was practised by the ancient Egyptians, who treated linen cloths with a mixture of lavender oil and Syrian asphalt, and dried them in sunlight. The cloths were used for wrapping mummies (Oster, 1969). Syrian asphalt was also used in the early 19th century for the preparation of lithographic printing plates, until replaced by the more convenient gelatin–dichromate process.

There were several attempts in the 1940s to formulate UV curing inks and other coatings based on natural drying oils such as tung oil or oiticica oil combined with materials such as carbon tetrabromide which were known to generate free radicals on irradiation (Wendt, 1948). These attempts were not particularly successful, and Carlick (1972) has suggested that it may have been the ozone from the mercury vapour lamps used, rather than the UV, which was responsible for the cure obtained. Interestingly, Klose (1968) found that a combination of short wavelength UV and ozone was effective in curing a range of conventional coating materials.

A step nearer to a usable UV curing system was made by Sites and Agruss (1962). They described formulations based on either drying oils such as tung oil or on unsaturated polyesters, in combination with available vinyl-type monomers such as styrene, divinyl benzene, acrylamide and methyl acrylate, using a variety of free radical generators such as sulphonyl chlorides, acyl chlorides, peroxides and organic titanates.

1.6.1.2 Unsaturated polyesters

The UV curing of coatings became a commercially viable process during the late 1960s, in Europe and in the United States. The impetus in Europe was the need for a more satisfactory method of applying fillers and other coatings to particle board, fibre board and other open grain wood products. Conventional methods used materials with high solvent contents which required multicoat applications and long curing cycles (Carlick, 1976). McCloskey and Bond (1955) had shown that unsaturated polyester resins dissolved in styrene could be crosslinked by UV if a suitable photoinitiator was incorporated. As this type of resin–monomer mix had been used for thermal curing with peroxide catalysts for many years it was logical to use the same system with a photoinitiator instead of the thermal catalyst. It was found to be quite satisfactory for filler coats. The rapid low temperature cure minimized the volatile loss and therefore gave better pore filling, and the particle board could be sanded immediately after curing (Deniger and Patheiger, 1969). Unsaturated polyester–styrene fillers are still in use, but the system proved less satisfactory for basecoats and topcoats, because of its slow cure speed, pronounced oxygen inhibition giving a tacky surface (Wicks, 1974), high monomer volatility (Younger, 1976) and lack of flexibility (Young, 1976; Schmidle, 1977). Monomer volatility could be reduced by the incorporation of waxes in the formulations (Schroeter and Moore, 1973). Wax was also found to reduce the inhibiting effects of atmospheric oxygen on the surface cure.

In Europe, UV curing allowed higher production rates and lower energy costs. In the United States, additional factors led to the adoption of UV curing, including Federal and state legislation restricting the emission of solvents into the atmosphere (Burachinsky, 1970) and local shortages of natural gas for hot air ovens (Libutti, 1974; Brenneman, 1976).

1.6.1.3 Acrylate chemistry

Early UV curing formulations, up to about 1970, were based on materials which were then commercially available, and used for other purposes. The lamps were usually medium pressure mercury lamps operating at up to $40\,W\,cm^{-1}$. The growth of UV curing technology since that date has been the result of the development of new resins and monomers specifically for incorporation into UV curing inks and coatings, and the introduction of higher powered lamps. This has enabled the full potential of the UV curing method to be exploited. It was realized that of all the double bonds that can be polymerized by a free radical process, that in the alkyl acrylate ester polymerizes most rapidly, and the rate is further enhanced by β-alkoxy or β-hydroxy substitution (Mesrobian, 1977). Acrylate esters polymerize at a rate which is an order of magnitude greater than is found with methacrylates which in turn polymerize faster than vinyl esters or vinyl aromatics, allyl esters or allyl ethers (Carlick, 1976; Schmidle, 1977). Resins were therefore synthesized with terminal or pendant acrylate groups, the backbone being derived from a polyester, polyether, epoxy, urethane or other suitable film-forming oligomer (Nishikubo *et al.*, 1974; Rybny *et al.*, 1975; Jackson *et al.*, 1976; Rowe, 1976; Young, 1976; Fuhr, 1977; Barclay, 1979).

The resin contributes viscosity, tack, pigment wetting and film-forming characteristics to the coating. Resins are normally very viscous, or even solid, and in order to reduce the viscosity to exactly the value required by the coating process it is necessary to use a reactive diluent. Well established monofunctional acrylate monomers such as ethyl acrylate and butyl acrylate, and even ethylhexyl acrylate, were found to be rather too volatile, the problems with volatility being aggravated by the very low odour threshold associated with the acrylate group (North *et al.*, 1976). Therefore a number of novel low volatility acrylate monomers were synthesized, e.g. phenoxyethyl acrylate (Pelgrims, 1978). However, it was realized that more rapid cure would be obtained if a reactive diluent containing two or more acrylate groups could be used and therefore polyfunctional acrylate ester reactive diluents of low volatility were synthesized by the reaction of polyols such as hexan-1,6-diol and trimethylolpropane with acrylic acid (Younger, 1976; O'Hara, 1978). The third reactive component in UV-curing acrylate formulations is the photoinitiator. Developments in photoinitiators, which involved not only the synthesis of new compounds, but also the discovery and exploitation of synergistic mixtures of existing compounds, led to more rapid cure, resistance to oxygen inhibition, and improved shelf life. It should be mentioned that the development of UV-curing inks and coatings was not being carried out in isolation. It was very much assisted by the intense interest of the academic photochemists in the photochemistry of ketones, and in particular the interaction of ketones with amines. Also, photopolymerizable systems were being developed for other purposes such as photoresists, and some cross-fertilization was able to take place. Photoinitiation will be discussed in detail in Chapter 3. UV-curing acrylate formulations usually contain a small amount of inhibitor such as a quinone to prevent premature gellation, and may contain a number of non-reactive components such as pigments, fillers, surfactants, waxes, lubricants, etc.

The new formulations, based on speciality materials, were marketed under various trade names by leading ink and coatings suppliers, and when used with efficient UV driers fitted with high intensity medium pressure mercury lamps, gave cure speeds and general performance which made them suitable for a wide range of printing and coating processes. A quite significant feature of the development was that for the first time it was possible to make the transition from a wet coating to a completely dry coating in a time comparable with that required for the coating process itself, and using a UV drier no larger and sometimes much smaller than the coating or printing machine. Since the introduction of acrylate-based formulations there has been a steady growth in the use of UV curing, a growth which would have been more rapid but for the worldwide economic recessions of the 1970s.

1.6.1.4 Health and safety

There have been gradual but not spectacular improvements in the performance of acrylate-based formulations since they were first introduced. A very significant influence has been the increasing concern for health and safety in the industrial environment. Acrylate monomers have been the subject of considerable scrutiny, particularly with regard to their potential for eye and skin irritation and skin sensitization (Broxterman, 1976; Mesrobian, 1977; Jenkinson, 1978; Nethercott, 1978; Reid, 1978; Barclay, 1979; Gelbke and Zeller, 1981; Tins, 1982). Several monomers, and in particular those with hydroxyl groups, e.g. hydroxethyl acrylate, have been found to be potent skin irritants and their use has been discontinued by many formulators. It is ironic that some of the most useful materials from a performance standpoint have been found to have the greatest hazards associated with their use (Huemmer, 1974), and a significant fraction of the formulators' development effort in the 1970s was aimed at finding acceptable alternatives to materials which came to be considered too hazardous to use. The result was that for several years the main improvement was in safety, rather than cure speed. However, there have been significant improvements in adhesion to difficult substrates, cure in thick pigmented films, lithographic properties, yellowing, odour and shelf stability.

1.6.1.5 Non-acrylate systems

While unsaturated polyester–styrene and acrylate-based systems have been the most widely used materials, some non-acrylate monomers, particularly vinyl acetate (Rybny *et al.*, 1978) and *N*-vinyl pyrrolidone (Lorenz, 1976) have been found to coreact well with acrylates, and are useful for viscosity reduction. The "thiol–ene" system (W. R. Grace), and the cationic systems using onium salt photoinitiators (American Can, General Electric, Minnesota Manufacturing and Mining, Imperial Chemical Industries) have many attractive features. These systems will be discussed in more detail in Chapter 3.

1.6.1.6 Pigmentation

An essential component in printing inks and coloured coatings is a pigment or mixture of pigments. The incorporation of pigments into UV-curing formulations presents much greater problems than is the case with formulations which cure by conventional methods. Pigments operate by selective absorption and reflection in the visible part of the spectrum. Unfortunately, all pigments absorb to a greater or lesser extent in the UV, and therefore will compete with the photoinitiator for the radiation from the lamp, and will limit the penetration of radiation into the ink or coating. The UV absorption spectra of pigments are poorly documented. A useful selection has been given by Hencken (1975, 1978). Measurement of a relevant absorption spectrum is complicated by the fact that pigments are dispersed rather than dissolved in the resin system and therefore behave as a discrete phase, with scattering taking place at the interface. This means that light transmitted through a thin pigmented film may emerge at an angle quite different from the angle of incidence. It is therefore desirable to collect all the light passing through the film and this is best done with an integrating sphere (Hauser *et al.*, 1978; Plews and Phillips, 1979). The effect of the presence of scattering pigments on the absorption by photoinitiator at different depths in a film has been calculated by Wicks and Kuhhirt (1975). It has been found experimentally that a very small concentration of a scattering pigment may actually enhance cure in thin films by increasing the path length through the film and therefore the likelihood of absorption by the photoinitiator (McGinniss, 1975; Hulme, 1976). The influence of the presence of organic pigments on cure has been studied (Bassemir, 1974; Bassemir and Bean, 1974; Watanabe, 1978; Plews and Phillips, 1979; Noomen, 1981), and in general there appears to be an inverse relationship between cure rate and the pigment absorption. The most important coloured pigments are those used in four-colour process sets, the ease of cure being given by the sequence yellow > magenta > cyan > black.

White pigments have received much more attention than coloured pigments, largely because of the activities of Dr B. E. Hulme and his coworkers at Tioxide Ltd. The aim has been to formulate a low viscosity opaque white coating which can be applied by a roller coater to give a cured film thickness similar to that obtained with conventional coatings. The latter contain large quantities of solvent so that the pigment concentration in the wet coating can be much lower than in the cured coating. This is not the case with a 100% reactive system (O'Hara, 1978), and therefore because the UV curing coating is *applied* at a much lower film weight (to achieve the same *cured* film weight), the pigment level must be much higher. In order to avoid problems with rheology, it is essential that the most effective white pigments are used. Although many have been studied (Pappas and Kuhhirt, 1975; Noomen, 1981), titanium dioxide is by far the best. There are two forms of TiO_2, anatase (refractive index 2·51) and rutile (refractive index 2·71). Rutile is favoured for UV curing coatings because it provides better opacity and better rheology for the coating (Hulme, 1975, 1976). When light strikes a coating pigmented with rutile, the pigment particles will either absorb the light or scatter it, depending on the wavelength. Below 380 nm rutile is strongly absorbing (Hird, 1976). At higher wavelengths rutile is strongly scattering, a result of its extremely high refractive index. This scattering provides the required opacity and appearance of whiteness. Success in curing white coatings pigmented with rutile has been achieved by the development of photoinitiator systems which respond to radiation between 380 and 420 nm (see Section 3.3.4.3), and the use of lamps with enhanced output in this region.

Although in formulations which cure by a free radical mechanism the primary effect of pigments on cure results from competition for radiation, secondary effects have been observed, with either acceleration or inhibition of cure by chemical interaction (Bassemir, 1974; Bridgeman, 1977; Chang, 1977). For instance, TiO_2 can act either as a photosensitizer, accelerating cure, or as a quencher of excited states, inhibiting cure (Pappas and Kuhhirt, 1975). Pigments often have a deleterious effect on the shelf stability of inks, causing premature gellation (Carlick, 1976). The formulation of cationic curing inks has proved to be very difficult because many commercial pigments contain components which act as powerful inhibitors of the polymerization reaction (see Section 3.4.4).

1.6.1.7 Litho and letterpress

UV curing printing inks were first used commercially on high speed rotary offset letterpress and lithographic printing machines. In letterpress printing the printing plate has a relief image, so that the application of the inking roller coats only the raised image area. The lithographic plate is planographic, and during its preparation the image area is made oleophilic (and hydrophobic), and the non-image area hydrophilic. A dampening roller applies water to the plate, which coats the non-image area and confines the application of ink to the image area. In offset presses the inked image is not transferred directly from the plate to the stock, but is first offset on to a rubber blanket. Letterpress (known as "dry offset" or "letterset" when used offset) and lithography have in common the need for a high viscosity paste ink, and ink formulations are substantially the same for both types of press. The material to be printed can be in the form of individual flat sheets, for instance of carton board or tinplate, or in the form of a continuous web, for instance of newsprint, or, particularly in the packaging industry, in the form of a preformed container, such as a pot or bottle.

When conventional inks are used, oxidative crosslinking is the most common drying process, coupled with evaporation and/or penetration of solvent. If the stock is very absorbent, penetration may be rapid enough to eliminate the need, on a sheet-fed press, for an accelerated drying process. Web presses, with the exception of those for newsprint, almost always have assisted drying, using hot air, flame impingement or infra-red. Penetration is poor into the high quality coated carton board often used in packaging, and if accelerated drying is not available the sheets must be sprayed with starch or wax "anti-set-off" powder to prevent set-off (transfer of ink between sheets) in the sheet stack. Oxidative drying is slow at ambient temperature, and even if penetration, evaporation of solvent or the use of anti-set-off spray eliminates the problem of set-off, full hardening may take hours or even days. When UV curing was applied to the lithographic printing of carton board full cure was achieved almost instantaneously, which eliminated the need to stack and allowed in line operations such as cutting, folding, creasing, glueing, foil blocking and overlaminating. As UV-curing inks do not contain volatile solvents, they offer better press stability and more consistent print quality. Inks can be left overnight in the ink ducts, which reduces downtime, and saves both ink and wash-up solvents. As cure takes only a fraction of a second it is possible to install UV drying units between the printing stations on a multicolour line, and therefore dry each colour before the next is applied ("dry trapping"). This allows higher quality multicolour half-tone work with the elimination of "dot gain" and "doubling" (Anderson, 1974). There is obviously no need for anti-set-off spray, the use of which accelerates the wear of machinery and leaves the print surface with a characteristic sandpapery feel, particularly unwelcome in cosmetics packaging. As Rosario (1977) has said, "a woman who buys a cream to give her a satin smooth skin is not likely to buy it in a rough carton". The dense crosslinking provided by the UV curing process gives the print better abrasion and chemical resistance, and lower residual odour and taint, particularly valuable for food cartons. UV curing has been particularly successful for the printing of difficult, non-absorbent substrates such as foil board and metallized paper (Mesrobian, 1977; Williams, 1981).

In the United States web-fed presses were an immediate growth area for UV curing. The hot air driers used with conventional inks were dependent on natural gas supplies, which in some areas were in short supply. The hot air driers caused many problems with paper webs, mainly resulting from the effect of the heat on the paper. Heat alters the moisture content and therefore the dimensions of paper stock. This affects web tension, the register of successive colours and the dimensional stability required for the accurate punching of holes and perforations on business stationery. With heat set, chill rollers are required to cool the web before it can be reeled or folded. Heat may cause fibre puffing, blistering of the paper and frequent web breaks (Chatterjee and Ramaswamy, 1975). UV curing eliminated all these problems, as there was minimal temperature rise of the web. UV curing provided an extra bonus when applied to narrow web label presses. The excellent chemical resistance of the UV cured ink eliminated the need for an overlacquer (Mesrobian, 1977). Web offset is used for printing newspapers and magazines, and there has been some limited use of UV curing in this area (Plumridge, 1974).

Another major use of UV curing has been for *metal decorating* in the packaging industry, for external application to two- and three-piece cans and to caps and closures. The metals involved are electrolytic tinplate, tin-free steel (TFS) and aluminium. When conventional inks are used, a major problem with printing on metal sheet is that it is only possible to print two, or at the most three, colours in one pass through a printing line, whereas with a permeable substrate up to six colours can be printed (Lott, 1976).

Many metal decorating designs require four colours, and this necessitates two passes through the line, and two stoving operations. Metal decorating with thermally cured inks is extremely energy intensive. Because metal is impermeable there is no possibility of penetration, and even more important, there is no access for oxygen into a sheet metal stack. This means that the print must be fully cured before it is stacked. This necessitates a curing cycle of about 10 min at 200 °C in a gas-fired wicket oven. The exact energy requirement depends on the chemistry of the ink, the solvent content, whether or not there is an overvarnish and the speed of the line. A typical figure for gas consumption is about 25 therms per hour, which is equivalent, in energy terms, to 750 kW h. Only a tiny amount of this energy is used to evaporate the small amount of solvent present in metal-decorating inks, and to raise the ink to the cure temperature. About half of the energy heats the air in the oven and is exhausted to the atmosphere, because it is necessary to introduce fresh air into the oven continuously to prevent the solvent concentration from approaching the explosive limit. Incinerators may be required to prevent exhausted solvent polluting the atmosphere. Almost all the rest of the energy heats the metal sheet, the wickets and the conveyor, and the mass of the oven, and is lost by radiation and convection. The gas is not the only source of energy required. Some 30–50 kW of electricity is needed to operate the conveyor system, fans, etc.

The energy requirement of UV sheet metal decorating depends on what is to be achieved. The first application of UV curing in Metal Box, in 1972, was for "interdeck setting", in which the process

(1) print colours 1 and 2 (heat set inks) on a Y-type press
(2) stove
(3) print colours 3 and 4 (heat set inks) on a Y-type press
(4) overprint varnish (heat set) on a roller coater
(5) stove

was converted to

(1) print colours 1 and 2 (UV curing inks)
(2) UV cure
(3) print colours 3 and 4 (heat set inks)
(4) overprint varnish (heat set)
(5) stove

This application of UV curing not only allows the entire sheet side to be decorated in one pass through the printing line, but also a stoving process with its attendant energy costs is saved. The purpose of UV curing in this instance is simply to dry the first two colours sufficiently to allow the third and fourth colours to be printed without loss of print quality. Two medium pressure mercury lamps rated at 80 W cm^{-1} suffice for this, each with a maximum consumption of about 9 kW.

When UV curing is used in a situation where metal sheet will be stacked immediately after the UV curing process, an operation known as "interpass setting" or "set stack", the cure requirements are more stringent, as all possibility of set-off, or even worse "blocking", must be avoided. UV driers for interpass setting normally incorporate four to six lamps at the end of the line.

Many beverage cans are now made in two pieces rather than three, and it is a formed container which is decorated. This necessitates quite different types of printing machine and UV curing systems (Libutti, 1974; Hamilton, 1976; Ury and Eastlund, 1977; LeFevre, 1979, 1980).

The printing of *common thermoplastic materials* such as polystyrene, polypropylene, polyethylene and poly(vinyl chloride) presents one of the most challenging problems to the printer. These materials have low energy surfaces which makes them difficult to wet, although this problem can often be alleviated by treatment with a flame or corona discharge. The materials are essentially impermeable to ink solvents, heat sensitive and easily distorted. When thermally curing inks are used they must be formulated from resins and solvents which allow rapid cure at a low temperature (e.g. 60–80 °C), and their costs are significantly higher than those of the inks formulated for paper and board, and for metal decorating. Furthermore, the volatile solvents used create problems with rheological stability. For these reasons UV curing has made a very significant impact on the dry offset printing of plastic pots, bottles, tubes, lids, etc., and it is likely that eventually almost all printing of this type will involve UV curing. The UV-curing inks, without volatile solvents, are much more stable on the press, and can be cured in a fraction of a second at ambient temperature.

Of particular interest is the "*on-mandrel curing*" of print on plastic containers (DeSaussure, 1978), as this demonstrates clearly the benefits of the very rapid cure which is possible. One of the principal problems associated with the printing of plastic containers is that of the high speed handling of lightweight containers. When they are printed they are held on mandrels on an indexing turret. The turret moves with intermittent motion, with stationary periods to allow delivery of the container, pretreatment (e.g. flame treatment or corona discharge), printing and take off. If there is an unused station between printing and take off, it is feasible to install a UV lamp and mandrel rotation, and to use the very short period (0·1–0·2 s) in front of the lamp to dry the print. Not only is the time available for irradiation very short, but only a fraction of a second elapses before the print must be scuff resistant. The demands are matched by the benefits. There is no need for a separate drier, with considerable savings in capital and operating costs, and in floor space. See also Section 14.8.5.

1.6.1.8 Flexography and gravure

There has been some limited use of UV curing for the very low viscosity "liquid" inks required for flexographic and gravure printing (Anderson, 1975; Fairbanks, 1976; Nowak and Rybny, 1977; Pelton, 1978; Rybny, 1978). UV curing allows a large reduction in the use of solvents, and the pollution they cause. However, there are problems involved in the formulation and cure of very low viscosity pigmented systems because of the volatility and toxicity of low molecular weight acrylate monomers (Bean, 1975; Carter, 1976; Gerhardt, 1976).

1.6.1.9 Screen printing

UV curing has been found to offer great benefits to the screen printer. Conventional screen inks contain 60–70% solvent. UV curing inks contain no volatile solvents and therefore do not dry in the screen mesh. It is possible to use finer mesh screens, and thereby obtain finer line and half-tone printing (Reed, 1976; Bradigan, 1977), with lower ink usage and lower ink wastage. There are problems involved in the printing of multicolour half-tone designs on large format display items with heat-cured screen inks because of heat-induced dimensional changes which affect the register of successive colours. This problem is avoided with UV curing. Screens will not normally need cleaning until the job is finished, and this saves time and ink and reduces screen wear. When a UV curing screen ink is used, the thickness of the applied film is retained after cure (apart from a small shrinkage due to the polymerization reaction), and this enhances the three-dimensional quality of the print.

UV screen printing is used for decorating sheet plastics, glass, metal, paper and board and moulded plastic containers. The rapid cure with UV allows screen printing to be used by the reel to reel label printer (Williams, 1981). UV screen printing is being increasingly used for the application of resists in printed circuit board manufacture (see Section 1.7).

1.6.1.10 UV curing coatings

Inks are printed on substrates primarily for the purpose of information and decoration, and the print is normally only required to maintain its own integrity. Coatings may be applied to improve appearance, but they are generally also required to improve the functional properties or even maintain the integrity of the substrate by for instance increasing abrasion resistance, reducing the possibility of staining or corrosion, reducing the coefficient of friction, providing a moisture barrier and so on. Therefore, although cure presents less of a problem with unpigmented coatings than with ink films of the same thickness, the demands on the cured film are generally greater.

A major application of UV curing coatings has been in wood finishing (Currier, 1974; Niehaus, 1974; Gruber, 1978; O'Hara, 1978; Dufour, 1979; Vrancken, 1980). Materials used include fillers for chipboard, which contain large proportions of inert transparent fillers such as calcite and barytes to lower cost and improve sandability, sealers for hardboard, opaque basecoats for printing wood grain, and matt or gloss topcoats.

Coatings can be applied to a substrate by a variety of methods including roller coating, curtain coating, electrocoating (McGinniss, 1982), screen printing and application through the dampening system of a lithographic printing machine. UV curing varnishes and lacquers are "100% solids", and therefore the entire weight applied is retained after cure. Very high gloss and a high quality appearance are possible, and UV curing lacquers have been used successfully as a cheaper alternative to film lamination, for coating record sleeves, book and magazine covers, greeting cards, playing cards, liquor and cosmetics cartons and label stock. UV curing varnishes have been used for such diverse and demanding applications as coating vinyl and vinyl asbestos floor tiles (Prane and Bluestein, 1976; Prane, 1980), flexible plastic sheet (Kurpiewski, 1981), leather (Knight and Marriott, 1978), wallpaper (Levy, 1976), aluminium ring pull can ends, skis, magnetic tape, wire (Ketley, 1975), optical fibres (Schonhorn *et al.*, 1976, 1977, 1979; Vazirani *et al.*, 1977; Stevens and Keough, 1978; Lawson *et al.*, 1980; Paek and Schroeder, 1981; Schlef *et al.*, 1982), and glass bottles and flashbulbs to reduce the risks of breakage (Schroeter *et al.*, 1978; Tozier and Shaffer, 1982).

It has been possible for formulate UV curing varnishes which can be applied "wet-on-wet" to conventional inks, so providing an immediate tack free surface and the possibility of in line operations without the extra ink costs of a total UV curing system (Strand, 1976; Magdelinskas and Wolinski, 1976; Yick, 1977; Pierce, 1978).

1.6.1.11 Problems associated with UV curing

The applications of UV curing which have been described have been chosen to illustrate how the basic advantages of the process can be realized in practice. However, the process is not without its problems, and these explain why UV curing inks and coatings have not replaced all other types, but rather have found success in certain specialized areas. As in any field of endeavour, the problems are there to be solved, and while they should be recognized as such it must be said that they can often be partly or wholly overcome by clever formulation and by more careful or modified methods of application.

(1) The major disadvantage is not a technical problem at all, but a

question of economics. UV-curing materials have often been quite a lot more expensive than their conventional counterparts, with a premium, based on price per kilogram, which is typically 50–100%, although it varies greatly with the application. It was hoped that as the process became more widely used and the market for the materials expanded, and as the cost of oil-based solvents increased, the extra cost of UV-curing materials would decrease, but this has not proved to be the case. However, the "mileage" of UV-curing materials will tend to be higher than that of conventional materials and so the premium based on price per square metre coated may be less daunting.

(2) Very rapid polymerization at ambient temperature of free radical curing materials involves a sudden shrinkage of the applied coating (McGinniss and Kah, 1977; Armbruster *et al.*, 1978; Rybny *et al.*, 1978; Dufour, 1980), and this results in adhesion on non-porous substrates which is generally poorer than is obtained with a slower higher temperature thermal cure. This unfortunately means that in metal decorating a "thermal bump" is required to anneal the stresses in the film and achieve the functional performance required.

(3) The very short delay between application of an ink or coating and cure allows little time for thorough wetting of the substrate, and for flow out, and therefore careful attention to rheology is required if adequate levelling and gloss is to be obtained (Anderson, 1974; Fairbanks, 1976).

(4) The chemical resistance of cured print is generally superior to that of conventional print, which in itself is an advantage, but it does make de-inking of paper more difficult (Vanderhoff, 1975, 1976).

(5) The can stability of UV-curing inks is generally poorer than that of inks which dry by oxidation (Williams, 1977).

(6) Because there is no solvent, higher pigment levels are required in wet coatings. This may cause rheological problems (O'Hara, 1978).

(7) The penetration of UV into inks and coatings is limited, particularly if the coating is heavily pigmented, and therefore UV curing is only suitable for relatively thin films. However, it is surprising what can be achieved with the right choice of photoinitiator and radiation source.

(8) As UV-curing materials are 100% solids it may be difficult to obtain very thin cured coatings without modification of coating machinery.

(9) The resins and monomers used in UV-curing formulations tend to be polar in nature, and therefore UV-curing inks are more difficult

to print by lithography (Bean, 1975; Mesrobian, 1977). The ink–water balance is more critical (Enchelmeyer, 1980).

(10) UV-curing monomers are often strong solvents. Printing plates, blankets and roller materials must be chosen carefully (Gruebel, 1974; Fadner, 1976).

(11) UV curing is a "line of sight" curing method, and therefore there must be no obstructions between the radiation source and the print or coating. In particular, gripper shadow must be avoided on sheet-fed presses.

1.7 Photoresists

(See Hepher, 1964; Woods, 1968; Bulloff, 1971; DeForest, 1975; Moreau and Viswanathan, 1975; Delzenne, 1976, 1979; Feit, 1978; Williams and Molaire, 1982.)

1.7.1 Principles

A "*resist*" is a coating which is applied to a surface, or to selected parts of a surface, to protect that surface against chemical or physical effects. If a photochemical reaction is used to alter the solubility of the resist material, the resist is known as a "*photoresist*". If irradiation brings about a *decrease* in solubility, the photoresist is "*negative working*", while if there is an *increase* in solubility the photoresist is "*positive working*". Most photoresist materials are organic, but inorganic types have been developed and have advantages for particular applications (Chang *et al.*, 1979).

Protection of *selected areas* of the substrate or support is an essential feature of the action of a photoresist. The photoresist material may be coated over the entire surface, and then selectively irradiated through a mask (usually a negative or positive photographic emulsion), to insolubilize (negative working) or solubilize (positive working) it selectively in the irradiated areas. The resist is then "developed" in a suitable solvent to remove the soluble material (Fig. 1.1). An alternative approach to obtaining spatial selectivity is to *print* the required image on the substrate, using a negative working resist ink. This method is suitable only when high resolution is not required. It is widely employed for the manufacture of printed circuit boards. The printing process used is almost always screen printing (Beeler, 1975; Berg, 1976; Peters, 1978).

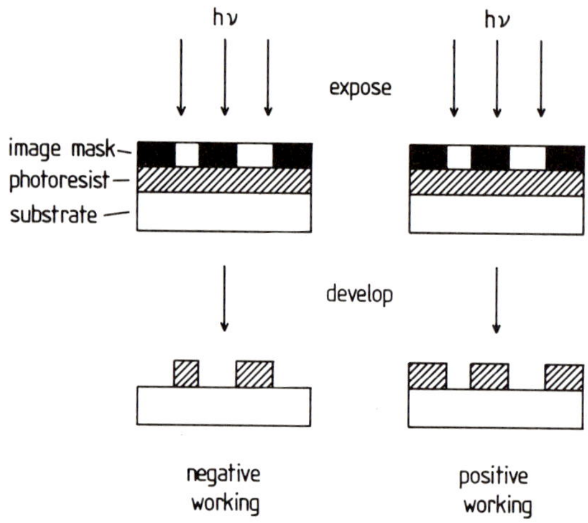

Fig. 1.1. Negative and positive photoresists. This schematic diagram shows contact between the image mask and the photoresist. Non-contact (projection) systems may also be used.

In some applications the insolubilized resist does not form an essential part of the final product. Its function is to form a mask which restricts the action of, for instance, an etchant, a plating solution or a solder bath to selected areas of the substrate, and after it has served this function it may be removed. Typical applications are the production of printed circuit boards and integrated circuits in the electronics industry, the etching of glass and the chemical milling of precision mechanical components and the processing of a deep etch lithographic printing plates.

In other applications the insolubilized photoresist remains as an essential part of the final product. These applications are mainly in the printing industry. For instance, the resist material provides the oleophilic image area for surface-type lithographic plates. Photoresist materials are used to make photopolymer letterpress plates. Selective insolubilization allows the creation of a relief image. The preparation of screens for screen printing involves the application of a photoresist stencil material to the mesh. Selective irradiation insolubilizes the photoresist in certain areas, to give, after development in a suitable solvent, a screen with the required pattern. The photoresist, by blocking the holes in the mesh, forms the non-image area.

1.7.2 Negative working photoresists

Most negative working photoresists are based on chemical systems which undergo photocrosslinking or photoinitiated chain polymerization on irradiation, and the increase in molecular weight reduces the solubility. For many years negative working photoresists were based on natural organic colloids such as albumen, animal and fish glues, casein and gelatin (all proteins) and gum arabic (a polysaccharide), made photosensitive by the addition of dichromates. In recent years superior synthetic materials have been introduced. Important photocrosslinking systems are those which involve the photocycloaddition of pendant cinnamate groups incorporated in polymer chains, and those which use bis-azides to crosslink cyclized rubbers. The photoinitiated chain polymerizations employed are generally similar to those used for the UV curing of inks and coatings, and involve the radical polymerization of oligomers and monomers with acrylate, methacrylate or allylic functionality; thiol–ene systems; and the cationic polymerization of epoxides. Further details of photocrosslinking and photoinitiated polymerization will be found in Chapter 3.

An alternative approach is to use a polymer having pendant solubilizing groups such as $—N_2^+X^-$, which are lost on irradiation (Jacobson and Jacobson, 1976, p. 176).

1.7.3 Positive working photoresists

The common type of positive photoresist consists of an acidic resin, which by itself would dissolve in aqueous alkali, combined with a solution inhibitor. The inhibitor is photosensitive, and on irradiation undergoes a series of reactions which converts it to an acidic hydrophilic material which is no longer capable of acting as a solution inhibitor. The acidic resin is normally either a phenol formaldehyde novolak, or an acrylic resin with pendant carboxylic acid groups derived from acrylic acid or methacrylic acid. The solution inhibitor is usually a quinone diazide, which is incorporated as a photosensitive group in a molecule large enough to have film-forming character (DeForest, 1975, p. 49). The quinone diazide, on irradiation, undergoes a Wolff rearrangement to give a ketene. The ketene may then react with water to give a carboxylic acid:

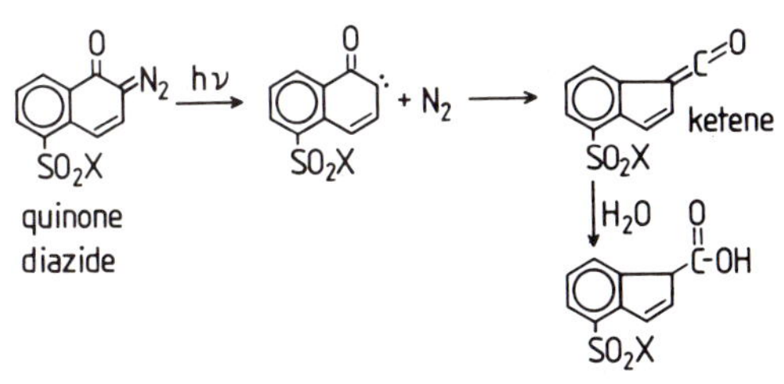

The above mechanism was suggested by Süs (1944). An alternative mechanism, suggested by Levine (1969), involves the ketene reacting with unreacted quinone diazide to form a lactone. Developing with aqueous alkali cleaves the lactone ring (DeForest, 1975, p. 52).

Positive resists tend to give better resolution than negative resists because image distortion phenomena are minimized (DeForest, 1975, p. 216). Another attractive feature is the use of aqueous solutions for processing.

1.7.4 Dry film photoresists

Photoresist materials contain non-reactive solvents, and after the coating process (e.g. roller coating, spin coating, dip coating, spraying) there is a solvent evaporation process to leave a solid dry film. This film of photoresist may be covered with a thin transparent plastic sheet to reduce contamination and oxygen inhibition, if this is relevant. When photoresists are used in microelectronics great care must be taken to eliminate particulate matter from the photoresist material.

The coating process is not always carried out by the user of the photoresist. The substrate may be supplied already coated. There is increasing use of dry film photoresists. The photoresist material is supplied as a thin dry film sandwiched between two thin plastic films, one of polyester and the other of polyethylene. Prior to use, the polyethylene film is removed and the photoresist is laminated to the substrate. The polyester film remains to provide protection until after the irradiation process.

1.7.5 Deep UV lithography

The optimum exposure wavelength for conventional photoresist materials usually lies between 330 and 450 nm. Radiation in this region has a low biological hazard, and is transmitted by glass, so that glass optical components can be used. Also, light sources with intense emission in this region are readily available. However, in the production of high density microcircuits, and diffraction gratings, and in other applications where a high resolution of fine detail is required, diffraction effects at these wavelengths become limiting factors which affect the minimum feature size and the maximum height to width aspect ratio. Electron beam and X-ray lithographic techniques have been developed to overcome this problem (Wittels, 1980). A simpler and cheaper alternative is to use conventional techniques modified for use at shorter wavelengths, <300 nm, which allows submicrometre details to be resolved (Lin, 1975). Transparent vitreous silica optics, aluminium reflectors and sources of intense short wavelength radiation are necessary. Photoresist materials suitable for use at short wavelengths are required. The most widely used "deep UV" photoresist has been poly(methyl methacrylate) (PMMA), which when exposed to radiation from a deuterium lamp, a Xe–Hg compact arc, or an excimer laser, is positive working, and can be developed in a variety of common solvents such as cellosolve acetate or methyl isobutyl ketone. PMMA has a peak absorption at 215 nm and is virtually transparent above 260 nm. Its absorption coefficient at 215 nm is low enough to allow deep penetration into resist films, and therefore the capability of replicating fine patterns with high aspect ratios (Mimura *et al.*, 1978). However, PMMA is not particularly sensitive, and long exposure times are required with non-laser sources. New materials based on the diazide/novolak system, with improved sensitivity and improved resistance to plasma etching, have recently been developed (Grant *et al.*, 1981; Iwayanagi *et al.*, 1981; Gipstein *et al.*, 1982).

References

Abdel-Hay, F. I., Barker, P. S. and Guthrie, J. T. (1980). *Makromol. Chem.* **181,** 2063–2070.

Alexander, I. J. and Scott, R. J. (1980). Eur. Pat. Appl. 14, 293 (Scott Bader Ltd).

American Dental Association Council on Dental Materials and Devices (1976). *J. Amer. Dent. Assoc.* **92,** 775–776.

Anderson, A. (1974). SME Tech. Pap. (Ser.) FC74-515.

Anderson, A. (1975). *Radiat. Curing* **2** (1), 17–18.

Anderson, N. K. (1968). *SPE J.* **24,** 81–84.

Anderson, W. T., Jr (1947). *Ind. Eng. Chem.* **39,** 844–846.

Anderson, T. F. and Voorhees, J. J. (1980). *Annu. Rev. Pharmacol. Toxicol.* **20,** 235–257.

Armbruster, D. C., Zeliznak, K. J., Rybny, C. R. and Vona, J. A. (1978). SME Tech. Pap. (Ser.) FC78-507; reproduced in *J. Radiat. Curing* **6** (4), 10–13, 16–19 (1979).

Arisman, R. K., Musick, R. C., Zeff, J. D. and Crase, T. C. (1980). *Proc. Ind. Waste Conf.* **35,** 802–808; *Chem Abs.* **95,** 29749.

Arnold, D. R. and de Mayo, P. (1971). *Chemtech.* **1,** 615–620.

Arthur, J. C. Jr (1980). *Dev. Polym. Photochem.* **1,** 69–92.

Bachmann, R. (1975). *Brown Boveri Rev.*, **62,** 206–209.

Barclay, R. Jr (1979). *Radiat. Curing* **6** (3), 4–11.

Barker, R. (1979). Electricity Council Research Centre (Memo) ECRC/M1233; *Chem. Abstr.* **91,** 78549.

Barker, R. (1981). Private communication.

Barker, P., Guthrie, J. T., Davis, M. J., Godfrey, A. A. and Green, P. N. (1981). *J. Appl. Polym. Sci.* **26,** 521–527.

Barker, P. S., Bottom, R. A., Guthrie, J. T. and Beddows, C. G. (1982). *Polym. Photochem.* **2,** 87–95.
Bassemir, R. W. (1974). *Amer. Ink Maker* **52** (12), 33, 34, 61, 62.
Bassemir, R. W. and Bean, A. J. (1974). Paper presented at 26th Annual Meeting of TAGA, St Paul, Minnesota, 13–15 May 1974; *TAGA Proc.* **1974,** 133–146.
Bayliss, C. E. and Waites, W. M. (1979). *J. Appl. Bacteriol.* **47,** 263–269.
Bean, A. J. (1975). SME Tech. Pap. (Ser.) FC75–310.
Beeler, J. (1975). *Circuits Manufacturing* **15** (7), 44–45.
Beightol, L. E. (1971). Brit. Pat. 1,245,937 (Solar Laboratories Inc.).
Berg, C. J. (1976). SME Tech. Pap. (Ser.) FC76-511.
Blogoslawski, W. J. and Alleman, D. W. (1979). *Ozone: Sci. Eng.* **1,** 55–60.
Bloomfield, J. J. and Owsley, D. C. (1982). *In* "Kirk–Othmer Encyclopedia of Chemical Technology", 3rd ed, Vol. 17, pp. 540–559. Wiley–Interscience, New York.
Boyle, I. T. (1980). *Scottish Med. J.* **25,** 1–3.
Boynton, H. G., Lewis, E. W. and Watson, A. T. (1959). *Ind. Eng. Chem.* **51,** 267–270.
Brack, K. (1976). SME Tech. Pap. (Ser.) FC76-500.
Bradigan, J. (1977). *Screen Print. Mag.* **67** (10), 96, 97, 100, 100A, 135, 136.
Braendli, G., Egger, H. and Hurlimann, R. (1981). *Brit. Pat.* 1,584,385 (Brown Boveri).
Brenneman, R. S. (1976). SME Tech. Pap. (Ser.) FC76-487.
Bridgeman, I. (1977). *Paint Manufacture* **47** (5), 29.
Brooks, G. T. (1974). "Chlorinated Insecticides". CRC Press, Cleveland.
Broxterman, W. E. (1976). SME Tech. Pap. (Ser.) FC76-535.
Bulloff, J. J. (1971). *Polym. Eng. Sci.* **11,** 405–415.
Buonocore, M. G. (1970). *J. Amer. Dent. Assoc.* **80,** 324–328.
Burachinsky, B. V. (1970). *Gravure* **16** (12), 8, 10, 29, 31; *Modern Lithography* **38** (11), 56–57.
Cannon, P. (1965), US Pat. 3,194,628 (General Electric Co.)
Carless, H. A. J. (1980). *Chem. Brit.* **16,** 456–459.
Carlick, D. J. (1972). *Modern Packaging* **45** (12), 64–67.
Carlick, D. J. (1976). *In* "Encyclopedia of Polymer Science and Technology", Suppl. 1, pp. 367–378. Wiley–Interscience, New York.
Carter, R. G. (1976). SME Tech. Pap. (Ser.) FC76-533.
Chang, Y. C. (1977). *Photogr. Sci. Eng.* **21,** 348–350.
Chang, M. S., Hou, T. W., Chen, J. T., Kolwicz, K. D. and Zemel, J. N. (1979). *J. Vac. Sci. Technol.* **16,** 1973–1976.
Chatterjee, P. C. and Ramaswamy, R. (1975). *Brit. Ink Maker* **17** (2), 76–79.
Christiaens, B. and Martens, G. (1973). US Pat. 3,748,243 (Solvay et Cie).
Clarke, N. and Knowles, G. (1982). *Effluent Water Treat. J.* **22,** 335–338, 340–341.
Clements, A. D. (1980). *Chem. Brit.* **16,** 464–467.
Colson, J. G. (1979). *In* "Kirk–Othmer Encyclopedia of Chemical Technology", 3rd ed, Vol. 5, pp. 808–818. Wiley–Interscience, New York.
Cotruvo, J. A. (1981). *Environ. Sci. Technol.* **15,** 268–274.
Cremer, R. J., Perryman, P. W. and Richard, D. H. (1958). *Lancet i*, 1094–1097.
Currier, G. M. (1974). SME Tech. Pap. (Ser.) FC74-523.
D'Anguiro, L. and Cremonesi, P. (1982). *Biotechnol. Bioeng.* **24,** 207–216.
Decker, C. (1983). *J. Appl. Polym. Sci.* **28,** 97–107.

DeForest, W. S. (1975). "Photoresist: Materials and Processes". McGraw-Hill, New York.
Delzenne, G. A. (1976). *In* "Encyclopedia of Polymer Science and Technology", Suppl. 1, pp. 401–443. Wiley–Interscience, New York.
Delzenne, G. A. (1979). *Makromol. Chem.* Suppl. 2, 169–188.
Deniger, W. and Patheiger, M. (1969). *Industrie-Lackier-Betrieb* **37,** 85–91.
DeSaussure, E. C. (1978). SME Tech. Pap. (Ser.) FC78-522.
Devgun, M. S., Cruickshank, A. J. M., Johnson, B. E. and Paterson, C. R. (1981). *Postgrad. Med. J.* **57,** 159–163.
Diffey, B. L. (1978). *Brit. J. Dermatol.* **98,** 703–705.
Diffey, B. L. (1982). "Ultraviolet Radiation in Medicine" Adam Hilger, Bristol.
Doede, C. M. and Walker, C. A. (1955). *Chem. Eng. (N.Y.)* **62** (2), 159–178.
Downes, A. and Blunt, T. P. (1877). *Proc. Roy. Soc.* **26,** 488–500.
Dufour, P. (1979). *J. Oil Colour Chem. Assoc.* **62,** 59–62.
Dufour, P. (1980). *Farbe Lack* **86,** 121–126.
Elad, D. (1969). *Org. Photochem.* **2,** 168–212.
Enchelmeyer, W. (1980). *Package Printing* **27** (10), 22, 68, 73.
Epstein, J. H. (1977). *In* "The Science of Photobiology" (Ed. K. C. Smith) pp. 175–207, Plenum, New York.
Fadner, T. A. (1976). SME Tech. Pap. (Ser.) FC76-496.
Fairbanks, D. F. (1976). SME Tech. Pap. (Ser.) FC76-530.
Feit, E. D. (1978). In "UV Curing Science and Technology" (Ed. S. P. Pappas), pp. 229–254. Technology Marketing Corporation, Stamford, Conn.
Fiessinger, F., Richard, Y., Montiel, A. and Musquere, P. (1981). *Sci. Total Environ.* **18,** 245–261; *Chem. Abstr.* **94,** 180415.
Fischer, M. (1978). *Agnew. Chem. Int. Ed. Engl.* **17,** 16–26.
Fremed, R. F. (1955). *Chem. Eng.* (N.Y.) **62** (2), 159.
Fry, L. (1977). *Brit. J. Dermatol.* **96,** 327.
Fuhr, K. (1977). *Polym. Paint Colour J.* **167,** 672–675.
Garey, J. F. and De Hart, R. (1979). *Ozone: Sci. Eng.* **1,** 201–207.
Garnett, J. L. and Leeder, J. D. (1977). *ACS Symp. Ser.* **49,** 197–220.
Garrison, L. R. (1980). *Radiat. Curing* **7** (3), 26–28, 30.
Gelbke, H. P. and Zeller, H. (1981) *J. Oil Colour Chem. Assoc.* **64,** 186–190.
Gemne, G., Hoffner, S. and Stenstroem, T. A. (1981). *Vatten* **37,** 265–274.
Gerhardt, G. W. (1976). SME Tech. Pap. (Ser.) FC76-531.
Gibson, G. E. and Bayliss, N. S. (1933). *Phys. Rev.* **44,** 188–192.
Gipstein, E., Ouano, A. C. and Tompkins, T. (1982). *J. Electrochem. Soc.* **129,** 201–205.
Goodeve, C. F. and Katz, S. (1939). *Proc. Roy. Soc. A* **172,** 432–444.
Governale, L. J. and Clarke, J. T. (1956). *Chem. Eng. Progr.* **52** (7), 281–285.
Grant, B. D., Clecak, N. J., Twieg. R. J. and Willson, C. G. (1981). *IEEE Trans. Electron Devices* **ED-28,** 1300–1305.
Green, G. E. and Stark, B. P. (1981). *Chem. Brit.* **17,** 228, 230, 232, 234.
Greubel, P. W. (1974). *Print. Plates Mag.* **60** (5), 3–7.
Gruber, G. W. (1978). In "UV Curing Science and Technology" (Ed. S. P. Pappas), pp. 171–184. Technology Marketing Corporation, Stamford, Conn.
Hamilton, C. R. (1976). SME Tech. Pap. (Ser.) FC76-506.
Hauser, P., Osterloh, R. and Jacobi, M. (1978) *Proc. XIVth FATIPEC Congress, Budapest*, 241–247.

Havinga, E. (1973). *Experientia* **29,** 1181–1193.
Hencken, G. (1975). *Farbe Lack* **81,** 916–923.
Hencken, G. (1978). *Amer. Ink Maker* **56** (3), 57, 58, 60, 64, 66, 68, 70.
Hepher, M. (1964). *J. Photogr. Sci.* **12**, 181–190.
Hérold, R. and Fouassier, J. P. (1980). *Angew. Makromol. Chem.* **86,** 123–141.
Hérold, R. and Fouassier, J. P. (1981a). *Makromol. Chem., Rapid Commun.* **2,** 699–702.
Hérold, R. and Fouassier, J. P. (1981b). *Angew. Makromol. Chem.* **97,** 137–152.
Hird, M. J. (1976). *J. Coat. Technol.* **48**, (620), 75–82.
Hirschkind, W. (1949). *Ind. Eng. Chem.* **41,** 2749–2752.
Ho, K. W. A. and Bohm, P. (1981). *Water Pollut. Res. J. Can.* **16,** 33–44.
Holick, M. F. (1981). *J. Invest. Dermatol.* **77,** 51–58.
Holick, M. F., MacLaughlin, J. A., Clark, M. B., Holick, S. A., Potts, J. T. Jr, Anderson, R. R., Blank, I. H., Parrish, J. A. and Elias, P. (1980). *Science* **210,** 203–205.
Horikawa, K., Wako, H. and Sato E. (1976). *Kogyo Yosui* (214) 21–24; *Chem. Abstr.* **87,** 140619.
Huemmer, T. F. (1974). *J. Radiat. Curing* **1** (3), 9–23.
Hulme, B. E. (1975). *Paint Manufacture* **45** (4), 12–16.
Hulme, B. E. (1976). *J. Oil Colour Chem. Assoc.* **59,** 245–252.
Hulme, P. and Turner, P. (1968). *Chem. Eng.* (*N.Y.*) **75** (7), 80–82.
Hutson, T. and Logan, R. S. (1972). *Chem. Eng. Progr.* **68** (5), 76–77.
Ito, Y. and Matsuda, S. (1969). *Ann. N.Y. Acad. Sci.* **147,** 618–624.
Iwayanagi, T., Kohashi, T., Nonogaki, S., Matsuzawa, T., Douta, K. and Yanazawa, H. (1981). *IEEE Trans. Electron Devices* **ED-28,** 1306–1310.
Jackson, R. J., Jones, P. D. and DeLaMare, H. E. (1976). SME Tech. Pap. (Ser.) FC76-494.
Jacobson, K. I. and Jacobson, R. E. (1976). "Imaging Systems". The Focal Press, London.
Jagger, J. (1958). *Bacteriological Reviews* **22,** 99–142.
Jenkinson, R. D. (1978). SME Tech. Pap. (Ser.) FC78-514.
Jewess, B. W. (1978). *Light. Res. Technol.* **10,** 184–188.
Johansen, E. S. and Myhrstad, J. A. (1978). *NIPH Ann.* **1,** 3–10.
Jonas, O. (1982). *Power* **126,** (9) 103–107.
Jubb, A. H. (1971). *Educ. Chem.* **8,** 23–25
Ketley, A. D. (1975). SME Tech. Pap. (Ser.) FC75-331.
Kilian, R. J. (1981). *Polym. Sci. Technol.* **14,** 411–417; *Chemtech.* **11,** 678–680.
Klose, R. (1968). *Industrie–Lackier–Betrieb* **36,** 234–242.
Knight, M. A. and Marriott, A. G. (1978). *J. Soc. Leather Technol. Chem.* **62** (1), 14–17.
Knudson, A. and Benford, F. (1938). *J. Biol. Chem.* **124,** 287–299.
Koller, L. R. (1965). "Utraviolet Radiation", 2nd edn. John Wiley & Sons, New York.
Kurpiewski, T. (1981). *Radiat. Curing* **8** (1), 19–27.
Lahl, U., Baetjer, K., Von Dueszeln, J., Gabel, B., Stachel, B. and Thiemann, W. (1981). *Water Res.* **15,** 803–814.
Lawson, D. E. M. (1978). *Proc. 4th Kellog Nutr. Symp.* 91–100; *Chem. Abstr.* **92,** 19280.
Lawson, K., Ansel, R. and Stanton, J. (1980). *Proc. Fifth Radiat. Curing Conf., Boston*, 431–459.

Lee, M. K., Yang, P. Y. and Wynveen, R. A. (1978). US Gov. Rep. LSI-ER-314-7-1; LSI-ER-314-7-5; *Chem. Abstr.* **91,** 216146; **91,** 216145.
Lee, N. E., Jolley, R. L., Denton, M. S. and Thompson, J. E. (1982). *Environ. Int.* **7,** 403–408.
LeFevre, P. H. (1979). SME Tech. Pap. (Ser.) FC79-720.
LeFevre, P. H. (1980). *Proc. Fifth Radiat. Curing Conf., Boston*, 325–335.
Legan, R. W. (1982). *Chem. Eng. (N.Y.)* **89** (2), 95–100.
Leitis, E., Zeff, J. D. and Smith, M. M. (1981). US. Gov. Rep. W81-04656, OWRT-C-90110-R(9436)(1); *Chem. Abstr.* **96,** 91077.
Lenka, S. and Nayak, P. L. (1982). *J. Appl. Polym. Sci.* **27,** 3625–3628.
Levin, R. E. and Parrish, J. A. (1975). *Light Des. Applic.* **5,** 35–43.
Levine, H. A. (1969). *ACS Div. Polym. Chem. Polym. Preprints* **10** (1), 377–384.
Levy, M. (1976). *Double Liaison—Chim. Peint.* **23,** 310–319.
Libutti, D. (1974). *Brewers Dig.* **46** (10), 52–54, 56.
Lightner, D. A., Wooldridge, T. A., Rodgers, S. L. and Norris, R. D. (1980). *Experientia* **36,** 380–382.
Lin, B. J. (1975). *J. Vac. Sci. Technol.* **12,** 1317–1320.
Lipowicz, M. (1981). *Chem. Eng. (N.Y.)* **88** (22), 40–41.
Lorenz, D. H. (1976). SME Tech. Pap. (Ser.) FC76-526.
Lott, A. D. (1976). *J. Oil Colour Chem. Assoc.* **59,** 141–145.
Lucas, G. M., Patsis, A. and Bolon, D. A. (1980). *Radiat. Curing* **7** (1), 4–9.
Lucey, J. F. (1972). *Pediatric Clinics of North America* **19,** 827–839.
Luckiesh, M., Taylor, A. H., Knowles, T. and Leppelmeir, E. T. (1949). *J. Franklin Inst.* **248,** 311–325.
Macur, G. J., Alpaugh, W. A. and Sharkness, J. E. (1980). *Proc. Ind. Waste Conf.* **35,** 809–816. *Chem. Abstr.* **95,** 29750.
McCloskey, C. M. and Bond, J. (1955). *Ind. Eng. Chem.* **47,** 2125–2129.
McDonagh, A. F. (1976). *Birth Defects, Orig. Artic. Ser.* **12** (2), 30–40.
McGinniss, V. D. (1975). *ACS Div. Org. Coat. Plast. Preprints* **35** (1), 118–123.
McGinniss, V. D. (1982). Can. Pat. 1,115,659 (SCM Corp.).
McGinniss, V. D. and Kah, A. (1977). *Polym. Eng. Sci.* **17,** 478–483.
Magdelinskas, M. M. and Wolinski, L. E. (1976). SME Tech. Pap. (Ser.) FC76-532.
Malaiyandi, M., Sadar, M. H., Lee, P. and O'Grady, R. (1980). *Water Res.* **14,** 1131–1135.
Mar, D. M. (1981). US. Pat. 4,273,558 (Envirotech Corp.).
Marcus, R. J., Kent, J. A. and Schenck, G. O. (1962). *Ind. Eng. Chem.* **54** (8), 20–28.
Matsuura, T. and Smith, J. M. (1970). *AIChE J.* **16,** 1064–1071.
Mauk, C. E., Prengle, H. W., Jr and Payne, J. E. (1976). US NTIS AD-A028306. *Chem. Abstr.* **86,** 26901.
Mellor, J. M., Phillips, D. and Salisbury, K. (1974). *Chem. Brit.* **10,** 160–166, 175.
Mesrobian, R. B. (1977). *Radiat. Phys. Chem.* **9,** 307–324.
Miron, J. and Skeist, I. (1978). *Adhesives Age* **21** (1), 35–38.
Mimura, Y., Ohkubo, T., Takeuchi, T. and Sekikawa, K. (1978). *Jap. J. Appl. Phys.* **17,** 541–550.
Miyama, H., Harumiya, N., Ito, Y. and Wakamatsu, S. (1968). *J. Phys. Chem.* **72,** 4700–4701.
Miyama, H., Fukuzawa, K., Harumiya, N., Ito, Y. and Wakamatsu, S. (1969). *J. Phys. Chem.* **73,** 4345–4347.

Moreau, W. and Viswanathan, N. (1975). *ACS Div. Org. Coat. Plast. Preprints* **35** (1), 108–117.
Mosher, M. W. and Bunce, N. J. (1971). *Canad. J. Chem.* **49,** 28–34.
Mullin, C. R. (1968). *In* "Kirk–Othmer Encyclopedia of Chemical Technology", 2nd ed, Vol. 15, pp. 331–354. Wiley–Interscience, New York.
Murray, B. D. (1976). SME Tech. Pap. (Ser.) FC76-498.
Myshkin, A. E. and Sakharov, V. N. (1982). *Russian Chem. Rev.* **51,** 40–50.
Nethercott, J. R. (1978). *Brit. J. Dermatol.* **98,** 541–552.
Niehaus, R. L. (1974). SME Tech. Pap. (Ser.) FC74-508.
Nishikubo, T., Imaura, M., Mizuko, T. and Takaoka, T. (1974). *J. Appl. Polym. Sci.* **18,** 3445–3454.
Noomen, A. (1981). *J. Oil Colour Chem. Assooc.* **64,** 347–354.
North, A. G., Orpwood, J. L. and Little, R. (1976). *J. Oil. Colour Chem. Assoc.* **59,** 9–18.
Nowak, M. T. and Rybny, C. B. (1977). *Amer. Ink Maker* **55** (12), 25, 26, 51.
O'Hara, K. (1978). *Proc. XIVth FATIPEC Congr., Budapest*, 485–498.
Ogiwara, Y., Torikoshi, K. and Kubota, H. (1982). *J. Polym. Sci., Polym. Lett. Ed.* **20,** 17–21.
Ohloff, G., Klein, E. and Schenck, G. O. (1961). *Angew. Chem.* **73,** 578.
Ohtsuka, N. and Seki, T. (1977). Japan Kokai 77 66,872; *Chem. Abstr.* **87,** 172215.
Oster, G. (1969). *In* "Encyclopedia of Polymer Science and Technology", Vol. 10, pp. 145–156. Wiley–Interscience, New York.
Paek, U. C. and Schroeder, C. M. (1981). *Appl. Opt.* **20,** 1230–1233.
Pape, M. (1975). *Pure Appl. Chem.* **41,** 535–558.
Pappas, S. P. and Kuhhirt, W. (1975). *J. Paint Technol.* **47** (610), 42–48.
Parrish, J. A., Fitzpatrick, T. B., Tanenbaum, L. and Pathak, M. A. (1974). *New Engl. J. Med.* **291,** 1207–1211.
Parrish, J. A., Anderson, R. R., Urbach, F. and Pitts D. (1978). "UVA: Biological Effects of Ultraviolet Radiation with Emphasis on Human Responses to Longwave Ultraviolet". John Wiley & Sons, Chichester.
Parrish, M. A. (1977). *J. Oil Colour Chem. Assoc.* **60,** 474–478.
Parsons, B. J. (1980). *Photochem. Photobiol.* **32,** 813–821.
Patel, A. S., Hovnanian, H. P. and Kozar, E. (1976). SME Tech. Pap. (Ser.) FC76-513.
Paulus, A. (1942). *J. Appl. Phys.* **13,** 300–304.
Pelgrims, J. (1978). *J. Oil Colour Chem. Assoc.* **61,** 114–118.
Pelton J. C. (1978). *Flexographic Techn. J.* **3** (1), 20–22.
Perkins, W. C. (1980). *Radiat. Curing* **7** (3), 4, 6, 8, 10, 12, 14, 16, 17, 24.
Peters, W. (1978). *Metalloberflaeche* **32,** 469–474.
Phillips, G. B. and Novak, F. E. (1956). *Appl. Microbiol.* **4,** 95–96.
Pierce, R. J. (1978). *Brit. Printer* **91** (1), 18–19.
Plews, G. and Phillips, R. (1979). *J. Coat. Technol.* **51** (648), 69–77.
Plumridge, R. (1974). *Brit. Printer* **87** (3), 49–51.
Prane, J. W. and Bluestein, C. (1976). SME Tech. Pap. (Ser.) FC76-527.
Prane, J. W. (1980). *Proc. Fifth Radiat. Curing Conf., Boston*, 3–11.
Prengle, H. W., Jr and Mauk, C. E. (1976). *Ozone/Chlorine Dioxide Oxid. Prod. Org. Mater. Proc. Conf.*, 302–320; *Chem. Abstr.* **90,** 126916.
Prengle, H. W., Jr and Mauk, C. E. (1978). *AIChE Symp. Ser.* **178,** 228–244.

Prengle, H. W., Jr, Mauk, C. E., Legan, R. W. and Hewes, C. G., III (1975). *Hydrocarbon Process.* **54** (10), 82–87.
Prengle, H. W., Jr, Mauk, C. E. and Payne, J. E. (1976). *Proc. Forum Ozone Disinfect.* 286–295; *Chem. Abstr.* **87,** 141028.
Reed, C. F. (1936). US Pat. 2,046,090
Reed, D. A. (1978). SME Tech. Pap. (Ser.) FC78-535.
Reed, K. J. (1976). *Point of Sale News* **26** (12), 24, 25, 27, 28.
Reid, J. B. (1978). SME Tech. Pap. (Ser.) FC78-511.
Richtzenhain, H. and Stephan, R. (1971). Ger. Offen. 2,026,671 (Dynamit Nobel AG).
Roffey, C. G. (1982). "Photopolymerization of Surface Coatings", John Wiley and Sons, Chichester.
Rogers, S., Marks, J., Schuster, S., Vella Briffa, D., Warin, A. and Greaves, M. (1979). *Lancet i,* 455–458.
Rosario, E. (1977). *Australian Lithographer, Printer and Packager* **11** (62), 34–35.
Roth, M. and Murphy, J. M. Jr (1979). US Gov. Rep. EPA/600/2-79/129; *Chem. Abstr.* **93,** 79002.
Rowe, W. (1976). SME Tech. Pap. (Ser.) FC76-495.
Russell, A. D. (1982). *In* "Principles and Practice of Disinfection, Preservation and Sterilization" (Eds A. D. Russell, W. B. Hugo and G. A. J. Ayliffe), pp. 534–547. Blackwell Scientific Publications, Oxford.
Rybny, C. B., Trebellas, J. C. and Vona, J. A. (1975). SME Tech. Pap. (Ser.) FC75-304.
Rybny, C. B. (1978). SME Tech. Pap. (Ser.) FC78-553.
Rybny, C. B., Armbruster, D. C. and Vona, J. A. (1978). *Mod. Paint Coatings* **68** (4) 59–65.
Sausville, J. W., Sisson, T. R. C. and Berger, D. (1972). *J. Illum. Eng. Soc.* **1,** 112–118.
Schenck, G. O. (1979). *Parfuem. Kosmet.* **60,** 433–443.
Schenck, G. O. (1981). Eur. Pat. Appl. 27,278; *Chem. Abstr.* **95,** 156292.
Schlef, C. L., Narasimham, P. L. and Oh, S. M (1982). *J. Radiat. Curing* **9** (2), 11–13.
Schmidle, C. J. (1977). *Amer. Ink. Maker* **55** (3), 41, 42, 44, 103.
Schonhorn, H., Kurkjian, C. R., Jaeger, R. E., Vazirani, H. N., Albarino, R. V. and DiMarcello, F. V. (1976). *Appl. Phys. Lett.* **29,** 712–714.
Schonhorn, H., Wang, T. T. and Vazirani, H. N. (1977). *J. Radiat. Curing* **4** (4), 20–21.
Schonhorn, H., Torza, S., Albarino, R. V., Vazirani, H. N. and Wang, T. T. (1979). *J. Appl. Polym. Sci.* **23,** 75–84.
Schroeter, S. H. and Moore, J. E. (1973). *In* "Nonpolluting Coatings and Coating Processes" (Eds J. L. Gardon and J. W. Prane), pp. 135–147. Plenum, New York.
Schroeter, S. H., Bolon, D. A. and Lucas, G. M. (1978). US Pat. 4,076,489 (General Electric Co.).
Schwarcz, A. (1976). SME Tech. Pap. (Ser.) FC76-497.
Shearon, W. H., Jr., Hall, H. E. and Stevens, J. E., Jr (1949). *Ind. Eng. Chem.* **41,** 1812–1820.
Siddiqui, S. A., McGee, K., Lu, W. C., Alger, K. and Needles, H. L. (1981). *Amer. Dyest. Rep.* **70** (7), 20–22.

Sisson, T. R. C. (1976). *In* "Photochemical and Photobiological Reviews," Vol. 1 (Ed. K. C. Smith), pp. 241–268. Plenum Press, New York.
Sites, B. L. and Agruss, M. S. (1962). US Pat. 3,041,203, US Pat. 3,047,422, US Pat. 3,050,413, US Pat. 3,051,591, US Pat. 3,052,568 (all assigned to Miehle-Goss-Dexter Inc.).
Song, P.-S. and Tapley, K. J., Jr (1979). *Photochem. Photobiol.* **29,** 1177–1197.
Sowell, R. R., Cuthrell, R. E., Mattox, D. M. and Bland, R. D. (1974). *J. Vac. Sci. Technol.* **11,** 474–475.
Stern, R. S., Thibodeau, L. A., Kleinerman, R. A., Parrish, J. A. and Fitzpatrick, T. B. (1979). *New Engl. J. Med.* **300,** 809–813.
Steuben, K. C. (1977). *Adhesives Age* **20** (6), 16–21.
Stevens, J. M. and Keough, A. (1978). SME Tech. Pap. (Ser.) FC78-551.
Strand, R. C. (1976). SME Tech. Pap. (Ser.) FC76-507.
Süs, O. (1944). *Justus Liebigs Ann. Chem.* **556,** 65–84.
Tabb, P. A., Savage, D. C. L., Inglis, J. and Walker, C. H. M. (1972). *Lancet ii,* 1211–1213.
Taubenest, R. and Ubersax, H. (1980). *Solid State Technol.* **23** (6), 74–79.
Thorington, L., Cunningham, L. and Parascandola, J. (1971). *Illum. Eng.* **66,** 240–250.
Ultraviolet Products Inc. (1977). Ultraviolet Products Application Bulletin A109.
Tins, J. A. (1982). *Radiat. Curing* **9** (3), 2, 4–6.
Tozier, J. E. and Shaffer, J. W. (1982). US Pat. 4,341,817 (GTE Products Corp).
Ury, M. G. and Eastlund, B. J. (1977). *Proc. IAS Annual Meeting, Los Angeles,* 357–362; *EEA* **81,** 13041.
Vanderhoff, J. W. (1975). *ACS Div. Org. Coat. Plast. Preprints* **35** (2), 406–413.
Vanderhoff, J. W. (1976). SME Tech. Pap. (Ser.) FC76-483.
Vaughan, W. E. and Rust, F. F. (1942). *J. Org. Chem.* **7,** 472–476.
Vazirani, H. N., Schonhorn, H. and Wang, T. T. (1977). *J. Radiat. Curing* **4** (4), 18–19.
Vig, J. R. and LeBus, J. W. (1976). *IEEE Trans. Parts. Hybrids Packag.* **PHP12,** 365–370.
Vrancken, A. (1977). *Farbe Lack* **83,** 171–179.
Vrancken, A. (1980). *Proc. Fifth Radiat. Curing Conf., Boston,* 13–26.
Wanmaker, W. L., van Gils, J. F. and ter Vrugt, J. W. (1975). *Light. Res. Technol.* **7,** 19–22.
Watanabe, K. (1978). *Graphic Arts Japan* **20,** 14–18.
Wayne, R. P. (1970). "Photochemistry". Butterworths, London.
Wendt, G. L. (1948). US Pat. 2,453,769, US Pat. 2,453,770 (Sun Chemical Corp.).
Wicks, Z. W. (1974). SME Tech. Pap. (Ser.) FC74-501.
Wicks, Z. W. and Kuhhirt, W. (1975). *J. Paint Technol.* **47** (610), 49–59.
Williams, C. H. (1977). *Brit. Printer* **90** (3), 25–26
Williams, C. H. (1981). Coates Inklings, (118).
Williams, J. L. R. and Molaire, M. F. (1982). *In* "Kirk–Othmer Encyclopedia of Chemical Technology", 3rd edn, Vol. 17, pp. 680–708. Wiley-Interscience, New York.
Wittels, N. D. (1980). *In* "Fine Line Lithography" (Ed. R. Newman) pp. 1–104. North Holland, Amsterdam.
Woods, B. (1968). *Trans. Inst. Met. Finishing* **46,** Printed Circuit Supplement, 131–136.

Yick, P. J. (1977). *Canad. Printer Publisher* **86** (11), 11–14.
Young, S. E. (1976). *Progr. Org. Coatings* **4,** 225–249.
Younger, J. R. (1976). *J. Oil Colour Chem. Assoc.* **59,** 197–201.
Zeff, J. D., Wark, J., Shuman, R., Farrell, F. C. and Alhadeff, E. S. (1976). US NTIS AD-AO38609; *Chem. Abstr.* **87,** 122318.

2 The Nature of Light

2.1 Early Concepts

Early experiments in optics led, in the 17th century, to two apparently opposing concepts of the nature of light. These were the corpuscular theory of Newton and the wave theory of Huyghens. The corpuscular theory was the more generally accepted until the end of the 18th century. During the early part of the 19th century the investigation of interference and diffraction phenomena, which could be explained only in terms of a wave theory of light, led to a temporary eclipse of corpuscular theory. Wave theory received further support in the 1860s, when Maxwell developed his electromagnetic theory of light, in which light is considered to have an electric field vector and a magnetic field vector at right angles to each other, both being transverse to the direction of propagation. Newton had shown that white light could be separated into its constitutent colours by a prism, and Herschel in 1800 and Ritter in 1801 had shown that solar radiation has components which extend beyond both ends of the region visible to the eye. Visible light is only a very small part of an electromagnetic spectrum which stretches from very short wavelength cosmic rays to very long wavelength radio waves (Fig. 2.1).

2.2 Planck's Radiation Law

During the latter part of the 19th century there was much interest in the radiation emitted by hot bodies, and in particular by a "black body"—i.e. one which absorbs all radiation falling on it, reflecting none and trans-

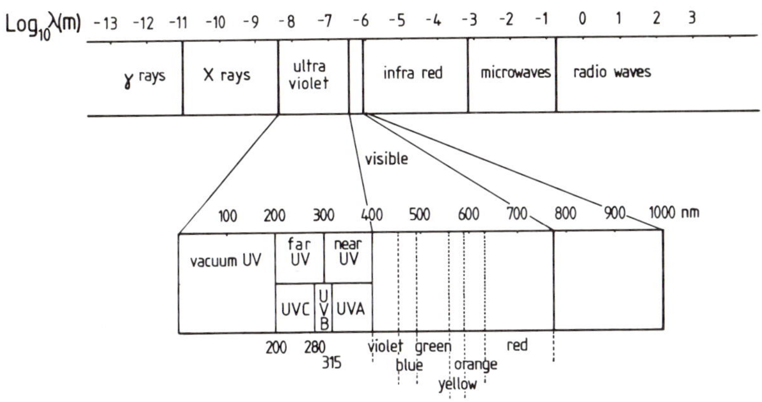

Fig. 2.1. The electromagnetic spectrum. The range of wavelengths between 100 and 400 nm has been divided into the regions UVA (315–400 nm), UVB (280–315 nm) and UVC (100–280 nm) by the Commission Internationale de l'Eclairage (1970).

mitting none. It was possible to predict from Maxwell's classical electromagnetic theory what the spectral distribution of thermal radiation should be. Unfortunately the predictions did not agree with the results of experiment.

The solution came from the revolutionary theory of Planck, who in 1900 postulated that a body radiating light of frequency ν would radiate energy only in multiples of $h\nu$, where h is a constant known as Planck's constant ($6{\cdot}626 \times 10^{-34}$ J s). Likewise a body can absorb radiation only in multiples of $h\nu$. By postulating this, and by using the Boltzmann energy distribution equation, Planck was able to derive a spectral distribution function for the radiation from a black body which agrees well with experiment (see Chapter 5).

Planck's quantum theory appeared at a time when phenomena were being discovered, for example the photoelectric effect, which could be explained only by a corpuscular theory of light. Einstein in 1905 extended Planck's theory and postulated that electromagnetic radiation itself is quantized and consists of discrete light quanta (photons) each of value $h\nu$. This neatly reconciled the previously paradoxical dual nature of light. Some years later, in 1923, de Broglie realized that matter in general has a dual nature and that any moving body has a wavelength associated with it, this being related to its momentum:

$$\lambda = h/mv.$$

2.3 Bohr's Theory

The concept of the atom as a dense positively charged nucleus surrounded by orbiting electrons is due to Rutherford. This model is not compatible with classical electromagnetic theory, because an orbiting electron has an acceleration towards the nucleus, and therefore should radiate, so losing energy, and spiral towards the nucleus. Bohr, in 1913, refined the Rutherford model by introducing the concept of discrete orbits (stationary states) in which no radiative loss would take place. This led to the Bohr theory of the hydrogen atom. The single electron could occupy certain discrete energy levels. Transition between a higher level (E_2) and a lower level (E_1) would result in a quantum of radiation being emitted, with a frequency given by

$$E_2 - E_1 = h\nu.$$

Bohr postulated that the energy of each discrete state is determined by a principal quantum number n. The energy of the electron in each state is a simple function of n, n having values 1, 2, 3, . . ., etc. The triumph of the Bohr theory lay in the prediction of the atomic spectrum of the hydrogen atom, and the explanation of the various empirical expressions which had been derived to link the wavelengths measured.

The Bohr theory was very successful in its treatment of the hydrogen atom, and hydrogen-like ions (e.g. He^+, Li^{++}), but was not capable of dealing with atoms with more than one electron.

2.4 The Schrödinger Equation

The Bohr theory thinks of the electron as a particle moving in a precise orbit with a precise momentum. De Broglie showed, however, that an electron in motion has a dual nature, with some of the properties of a wave. Schrödinger in 1926 combined the de Broglie equation with the classical equation for harmonic motion to give the Schrödinger equation for the hydrogen atom:

$$\frac{\delta\psi^2}{\delta x^2} + \frac{\delta\psi^2}{\delta y^2} + \frac{\delta\psi^2}{\delta z^2} + \frac{8\pi^2\mu(E + e^2/r)}{h^2}\psi = 0.$$

This equation is the basis of wave mechanics. ψ is called the wave function, and has the significance that $\psi^2\delta\tau$ is the probability of finding the electron in a small volume element in the atom $\delta\tau$, situated at space coordinates x, y

and z. μ is the reduced mass of the nucleus and electron. E is the total energy, and e^2/r the Coulombic energy of interaction, i.e. the potential energy.

One can place, *a priori*, certain restrictions on ψ, based on the significance of ψ^2. It must be a continuous function; it must be finite; it must be single valued; and it must be 0 when $r = \infty$. When the second-order differential equation is solved for ψ, it is found that valid solutions, i.e. ones which meet the restrictions, occur only at certain values of the total energy E, called "eigenvalues". The calculated eigenvalues for the electron in the hydrogen atom correspond exactly to the stationary state energies calculated from the Bohr theory. However, the principal quantum number n which appears does so not as a result of an arbitrary assumption, but as a natural result of the equation of wave motion, the integer n being related to the number of nodes in the wave structure.

2.5 Quantum Numbers

The solution of the Schrödinger equation for the hydrogen atom requires not just one, but three quantum numbers to describe the possible orbitals. These are n, l and m.

n, the principal quantum number, can have values of 1, 2, 3, etc., and can be thought of as defining the mean distance of the electron from the nucleus.

l and m arise from the quantization of the angular momentum of the electron. The orbital angular momentum, according to wave mechanics, is a vector quantity which is quantized and can have the values $\sqrt{l(l+1)}h/2\pi$, where l, the azimuthal quantum number, can have values of 0, 1, 2, . . ., $(n-1)$. l defines the shape of the orbital. When n is 1, l can only be 0. This indicates that the total orbital angular momentum is zero, which implies that the orbital must be spherically symmetrical about the nucleus. When $n = 2$, l can be 0 or 1. When $l \neq 0$, the orbital is asymmetric, and this causes a significant change in the total energy of the atom. The values of l give rise to the familiar s, p, d and f orbitals. A major contribution of wave mechanics has been to determine the shapes of these orbitals—these shapes being at the base of modern concepts of chemical bonding.

If the total angular momentum is not zero, i.e. $l \neq 0$, the atom will have a magnetic moment given by the expression

$$\sqrt{l(l+1)}eh/4\pi m_e c.$$

If the atom is placed in a strong magnetic field, the interaction of the field with the magnetic moment of the electron will cause the orbital angular momentum vector to precess about the direction of the external field. However, only certain orientations of the orbital angular momentum vector relative to the external field are allowed. The component in the direction of the external field is quantized and must be an integral number of times $h/2\pi$, i.e. $mh/2\pi$, where m is the magnetic quantum number. m may have integer values between $-l$ and $+l$. m can be thought of as defining the orientation of the orbit relative to some chosen axis. In the absence of an external field orbits having the same values of n and l, but different values of m, are degenerate (i.e. they have the same energy).

The three quantum numbers which arise from the Schrödinger equation are not sufficient to explain the fine structure of atomic spectra. To do this it is necessary to postulate that the electron spins about its own axis, so contributing to the total angular momentum of the system. This contribution is quantized and can have values $\sqrt{s(s+1)}h/2\pi$ where $s = \pm\frac{1}{2}$. s is the fourth quantum number, the spin quantum number. Each orbital as determined by n, l and m can accommodate two electrons, one having $s = \frac{1}{2}$ and the other $s = -\frac{1}{2}$. In the absence of an external field the energies are the same.

2.6 Atoms with Many Electrons

The application of the Schrödinger equation becomes difficult for atoms containing many electrons, and approximation methods are required. The assignation of electrons to available orbitals follows simple rules. The Pauli exclusion principle states that no two electrons in an atom may have the same four quantum numbers. Electrons having the same value of n are said to occupy the same shell. The maximum number of electrons which can occupy the same shell is $2n^2$. The orbitals are normally filled up to give the minimum total energy, i.e. low energy orbitals are occupied before high energy orbitals. Orbitals with the same values of n and l are occupied each with a single electron before any pairing takes place—so maintaining the maximum number of unpaired electrons (Hund's rule).

2.7 Coupling of Momenta

In an atom containing many electrons, the total orbital angular momentum for each electron may add vectorially to that of others. The situation is fortunately simplified by the fact that the sum for all the electrons in fully occupied shells is zero, and therefore only electrons in incomplete shells need be considered. The total orbital angular momentum of the atom is $\sqrt{L(L+1)}h/2\pi$, where L is the vector sum of the l values of the individual electrons. L is restricted by quantum conditions to integral values. Similarly the spin quantum numbers of the individual electrons may combine to give a sum S, which is also quantized, and subject to the restriction that if there is an odd number of electrons S must be an odd multiple of $\frac{1}{2}$, and if there is an even number of electrons S must be an integer. The total spin angular momentum is $\sqrt{S(S+1)}h/2\pi$.

An interaction or coupling takes place between the net orbital and net spin angular momenta. In most cases this is of the Russell–Saunders or L–S type. The L and S vectors combine to form a resultant J, with the restriction that if S is an integer J must be an integer, and if S is an odd multiple of $\frac{1}{2}$, J must also be an odd multiple of $\frac{1}{2}$. The total angular momentum of the atom is then $\sqrt{J(J+1)}h/2\pi$.

2.8 Term Symbols

The emission of a photon by an atom is a result of a change in the electronic state of an atom. The energy of a given electronic state is determined by n, L, S and J for the atom. These quantities are commonly combined in a "term symbol" which has the form

$$n^{(2S+1)}L_J.$$

Note that n is the principal quantum number for the electron about to undergo a transition. $2S+1$ is known as the "multiplicity".

Only certain transitions are possible. These are determined by selection rules:

(1) There are no restrictions on changes in n.
(2) $\Delta L = \Delta l = \pm 1$ when there is only one optical electron; $\Delta L = 0$ or ± 1 when there is more than one optical electron.
(3) $\Delta J = 0$ or ± 1, except that $J = 0 \nrightarrow J = 0$.
(4) $\Delta S = 0$.

Both these selection rules and the form of the term symbols are strictly relevant only for Russell–Saunders coupling. In heavy elements, such as mercury, a different form of coupling, known as spin–orbit or jj coupling predominates. In this case the orbital and spin angular momenta for each individual electron couple to form a resultant j, and then the individual j

values couple to form a resultant J. L and S are therefore not strictly meaningful in this case. However, the form of the term symbols is still, by convention, maintained as before.

2.9 Energy Level Diagrams

The main reason for this brief and rather superficial treatment of atomic physics is to allow the reader to understand the energy level or Grotrian diagrams for the atoms which are important in artificial radiation sources. Electronic transitions in atoms are the predominant source of useful UV radiation, and energy level diagrams are the key to understanding the wavelengths emitted, particularly by gas discharges.

The energy level diagram for mercury is shown in Fig. 8.22. Note that the electronic states are divided into singlets ($S = 0$) and triplets ($S = 1$). This is because there are two electrons in the valence shell. In contrast, sodium, with a single valence electron, has a simpler energy level diagram, with all states being doublets. The transitions which give rise to some of the more important wavelengths emitted by mercury lamps are shown on the diagram and it will be noticed that some of these are forbidden by the Russell–Saunders selection rules. Certain electronic states exist in atoms from which no radiative transition is possible—these are known as "metastable states".

Although we are mainly concerned with the energy levels in neutral atoms, in certain circumstances, such as in a gas discharge with a very high plasma temperature, there may be a substantial concentration of ions and radiation will be emitted as a result of electronic transitions within the ions. The energy levels in an ion will probably be quite different from those in the parent neutral atom. Hg^+ has only one valence electron and therefore all the states are doublets. The spectrum of neutral mercury is by convention referred to as Hg(I), that of singly ionized mercury Hg(II), that of doubly ionized mercury Hg(III), etc.

2.10 Mechanisms of Photon Production

There are three important mechanisms by which photons in the UV and visible are produced:

(1) Radiative transitions, spontaneous or stimulated, between electronic energy levels in atoms and molecules.

(2) Two-body radiative recombination (mainly electron/ion).
(3) Electron retardation (*Bremsstrahlung*)—when an electron is caused to deviate from its normal path by, for instance, passing close to a positive ion, classical electromagnetic theory requires that radiation be emitted. This is also the source of synchrotron radiation—produced by electrons travelling at high speed in circular paths.

The first source of artificial light was undoubtedly fire, i.e. the chemical combustion of organic matter. The production of light in the combustion process is complex because although some may arise from excited species directly formed by the chemical reactions taking place, much is a result of exothermicity which causes incandescence. The generation of light directly by chemical reactions is fairly uncommon and is known as "chemiluminescence". When this happens in nature, as in the firefly, it is known as "bioluminescence". When the energy is supplied by light which is absorbed by the body, the re-emission of the absorbed energy may be called "photoluminescence". Fluorescence and phosphorescence are forms of photoluminescence. Dixon and Hughes (1975) obtained a measurable light output from a Pyrex sphere containing a mixture of mercury and nitrogen simply by rotating it. This is an example of "triboluminescence". Changes of state are sometimes accompanied by light emission, e.g. "crystallo-luminescence". If a body emits light simply because it is at an elevated temperature, the light is known as "temperature radiation".

For most viable light sources the source of energy is electricity. Whenever electricity is the energy source, the primary process is the acceleration of charged particles, particularly electrons, in an electric field.

2.11 Spontaneous and Stimulated Emission

Let us consider two energy states for an atom, E_1 and E_2. E_2 is at a higher level than E_1. There are three possible processes involving radiation which can connect these two levels:

(1) If an atom in state E_1 is placed in a radiation field of wavelength λ such that $E_2 - E_1 = hc/\lambda$, then there is the possibility that the atom will absorb a photon and be excited to level E_2. This is simple *absorption*.
(2) If an atom in state E_2 is placed in a radiation field of wavelength λ such that $E_2 - E_1 = hc/\lambda$, then there is the possibility that a photon will stimulate the atom to emit a photon, coherent with the

impinging photon, and drop to level E_1. This is *stimulated* or *induced emission*.

(3) An atom in state E_2 can spontaneously emit a photon and drop to level E_1. This is *spontaneous emission*.

If E_2 and E_1 are separated by several electron volts, and a collection of atoms is in thermal equilibrium, then the Boltzmann distribution law indicates that the number of atoms in level E_1 will greatly exceed the number in level E_2. This means that in normal circumstances an impinging photon is more likely to be absorbed than to cause stimulated emission. Also, the lifetimes of non-metastable states are so short that unless the photon flux density is very high, atoms in level E_2 are more likely to lose energy by spontaneous emission than by stimulated emission. However, stimulated emission is the source of laser radiation. The conditions for stimulated emission are made favourable by arranging a *population inversion*. Stimulated emission will be discussed in greater detail in Section 12.1.

2.12 Units

Useful units of wavelength (λ) are the nanometre (nm) = 10^{-9} m and the micrometre (μm) = 10^{-6} m. In the literature you may also come across other units, including the ångstrom (Å) = 10^{-10} m, the millimicron (mμ) = 10^{-9} m and the micron (μ) = 10^{-6} m. These last three units are not recommended SI units.

Frequency may be expressed in hertz (cycles per second) or wavenumber, $1/\lambda$ cm^{-1}. Wavenumber is sometimes given the unit kayser (K); this is not an SI unit.

The basic unit of energy is the joule (J).

The energy of a photon is given by $E = h\nu = hc/\lambda$.

Planck's constant $h = 6{\cdot}626 \times 10^{-34}$ J s.

The velocity of light (c) = $2{\cdot}998 \times 10^{8}$ m s^{-1}.

Therefore, E (per photon) = $1{\cdot}986 \times 10^{-16}/\lambda$ J if λ is expressed in nanometres.

A "mole" of photons, i.e. Avogadro's number ($6{\cdot}022 \times 10^{23}$) of photons, is known as an "einstein". Therefore, E (per einstein) = $1{\cdot}196 \times 10^{8}/\lambda$ J, if λ is expressed in nanometres.

One joule is the energy of $5{\cdot}034 \times 10^{15}\lambda$ photons (λ in nanometres). Therefore one watt is equivalent to $5{\cdot}034 \times 10^{15}\lambda$ photons, or quanta, per second.

An *electron volt* is the kinetic energy which an electron acquires when it is accelerated in an electric field through a potential difference of one volt. One electron volt is equivalent to $1{\cdot}602 \times 10^{-19}$ J.

References

Commission Internationale de l'Eclairage (1970). "International Lighting Vocabulary", CIE, Paris.
Dixon, A. J. and Hughes, J. R. (1975). *J. Phys. E* **8,** 425–426.

3 Photochemistry and Photopolymerization

3.1 Basic Concepts of Photochemistry

Photochemistry is the branch of chemistry concerned with the action of light (mainly visible and ultraviolet) on chemical systems. It is essentially the chemistry of atoms and molecules in *electronically* excited states.†

3.1.1 The Grotthus–Draper and Stark–Einstein laws

There are two basic laws of photochemistry. The first, known as the *Grotthus–Draper law*, states that only light which is *absorbed* is capable of initiating a photochemical process. The second, the *Stark–Einstein law*, is derived from quantum theory and states that when an atom or molecule undergoes a photochemical change it is as a result of the absorption by the atom or molecule of a single photon. Because excited states normally have very limited lifetimes, it is seldom that an atom or molecule in an excited state will have time to absorb a further photon to give a still higher energy state. However, there is evidence that this may happen under conditions of intense irradiation. It is even possible for a molecule to absorb two photons simultaneously (biphotonic absorption) under extreme conditions (Wayne, 1970, pp. 3, 154). Despite these apparent deviations, the essential point of the law remains—molecules absorb radiation in discrete quanta, and the number of excited molecules produced is proportional to the number of photons absorbed.

† But see Section 12.5.3.

3.1.2 Primary and secondary processes; reciprocity and reciprocity failure

The absorption of light by molecules and the formation of excited states lies in the province of spectroscopy. Photochemistry is concerned primarily with the fate of the excited state once formed, i.e. with the nature of the process

$$A^* (+ B) \longrightarrow C + D + \ldots$$

where A^* is a molecule in an electronically excited state, and C, D, etc., are products of the *primary* photochemical reaction. The mechanism by which C, D, etc., are formed may involve A^* reacting with a ground state molecule B. The process shown above falls into the category "excitation/chemical change". It is likely that there will be competitive photophysical processes by which A^* returns to the ground state. In general the lifetimes of excited states are short, and the photon flux densities available from technologically important incoherent light sources are sufficiently low, to ensure that the instantaneous concentration of excited molecules is small, so that reactions *between* excited molecules are unlikely.

In some branches of photochemistry, notably those concerned with photography and photobiology, there is a third law quoted known as the *Bunsen–Roscoe law*, or the *law of reciprocity* (Kosar, 1965, p. 5; Harm, 1980, p. 15). This states that the quantity of the reaction products of a photochemical reaction is proportional to the product of the intensity of the incident radiation and the time of exposure. The law is based firstly on the assumption that the total number of excited molecules formed is proportional to the integrated irradiance, i.e. the incident dose, and secondly on the assumption that the rate of the photochemical process is proportional to the concentration of excited molecules. It is found in practice that these assumptions are not always fully justified. Possible reasons are as follows:

(1) The relationship between *incident* intensity and *absorbed* intensity may vary during a reaction if the products absorb radiation and compete with the reactant molecule for radiation. This is known as the "*inner filter effect*". Although the incident intensity may remain constant, the rate of excitation of reactant falls. This is a problem to avoid if possible when using chemical dosimeters and actinometers (see Chapter 4).

(2) Sometimes the primary photochemical reaction, the rate of which may be proportional to the concentration of excited molecules, is in competition with another process which has a different dependence on the concentration. An example is the influence of atmospheric

> oxygen on excited photoinitiator molecules in thin polymerizing coatings. Many photoinitiators produce radicals from a triplet state, and the process by which radicals are produced must compete with the action of any triplet state quencher which is present. Oxygen is a triplet state quencher and in this case the conditions are such that the rate at which quenching takes place is more dependent on the rate at which oxygen can diffuse to and into the surface, than on the concentration of triplet photoinitiator. Production of radicals is therefore favoured by a high intensity of irradiation.

A practical implication of the Bunsen–Roscoe law is that the quantity of the reaction products formed is dose dependent, i.e. a high intensity of irradiation for a short period gives the same result as a lower intensity for a longer period, and continuous irradiation gives the same result as intermittent irradiation, as long as the irradiance integrated with respect to time remains constant. If this aspect of the Bunsen–Roscoe law is not obeyed, the system is said to exhibit "*reciprocity failure*".

In many situations of technological importance, the *overall* change taking place does not obey the law of reciprocity because the products of the primary photochemical reaction act as intermediates which engage in *secondary non-photochemical* processes to produce final products. There may be a number of competing secondary reactions and the *ratio* of final products may depend on the concentration of products of the primary reaction. A classic example lies in the kinetics of photoinitiated free radical polymerization, which will be discussed in Section 3.2.4. Formation of polymer (in the absence of oxygen) is favoured by a low intensity of irradiation, while formation of radical termination products is favoured by a high intensity of irradiation.

Although reciprocity and reciprocity failure is a subject ignored by the authors of many of the standard texts on photochemistry, it is of great relevance to the design of equipment for the technological application of photochemical reactions, and in particular to the choice of radiation source.

3.1.3 Absorption of light; the Beer–Lambert law

The absorption of monochromatic radiation by a homogeneous substance (e.g. molecules in solution) is governed by the Beer–Lambert law, which states that when a beam of incident intensity I_0 passes through a layer of thickness d, its intensity is reduced to I, where

$$I = I_0 10^{-\varepsilon c d}.$$

ε is the extinction coefficient, and c the concentration of the absorbing species. If d is in centimetres, and c is the *molar* concentration (in moles per litre), ε is the *molar extinction coefficient*, and has the units litres per mole-centimetre.

The *absorbance* or *optical density* of a layer is defined as

$$A = \log_{10} I_0/I = \varepsilon cd.$$

The *transmittance* T is defined as I/I_0, and therefore

$$T = 10^{-A}, \qquad A = -\log T.$$

The Beer–Lambert law may be derived theoretically if it is assumed that the rate of loss of photons is proportional to the rate of bimolecular collisions between photons and the absorbing species (Wayne, 1970, p. 25). If there is no molecular interaction, ε is independent of c, but varies with wavelength, temperature and solvent.

If there is more than one absorbing species, the total absorbance is the sum of the absorbances of the individual components i.e. $A = \sum \varepsilon cd$, provided that the components do not interact.

If the material is heterogeneous, and there are discrete phases with different refractive indices, reflection and scattering will complicate matters.

3.1.4 *Molecular excitation and its consequences*

3.1.4.1 Molecular orbitals

Molecules are formed by the joining of atoms by chemical bonds. The simplest description of bond formation and the electronic structure of molecules is given by *molecular orbital theory*. Molecular orbitals are derived from the *linear combination of atomic orbitals* (LCAO). It is assumed that only valence shell electrons are involved, and that inner shell electrons remain in their atomic orbitals. When a single bond is formed between two atoms, the atomic orbitals of the individual atoms combine to form a molecular orbital which embraces both nuclei and accommodates both electrons. When two atomic orbitals A and B represented by wave functions ψ_A and ψ_B combine, two different molecular orbitals are possible, represented by wave functions ψ_1 and ψ_2, depending on the method of combination: either

$$\psi_1 = \psi_A + \psi_B,$$

in which case the atomic orbitals are combined *in phase*, or

$$\psi_2 = \psi_A - \psi_B,$$

and the atomic orbitals are combined *out of phase*. For efficient combination the individual atomic orbitals must have similar symmetry, comparable energies and the ability to overlap. The in-phase combination ψ_1 gives a molecular orbital which has an energy less than that of the atomic orbitals, and this molecular orbital is known as a *bonding* orbital. Conversely, the out-of-phase combination ψ_2 gives a molecular orbital which has an energy higher than that of the atomic orbitals, and this is known as an *antibonding* orbital. Normally when a bond is formed the two electrons from the atomic orbitals occupy the bonding molecular orbital, and the internuclear distance adjusts to minimize the energy. The total number of molecular orbitals formed is equal to the number of atomic orbitals participating, but normally only the bonding orbitals are occupied, each by a pair of electrons having opposite spins.

When carbon atoms are incorporated in molecules it is normal for all the valence electrons to be used in bond formation. In the case of elements in Groups VA, VIA and VIIA of the periodic table there are "lone pairs" of electrons which do not participate. These non-bonding electrons are known as "n" electrons.

If the atomic orbitals which combine to form a molecular orbital are symmetrical about the internuclear axis, the molecular orbital is known as a σ type if bonding, and σ^* if antibonding. Atomic orbitals of the s type always combine in this way, and p orbitals can do so. In the case of carbon, the hybridization of s and p orbitals to give sp, sp^2 and sp^3 hybrids yields atomic orbitals which again combine to give σ and σ^* types. σ-type molecular orbitals provide the main framework of organic molecules. The other types of molecular orbital which are important in organic chemistry are those produced by the lateral overlap of two parallel p orbitals, and these are called π and π^* molecular orbitals. The ways in which σ, σ^*, π and π^* orbitals are formed are shown in Fig. 3.1. Note that when a molecular orbital is antibonding, there is a node between the nuclei. The node represents a point where the wave function is zero, and therefore where the probability of finding an electron is also zero. A single bond, for instance between two carbon atoms, is formed by a σ orbital, while a double bond is a combination of a σ orbital plus one π orbital, and a triple bond is a combination of a σ orbital and two π orbitals. Electrons in σ orbitals can be considered to be localized to the volume between and around the nuclei of the atoms forming the bond. When double bonds are conjugated, as in the benzene ring, the π orbitals merge, and therefore electrons in the π orbitals are delocalized and can move freely around the conjugated π system.

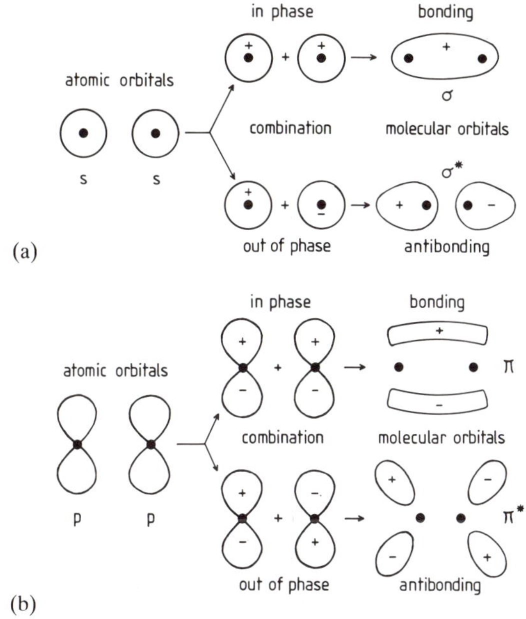

Fig. 3.1. (a) Combination of s atomic orbitals to form σ (bonding) and σ* (antibonding) molecular orbitals. (b) Combination of p atomic orbitals to form π (bonding) and π* (antibonding) molecular orbitals.

3.1.4.2 Electronic excitation in molecules

Electronic excitation takes place when an electron in a bonding molecular orbital, or an n atomic orbital, is raised into an antibonding orbital. In most organic molecules σ orbitals have the lowest energy, with the energies of π, π* and σ* orbitals being progressively higher. The energy of π orbitals is strongly influenced by the degree of conjugation. The energy of n orbitals may lie above or below that of the π orbitals. The most common transitions are $n \rightarrow \pi^*$, $\pi \rightarrow \pi^*$, $n \rightarrow \sigma^*$ and $\sigma \rightarrow \sigma^*$. Depending on which transition takes place, a variety of molecular excited states is possible. An excited state derived from an $n \rightarrow \pi^*$ transition is known as an n, π* state, that from a $\pi \rightarrow \pi^*$ transition as a π, π* state, and so on. The excited molecule contains two unpaired electrons. If the spins are opposed the state has a multiplicity of one, and is a *singlet*, while if the spins are parallel the state has a multiplicity of three and is a *triplet*. In almost all organic molecules the ground state is a singlet, and therefore the various singlet states can be designated, in order of increasing energy, S_0, S_1, S_2, etc., while the equivalent triplet states are T_1, T_2, etc. Triplet states have lower energies than the equivalent singlet states.

Electronic energy is not the only type of quantized energy possessed by a molecule. Unlike an atom, molecules can have vibrational and rotational energy, and it is possible that when a transition takes place there is a change not only in the electronic state but also in the vibrational and rotational states. This explains why the absorption spectra of molecules consist of bands rather than lines.

3.1.4.3 The Jablonski diagram

Transitions between energy states in molecules may be shown on an energy level or "Jablonski" diagram (Jablonski, 1935), a schematic example of which is shown in Fig. 3.2. Note that each electronic energy state has a

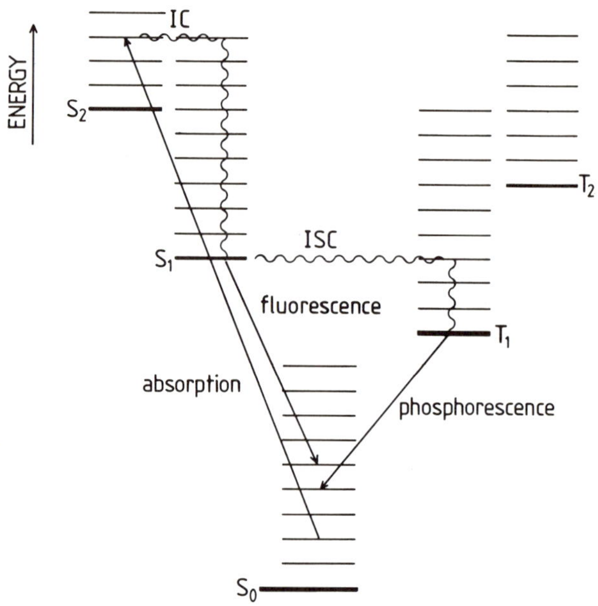

Fig. 3.2. The Jablonski diagram. This shows the singlet and triplet levels, each of which has many vibrational levels associated with it, and the possible transitions. IC, internal conversion; ISC, intersystem crossing; wavy lines represent non-radiative transitions; straight lines represent radiative transitions.

series of vibrational states associated with it. Transitions are of two main types, those which involve the absorption or emission of radiation, and those which are radiationless. Radiationless electronic transitions involve moving to a lower electronic energy state at a higher vibrational level. Excess vibrational energy can be readily dissipated to other molecules. A radiationless transition between states of the same multiplicity is known as "*internal conversion*" (IC), while when a change in multiplicity occurs the radiationless transition is known as "*intersystem crossing*" (ISC). ISC is favoured if the singlet state is n, π^*, as the energy gap between singlet and triplet is then small (Porter and Suppan, 1964).

As with atomic orbitals there are selection rules which govern the possible transitions between excited states. Perhaps the most significant selection rule is that transitions which involve the absorption or emission of radiation can only take place between states of the same multiplicity. This implies that for almost all organic molecules the immediate result of absorption is the formation of an excited singlet state. If this state is S_2, S_3, etc., it will in all probability decay rapidly to either the lowest excited singlet state (S_1) or the lowest triplet state (T_1) by internal conversion and/or intersystem crossing (Kasha, 1950). Therefore, the primary photochemical or photophysical process by which the molecule loses its energy almost invariably occurs from one or other of these lowest excited states. This explains why the nature of the primary process(es) is not dependent on the initial excited state produced and therefore is independent of the wavelength of the light used to excite the molecule.

Process	First-order rate constant (s^{-1})	
	General range	Benzophenone (in benzene)
Absorption	10^{15}–10^{16}, very fast	—
$S_2 \xrightarrow{IC} S_1$	10^{11}–10^{14}, very fast	10^{12}
$S_1 \xrightarrow{IC} S_0$	10^{6}–10^{12} } in competition	—
$S_1 \xrightarrow{\text{fluorescence}} S_0$	10^{6}–10^{9} } in competition	10^{6}
$S_1 \xrightarrow{ISC} T_1$	10^{4}–10^{12} } in competition	10^{10}
$T_1 \xrightarrow{ISC} S_0$	10^{-1}–10^{5}, slow	10^{5}
$T_1 \xrightarrow{\text{phosphorescence}} S_0$	10^{-2}–10^{4}, slow	10^{2}
Vibrational relaxation	10^{12}, very fast	10^{12}

The extinction coefficient for the absorption of radiation depends on the nature of the transition. A large extinction coefficient is favoured by the two participating orbitals occupying the same region of space. Thus large extinction coefficients are associated with $\pi \rightarrow \pi^*$ and $\sigma \rightarrow \sigma^*$ transitions, and small extinction coefficients with $n \rightarrow \pi^*$ transitions.

The excited state S_1 of an isolated molecule has a very short lifetime because the process $S_1 \rightarrow S_0 + h\nu$ is facile. This process is fluorescence. A chemical reaction of S_1, or indeed intersystem crossing, needs to be very rapid to compete with this radiative decay mechanism. The process $T_1 \rightarrow S_0 + h\nu$, on the other hand, is forbidden, or to be more precise has a low probability. The lowest triplet state therefore has a relatively long lifetime, because the radiative decay (phosphorescence) is so slow. Therefore intersystem crossing to a triplet state favours a non-radiative decay mechanism, and most technologically significant photochemical processes are derived from the chemical reactivity of the triplet state.

3.1.4.4 Dissipative pathways

The mechanism by which the S_1 or T_1 state loses its energy may involve either physical decay or chemical change. In either case the process may be first order (i.e. unimolecular) or second order (i.e. bimolecular). The principal dissipative pathways are shown in Table 3.1. More comprehensive schemes are given in the standard textbooks on photochemistry (e.g. Calvert and Pitts, 1966, p. 196).

Table 3.1. Classification of primary processes

Unimolecular processes		Bimolecular processes	
Photophysical	Photochemical	Photophysical	Photochemical
Luminescence, i.e. fluorescence, phosphorescence	Dissociation Isomerization	Physical quenching Intermolecular energy transfer	Dimerization Hydrogen abstraction Electron transfer etc.

In *physical quenching* the excitation energy is lost to another molecule in translational or vibrational excitation of that molecule (Wayne, 1970, p. 9). *Intermolecular energy transfer* causes the electronic excitation of another molecule, i.e. $A^* + B \rightarrow A + B^*$. B^* can then take part in photophysical and photochemical processes.

3.1.5 Quantum yield

The efficiency of a primary photophysical or photochemical process is measured by the *quantum yield* for that process. If molecule A absorbs a photon to give an excited state A^*, which can then lose its energy by several different processes, the quantum yield for any *one* of these processes is defined as the number of molecules undergoing that process divided by the number of photons absorbed. This is the definition of *primary quantum yield* and as a consequence of the Stark–Einstein law, it has a maximum value of unity.

Quantum yield may also be defined in terms of the *ultimate products* of a photochemical reaction, and the situation now becomes more complicated, particularly if the primary photochemical process is followed by other non-photochemical reactions. If absorption of a photon leads to the initiation of a chain reaction, e.g. an addition polymerization, and if the *overall quantum yield* is defined in terms of the number of monomer addition reactions which result from the absorption of each photon, the calculated value may be very high indeed.

3.2 The Chemistry of Photopolymerization

3.2.1 *Classification of photopolymerization: types 1 and 2*

The term "photopolymerization" is used in this book to describe any process by which "low" molecular weight materials are linked together to form a larger structure under the influence of light. Delzenne (1979a,b) has suggested a very helpful scheme for classifying such reactions. Photopolymerization reactions are first divided into two major classes, Type 1 and Type 2. In *Type 1* photopolymerization each increase in molecular weight requires its own photochemical activation, e.g. for homopolymerization

$$M_n \xrightarrow{h\nu} M_{n+1}$$

Practical applications of Type I photopolymerization usually involve the formation of crosslinks between pre-existing polymer chains.

In *Type 2* photopolymerization a photoinitiator (PI) absorbs a photon to give an excited state (PI*), which then by some mechanism forms an intitiator entity I, which interacts with monomer to form polymer by a chain reaction, i.e.

$$\mathrm{PI} \xrightarrow{h\nu} \mathrm{PI}^*$$
$$\mathrm{PI}^* \longrightarrow \mathrm{I}$$
$$\mathrm{I} + \text{monomer} \longrightarrow \text{polymer}$$

Whereas in Type 1 photopolymerization an excited state of the monomer is involved, in Type 2, often called "*photoinitiated polymerization*", the polymerization involves only ground state monomer, and is not *per se* a photochemical process.

Because the quantum yield in terms of polymerization steps has the potential of being much greater in Type 2 than in Type 1, it is Type 2 photopolymerizations which have found the greater practical application.

3.2.2 *Photocrosslinking reactions*

Perhaps the most technologically significant application of Type 1 photopolymerization is the incorporation in pre-polymer chains of pendant cinnamate groups, which dimerize by cyclobutane ring formation to give a crosslink (Minsk *et al.*, 1959; Hepher, 1964; Delzenne, 1976):

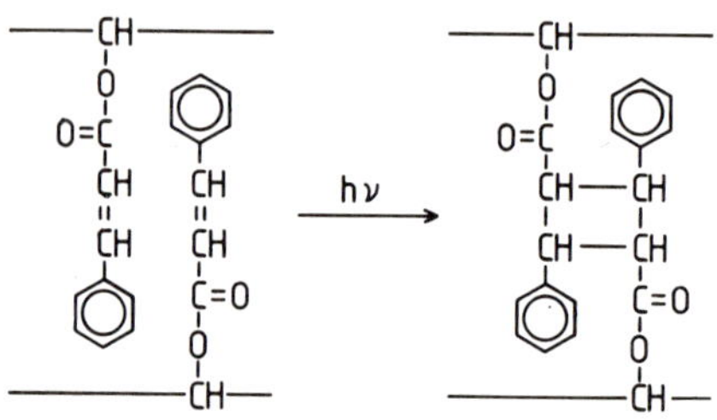

Cinnamic acid and its esters exist in two isomeric forms, *cis* and *trans*. Only the *trans* isomer dimerizes, but the *cis* isomer is converted to the *trans* isomer by a photoisomerization reaction. Dimerization can take place from either the singlet or the triplet state of the cinnamate chromophore (Delzenne, 1979b). The most common system uses poly(vinyl alcohol) esterified with cinnamic acid. The cinnamate group is sensitive to wavelengths below 320 nm. It can be sensitized to wavelengths above 320 nm by using a suitable photosensitizer which is capable firstly of absorbing radiation at longer wavelengths, and secondly of populating the triplet state of the cinnamate group by intermolecular energy transfer (see Table 3.1). Sensitization well into the visible is possible. Another crosslinking mechanism yielding a cyclobutane ring is the photosensitized dimerization of the norbornene ring (Osborn, 1976).

Type 1 photopolymerization is important in photoresist technology (Hepher, 1964; Kosar, 1965; Delzenne, 1971, 1976, 1979c; Green and Stark, 1981). Another important example, historically, is the insolubilization of gelatin or other colloidal proteins by dichromates. More recent systems have been reviewed by Butler and Ferree (1975). Azides (Reiser, 1980) have been the basis of successful photoresist formulations for the semiconductor industry. Difunctional azides photodecompose with the elimination of nitrogen to form highly reactive nitrene radicals which can crosslink polymers by a variety of mechanisms including insertion into carbon–carbon double bonds and carbon–hydrogen bonds, or by hydrogen abstraction. The most common polymer components are cyclized rubbers such as cyclized polyisoprene.

3.2.3 Active centres in photoinitiated polymerization: radicals and ions

Type 2 photopolymerizations may be subdivided into Type 2A and Type 2B. In Type 2A, both the initiating species and the propagating species are radicals, and *free radical* polymerization occurs. In Type 2B, the initiating and propagating species are cations, and *cationic* polymerization occurs. Note that both types are examples of *addition* or *chain growth* polymerization, as opposed to *condensation*, or *stepwise* polymerization, and there are no small molecules like water eliminated.

Addition polymerization involves monomers which contain either multiple bonds or strained rings. The multiple bonds may be —C=C—, —C=O or —C≡C—, although the olefinic double bond is by far the most important. Although Delzenne includes only free radical and cationic polymerization in his scheme, some olefinic double bonds will polymerize by a third mechanism which involves *anionic* active centres. However, no technologically feasible method of photoinitiating anionic polymerization has been devised.

In photopolymerization most useful monomers which contain olefinic double bonds have them in the form of a 1,2-ene group, i.e. CH_2=CXY, where X and Y are substituents. During polymerization the π bond part of the double bond is destroyed and polymerization occurs by the formation of two new σ bonds. Mechanisms of free radical and cationic polymerization of terminal olefinic groups are shown in Fig. 3.3.

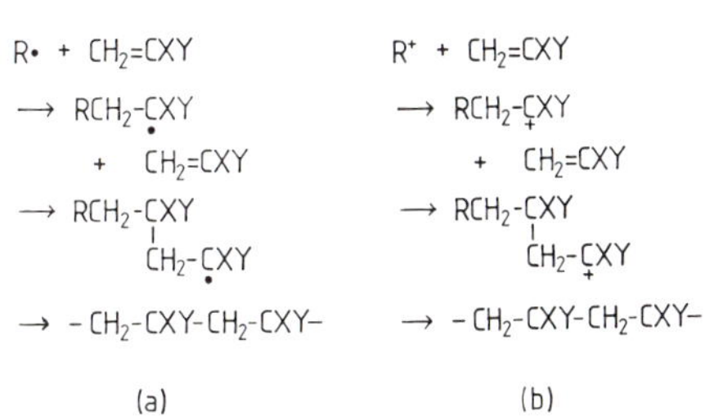

Fig. 3.3. Initiation and propagation of vinyl polymerization by (a) a free radical mechanism and (b) a cationic mechanism.

The mechanism by which a vinyl group will polymerize depends on the nature of the substituent groups attached to it. Free radical polymerization is possible if the substituents help to stabilize a propagating free radical. Such substituents are either electron withdrawing (e.g. Cl, COOR) or conjugating (e.g. Ar) (Billingham and Ledwith, 1974). Likewise, cationic polymerization is facilitated if the substituents can stabilize a propagating carbonium ion. Such substituents are electron releasing (e.g. alkyl, OR,

Ar). Note that aromatic groups can assist free radical polymerization by conjugating with the vinyl group, and cationic polymerization by being electron releasing. Hence styrene can be polymerized by either mechanism. The vinyl groups of most value in photopolymerization are shown in Table 3.2. The acrylate ester is the most reactive vinyl group for free radical polymerization. Vinyl ethers polymerize very rapidly by a cationic mechanism, but various practical problems have limited their use. Strained ring compounds polymerize by ring opening and always by a cationic mechanism. This will be discussed in Section 3.4.

Table 3.2. Examples of vinyl monomers CH_2CXY

X	Y		Mechanism
H	COOR	Acrylate ester	Radical
CH_3	COOR	Methacrylate ester	Radical
H	C_6H_5	Styrene	Radical, cationic
H	$OCOCH_3$	Vinyl acetate	Radical
H	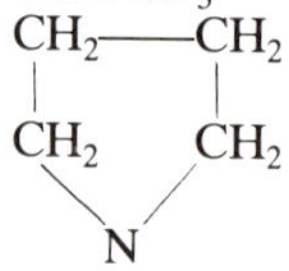	*N*-Vinyl pyrrolidone	Radical
H	OR	Alkyl vinyl ether	Cationic

3.2.4 *The kinetics of photoinitiated free radical polymerization*

Free radical polymerization is almost exclusively the addition polymerization of the olefinic double bond. The kinetics of the photoinitiated polymerization of vinyl monomers has been reviewed in depth by Oster and Yang (1968). The first step is the formation of an initiator radical $R^{\cdot}$, which can combine with a monomer molecule to form a new radical:

$$R^{\cdot} + M \longrightarrow RM^{\cdot} \qquad \textit{initiation}.$$

Initiation can also take place by hydrogen abstraction. The new radical can then add to another monomer molecule:

$$\left.\begin{array}{l} RM^{\cdot} + M \longrightarrow RM_2^{\cdot} \\[2ex] RM_n^{\cdot} + M \longrightarrow RM_{n+1}^{\cdot} \end{array}\right\} \qquad \textit{propagation}.$$

Competing with initiation and propagation are *chain transfer* and *termination* reactions, both of which limit the molecular weight which can be achieved. Chain transfer occurs when the growing polymer radical abstracts a hydrogen or other atom, or other species, from a compound present in the system, which may be a monomer, initiator, solvent or perhaps a material specifically added for this purpose (i.e. a chain transfer agent):

$$RM_n^{\cdot} + XA \longrightarrow RM_nX + A^{\cdot}$$

Termination occurs when two radicals are destroyed by mutual interaction. This is usually by combination, but can also be by disproportionation (Pappas and Chattopadhyay, 1975).

The rate of free radical polymerization is given by

$$R_p = k_p \left(\frac{R_i}{2k_t}\right)^{1/2} [M], \quad (3.1)$$

where k_p is the propagation constant, k_t is the termination constant, R_i is the rate of production of primary radicals, and [M] is the monomer concentration.

In the case of photoinitiated polymerization,

$$R_i = 2\phi I_a, \quad (3.2)$$

where I_a is the intensity of light *absorbed* and ϕ is the quantum yield for the production of primary radical pairs. The maximum possible value of ϕ is unity. Note that while R_i, being the rate of the primary photochemical reaction, is independent of temperature, k_p and k_t, being rate constants for secondary, thermal reactions, are temperature dependent.

If the monomer is in the form of a section of thickness b, then if the photoinitiator is present in concentration [PI] and has an extinction coefficient ε, the intensity of light absorbed will be related to the intensity of light incident on the surface of the section, I_0, by the expression

$$I_a = I_0(1 - 10^{-\varepsilon[PI]b}). \quad (3.3)$$

Combining expressions (3.1), (3.2) and (3.3) gives

$$R_p = k_p \left(\frac{I_0\phi(1 - 10^{-\varepsilon[PI]b})}{k_t}\right)^{1/2} [M]. \quad (3.4)$$

This expression† indicates that the rate of polymerization is proportional to $k_p/k_t^{1/2}$. Termination reactions require little activation energy, and are

† If the reader refers to the excellent review of Oster and Yang (1968), and to later sections of this chapter, he will appreciate that the derivation of expression (3.4) is based on a rather simplified model of photoinitiated free radical polymerization. The underlying assumptions may not in practice always be justified.

diffusion controlled. k_t therefore increases rapidly as the viscosity is reduced. The result is that $k_p/k_t^{1/2}$ increases as the viscosity is increased. Therefore, in a system in which polymerization increases the viscosity, the rate of polymerization may increase in the early stages, until depletion of monomer reverses the trend. This effect is known as "autoacceleration" or the "Trommsdorff effect" (Norrish and Smith, 1942; Trommsdorff *et al.*, 1948). Another conclusion from the expression is that the rate of polymerization is proportional to the square root of the light intensity. Therefore there should be reciprocity failure, as a result of the nature of the secondary reactions. A greater degree of polymerization should be achieved by using a low intensity of irradiation for a long period than a higher intensity for a correspondingly shorter period. This has a strong influence on the choice of radiation source, and creates problems when xenon flashlamps are used (Phillips, 1978).

It is important to realize that R_p is a measure of the total polymerization occurring in a section of thickness b, and [M] of the total monomer present. The degree of polymerization will vary in the section, because of the attenuation of the intensity of the light (Holman and Rubin, 1978). The situation will be complicated in many instances by the presence of components, other than the photoinitiator, which absorb light.

Although expression (3.4) is of considerable help in understanding the kinetics of the polymerization process, the situation is much complicated by the effects of oxygen inhibition unless steps are taken to exclude air. Molecular oxygen is a ground state diradical, and can interfere with the polymerization process in two ways. It can combine with radicals in the polymerizing system, producing unreactive peroxy radicals:

$$RM_n^{\cdot} + O_2 \xrightarrow{k_I} RM_nO_2^{\cdot}$$

k_I is $\sim 10^5 k_p$ (Wight, 1978), and therefore, if the concentration of oxygen is high, $k_I[O_2] \gg k_p[M]$, and there will be pronounced oxygen inhibition. Sometimes, if there is oxygen dissolved in the monomer, an induction period will be observed during which the oxygen is consumed. Oxygen is also an efficient triplet state quencher, and can therefore interfere with the mechanism of radical production. When a thin film is polymerized in air, the degree to which oxygen can interfere depends on the rate at which it can diffuse to and into the surface. The more rapid the polymerization the less opportunity oxygen has to influence the reaction, and therefore increasing the light intensity decreases the effect of oxygen inhibition (Rubin, 1974; Plews and Phillips, 1979). Other methods of reducing oxygen inhibition involve the use of amines to regenerate active radicals by a chain reaction (see Section 3.3.2.2), use of wax barrier coats (Bolon and Webb, 1978), conversion of triplet oxygen to singlet oxygen by dye sensitization, followed by scavenging of the singlet oxygen (Decker *et al.*, 1979), and, most logically, carrying out the polymerization in an inert atmosphere.

In general there is an *upper* limit to the photon flux density which can be employed, as a result of the Trommsdorff effect, this limit depending on the viscosity of the formulation and the photoinitiator concentration, and a *lower* limit below which oxygen inhibition becomes serious if cure takes place in air. It is convenient that the optimum range of photon flux density is approximately that given by a medium pressure mercury discharge lamp when as much as possible of the radiation is focused to a narrow beam by using a semi-elliptical reflector.

Figure 3.4 shows the cure of a UV curing coating with a fixed dose of radiation. The variables are film thickness, photon flux density, and atmosphere—air or nitrogen. A medium pressure mercury lamp was used (input power 80 W cm^{-1}) angled across a conveyor moving at a nominal 1 m s^{-1}, with either a focused beam or a diffuse beam. The constancy of the dose was checked with a "Light Bug" (see Section 4.4), and minor

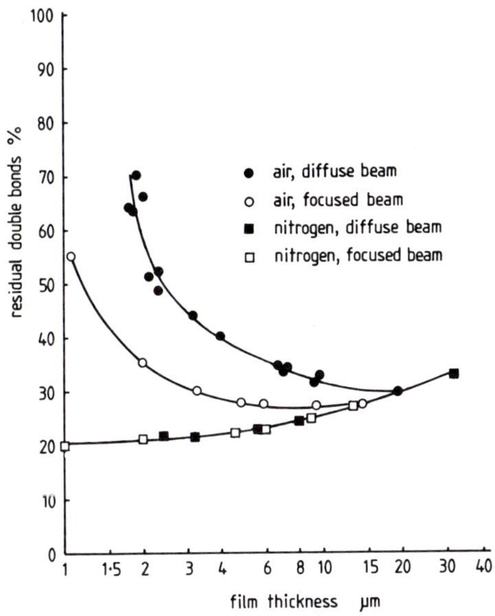

Fig. 3.4. Influence of film thickness, photon flux density and curing atmosphere on cure of acrylate-based coating using fixed dose of radiation. Film coated on NaCl disc. Cure measured by infra-red spectroscopy using 808 cm^{-1} absorption of acrylate group (CH_2 twist). Coating composition: aliphatic urethane acrylate resin, 68·7%; trimethylolpropane triacrylate, 26·3%; α,α-diethoxyacetophenone, 5·0%. (Figure adapted from Plews and Phillips, 1979.)

adjustments were made to the conveyor speed to maintain constancy. When the film was very thin, cure in air, as measured by infra-red spectroscopy, was poor, and even worse when the photon flux density was reduced. Cure under nitrogen, in a specially constructed cell with a Spectrosil® fused silica window, was very good, and independent of the flux density. The effects of both flux density and cure atmosphere were reduced as the film thickness increased. Ultimately, cure in thick films was reduced because of attenuation of the radiation.

Work by Bassemir and Bean (1974) with a set of process coloured inks (magenta, yellow, cyan and black) has shown that in the region of intensities available with medium pressure mercury lamps, increasing intensity causes a rapid increase in cure rate. Their results indicated a threshold of intensity, the darker colours having higher thresholds.

The rate of polymerization in UV-cured acrylate-based coatings is very rapid (Phillips, 1978). Figure 3.5 shows the results of a crude experiment in which a formulation in the form of a thin film on a sodium chloride disc was irradiated *in situ* in an IR spectrophotometer with a 500 μs flash of light from a xenon flash tube. The transmission of the film at an absorption peak of the acrylate group was monitored with a fast running chart recorder. It can be seen that the polymerization initiated by a single flash is complete within a small fraction of a second. Second and third flashes give further polymerization.

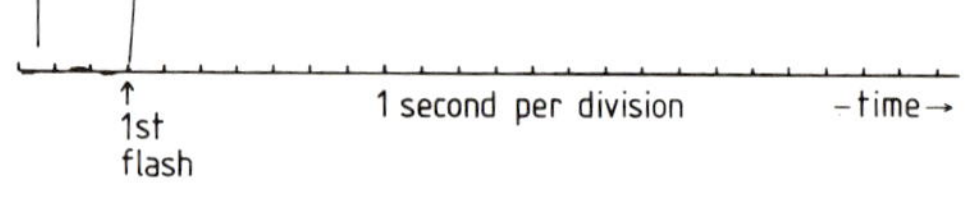

Fig. 3.5. Flash curing of acrylate-based coating *in situ* in an infra-red spectrophotometer. Coating composition: urethane acrylate resin, 57%; trimethylolpropane triacrylate, 38%; α,α-dimethoxy-α-phenylacetophenone, 5%.

3.2.5 Problems inherent in free radical polymerization

The early technological applications of photoinitiated polymerization were based exclusively on free radical mechanisms, and it is still true that the great majority of applications involve radicals as active centres. This approach has proved to be very successful, and its initial adoption was no doubt influenced by the fact that the mechanisms involved in free radical polymerization had been well established for many years. Also, photochemical methods of generating radicals which could act as initiating species were readily available. However, there are inherent problems associated with radical polymerization as applied to surface coatings. Most of these problems are related either to the nature of the termination reactions, or to the types of monomer which are required. Oxygen is a terminating species which is often difficult or expensive to eliminate, and even with carefully formulated coatings tends to detract from the properties of the cured surface. The radicals that are necessary for initiation and propagation can also terminate, and therefore there is no way in which one can avoid the presence of terminating species. All one can hope to do is to minimize the rate of termination relative to the rate of propagation. Radicals are inevitably shortlived species, and therefore the polymerization reaction ceases very soon after the irradiance ceases. Radical recombination reactions may cause yellowing. The Trommsdorff effect may cause problems with low viscosity formulations, and the radical concentration and the irradiance must be kept within certain limits if satisfactory cure is to be obtained. Free radical polymerization always involves monomers with double bonds and it is inevitable that some rapid shrinkage occurs during the formation of the polymer network which may detract from the adhesion of the coating to the substrate.

3.2.6 "Living" polymerization

During the last 10 years there has been much effort devoted to the development of curing systems based on *photoinitiated ionic chain polymerization.* There are potential terminating species for all types of active centre polymerizations, but in principle it is possible to avoid their presence when the active centres are ions. If termination species are eliminated the kinetics are simplified, and perhaps the most significant aspect is that a "living" polymer results, with propagation continuing long after radiation ceases. This is of great value when polymerization is taking place at ambient temperature because a stage is reached when the rate of propagation is slowed down by steric factors. In the case of radical polymerization, it is often difficult to eliminate the last 20–30% of functional groups unless irradiation takes place at an elevated temperature to allow more mobility in the system. It is also found that in general, a stoving operation subsequent to UV curing of a free radical system does not actually reduce the number of polymerizable double bonds unless a thermal mechanism for radical production is available. In a "living" polymer system, propagation continues, albeit slowly, after a densely crosslinked system has been formed, until in theory every functional group has been eliminated. The progress towards complete functional group conversion can be accelerated by raising the temperature after irradiation.

3.2.7 The rationale for cationic polymerization

In practice anionic polymerization is not suitable for use in surface coatings because it is even more prone to oxygen inhibition than radical polymerization, and it is very sensitive to other adventitious impurities (Crivello *et al.*, 1977). Cationic systems are much more promising. There is no apparent oxygen inhibition, and if nucleophilic species other than the monomer can be eliminated from the formulation, a "living" polymer system should result. Another major attraction of cationic polymerization is that it is not restricted to monomers containing polymerizable double bonds, but can use monomers containing strained rings such as cyclic ethers and lactones. Some of these monomers, particularly those containing epoxide groups, are very attractive to the formulator of surface coatings. Ring opening involves less shrinkage than double bond polymerization, and it is even possible to arrange for a slight expansion which may allow the coating to lock into irregularities in the substrate surface (Bailey, 1977). Whereas the formulator of free radical systems has needed to resort to specially designed polyfunctional monomers and resins, the formulator of cationic systems has had available a wide range of epoxides from low viscosity aliphatic and cycloaliphatic types to the higher molecular weight DGEBA and epoxy novolak resins, all of which are commercially available and have been widely used for other purposes for many years (Lee and Neville, 1967). Suitable monomers and their relative reactivities have been described by Crivello *et al.* (1977). Even low viscosity, low molecular weight examples tend to have a low odour, in contrast to the strong odour of many acrylate esters, or thiol–ene systems (Aloye, 1976).

If one has a propagating cation it is necessary for overall neutrality to have a counter-anion, which must be non-nucleophilic. Propagation takes place most rapidly if the propagating cation is a free ion without a close association with the anion. A high dielectric constant for the medium helps.

3.3 Free Radical Photoinitiators

3.3.1 System components

During the photoinitiation process, light is absorbed to produce a chemically active excited state, which by some mechanism leads to the production of radicals which are capable of initiating polymerization. A photoinitiator or photoinitiating system is required, and this may consist of up to three distinct components:

(1) A molecule which has an electronically excited state which by some chemical reaction can produce initiating radicals. This molecule is normally referred to as the *photoinitiator.*
(2) Although in most cases the electronically excited state of the photoinitiator is produced by absorption of a photon by the photoinitiator, it is sometimes possible to populate the excited state by triplet energy transfer from another molecule with a more useful absorption spectrum. This molecule is known as a *photosensitizer.* If a photosensitizer is used it may be appropriate to call the photoinitiator a *coinitiator* (Pappas, 1978).
(3) Sometimes the photoinitiator produces radicals by a bimolecular reaction (usually hydrogen transfer) with another molecule. This may be referred to as the *synergist, accelerator*, or *photoactivator.*

It can be seen that photoinitiators can be divided into two main classes, firstly those that give radicals by a unimolecular process, and secondly those that give radicals by a bimolecular reaction.

3.3.2 Criteria

The choice of a photoinitiator or photoinitiating system depends on a number of criteria:

(1) The absorption spectrum of the absorbing species must harmonize with the spectral output of the radiation source used, taking into account competition from pigments and other components, and allowing for the depth of penetration required.
(2) The excited state must be *chemically* active, i.e. lose its energy by a chemical reaction producing radicals, rather than by a physical decay process.

(3) The radicals produced must be effective in initiating polymerization of the monomer.
(4) The photoinitiator must be *ineffective* at initiating polymerization thermally, to allow long term stability in the dark.
(5) Neither the photoinitiator nor its by-products should be appreciably toxic.
(6) The photoinitiator and its by-products should be odourless and colourless.
(7) The by-products should preferably not compete for radiation with the photoinitiator—in other words the chromophore should be destroyed.
(8) Residual quantities, and by-products, should not detract from the physical properties of the cured film.
(9) It should be cheap and available in commercial quantities.
(10) Last but not least, it must be soluble in the monomer system. In the case of lithographic inks it should be insoluble in water (Fields, 1978). There has recently been an interest in water-soluble photoinitiators which can be used with water-soluble monomers such as *N*-vinyl pyrollidone and methacrylic acid for grafting on to textiles (Barker *et al.*, 1981).

Over the years a large number of compounds have been suggested for use as photoinitiators (Rabek, 1968) and there is an extensive patent literature. Relatively few have been commercially successful, largely because of the difficulty of meeting the criteria which have been listed. Almost all of those which have achieved lasting success have been aromatic carbonyl compounds (Ledwith, 1976a). These have absorption spectra which allow them to respond well to the wavelengths emitted by the medium pressure mercury lamp, and their excited states are often very reactive. The rate constant for intersystem crossing is large ($>10^8\,s^{-1}$) for all ketones and therefore the formation of the metastable triplet state is favoured (Wagner, 1976). Primary triplet reactions are often reversible. Their products are radicals, biradicals or charge transfer complexes, so that the quantum yield may be lower than would otherwise be expected.

Photoinitiators may be divided into two main groups depending on whether the process by which radicals are produced is unimolecular or bimolecular. Unimolecular radical production implies a fragmentation process. Bonds broken may be C—C, C—X, C—S, S—S, C—O, N—O, S—X (X = halogen).

3.3.3 Unimolecular radical production

Examples of aromatic carbonyl compounds which produce radicals by a unimolecular process and have become commercially successful, are shown in Fig. 3.6. Most but not all fragment by α or Norrish Type I cleavage:

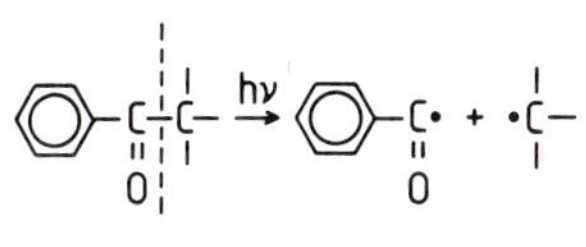

3.3.3.1 Benzoin ethers

The photolysis of benzoin has been known since the early years of academic photochemistry (Ciamician and Silber, 1901). Benzoin ethers as photoinitiators for the polymerization of acrylates were the subject of a Dupont patent (US Patent 2,448,828) as long ago as 1946, and the subsequent development and commercialization of the Dupont Dycril photopolymer plate in the 1950s (US Patent 2,760,863, 1952) led to what was primarily the first major use of photoinitiated polymerization. Benzoin ethers have been the most extensively studied and used initiators (Mleziva and Cermak, 1973; Delzenne, 1979a).

Benzoin and its derivatives fragment by α cleavage (Chinmayandam and Melville, 1954; Kornis and de Mayo, 1964; Closs and Paulson, 1970; Ledwith *et al.*, 1972). Benzoin itself is a rather poor photoinitiator, and has a low solubility in resin–monomer systems, and therefore derivatives having the general structure

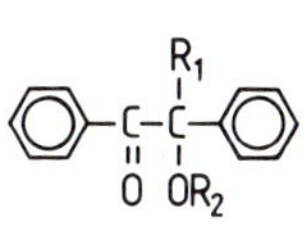

have been examined. The relative merits of a wide range of derivatives were assessed by workers at Bayer in Germany, with the emphasis on their use in styrene–unsaturated polyester systems (Heine, 1972; Heine *et al.*, 1972; Heine and Traenchner, 1975; Rudolph *et al.*, 1975). It was found that when $R_1 = H$, the order of merit for varying R_2 was alkyl $\geqslant$ aryl $>$ H $>$ COR. Alkyl ethers are therefore most effective, and alkyl esters least effective. The rate constant (k_α) for α cleavage for the alkyl ethers is very high, $>10^{10}\,s^{-1}$, whereas for benzoin it is $\sim 10^9\,s^{-1}$, and for benzoin acetate only $\sim 5 \times 10^7\,s^{-1}$ (Heine and Traenchner, 1975). This variation has been associated with the ability of the benzylic substituent to stabilize an adjacent positive charge, and so weaken the bond. Hence substituents which are electron donating, such as alkyl ethers, are more

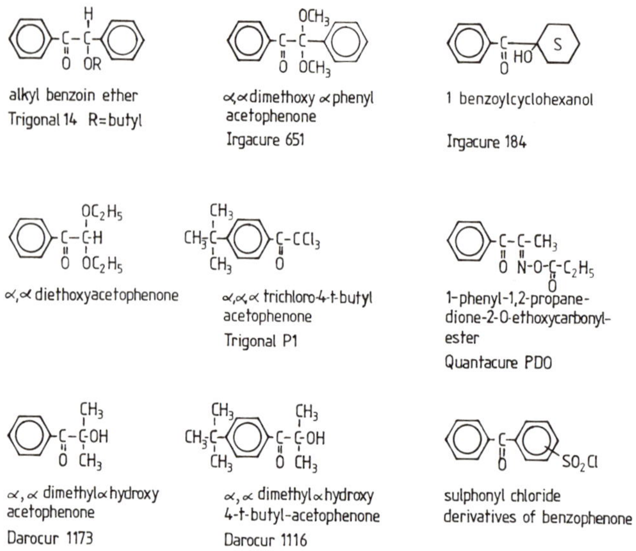

Fig. 3.6. Aromatic carbonyl compounds which act as photoinitiators by a unimolecular process. Darocur is a trademark of E. Merck; Irgacure is a trademark of Ciba Geigy; Trigonal is a trademark of Akzo; Quantacure is a trademark of Ward Blenkinsop.

effective than those which are electron accepting, such as alkyl esters. The effectiveness of the benzylic substituent can be related quantitatively to the Hammett σ factor (McGinniss, 1979). There are differences in reactivity between the various alkyl ethers. There are four butyl isomers, and Osborn and Sandner (1974) have shown that those derived from primary alcohols (*n*-butyl and isobutyl) are more effective than the one derived from a secondary alcohol (*s*-butyl) with *t*-butyl being least reactive.

The primary radicals formed from benzoin ethers are benzoyl and benzyl ether:

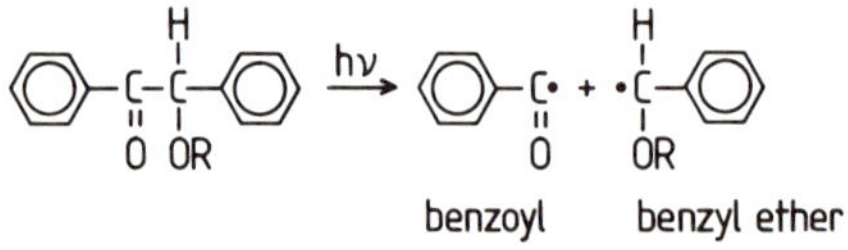

The quantum yield for primary radical formation is independent of wavelength (Carlblom and Pappas, 1977), and is almost unity. However, there may be some recombination of primary radicals due to the "cage effect" (Osborn and Sandner, 1974). In acrylates and methacrylates the benzoyl and benzyl ether radicals have been found to be equally effective in initiating polymerization (Pappas and Chattopadhyay, 1975), but in styrene the benzoyl radical has been found to be more effective (Heine and Traenchner, 1975; Fuhr, 1977; Hageman *et al.*, 1979). The reason for this apparent discrepancy has been explained by Pappas (1981) in terms of the differences between the experimental conditions used. Under the conditions of UV curing (low monomer concentration, high radical concentration) the benzoyl radical is primarily responsible for initiation, and the benzyl ether radical for chain termination. However, Osborn and Sandner (1974) suggested that some secondary fragmentation of the benzyl ether radical could take place, giving benzaldehyde and an alkyl radical:

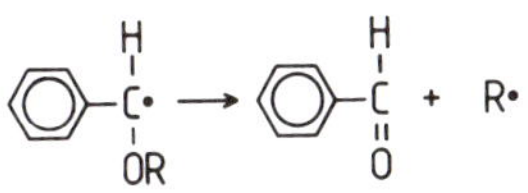

and that this could explain the differences in reactivity between the various benzoin ethers. Benzaldehyde is found as a by-product but could also be formed by hydrogen abstraction by benzoyl radicals (Berner *et al.*, 1978). Hydrogen abstraction is a mechanism of initiation (Carlblom and Pappas, 1977).

There has been considerable debate as to whether the α cleavage in benzoin ethers takes place from the excited singlet or the excited triplet state. The effectiveness of benzoin ethers in styrene, a powerful triplet quencher, led the workers at Bayer to conclude that splitting takes place from the singlet state. The rate of α cleavage is comparable with the rate of intersystem crossing in aromatic ketones. However, benzoin ethers can be sensitized by energy transfer from triplet sensitizers, which suggests that splitting can take place from the triplet state (Pappas and Chattopadhyay, 1973). Perhaps one should agree with Kuhlmann and Schnabel (1977) that splitting can take place from either state.

Benzoin ethers have been found to give poor can stability, although the larger the alkyl group the less the problem of premature gellation (Heine and Traenchner, 1975). The poor stability has been associated with the α hydrogen atom, and a possible mechanism for premature initiation involving hydroperoxide formation has been suggested by Pappas (1978). Various additives have been suggested to improve can stability (Pryce, 1976).

3.3.3.2 α,α-Dimethoxy α-phenyl acetophenone

A logical approach to improving the performance of a benzoin ether is to replace the α hydrogen by an alkoxy group. This not only provides a dramatic improvement in can life, but also enhanced reactivity. The success of this approach is exemplified by the excellent performance of α,α-dimethoxy α-phenyl acetophenone (DMPA), marketed commercially by Ciba Geigy as Irgacure 651 (Berner *et al.*, 1978). The enhanced reactivity as compared with benzoin ethers is partly due to the extra electron-donating alkoxy group, and partly due to the well established secondary fragmentation of the substituted benzyl radical following the initial α cleavage, yielding a highly reactive methyl radical (Sandner and Osborn, 1974):

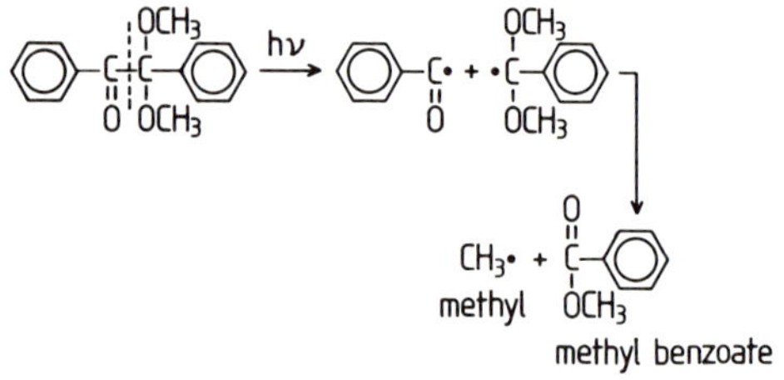

The methyl benzoate formed may leave the coating with a residual odour. Chemically induced nuclear polarization (CIDNP) has been used to identify mechanisms and by-products (Borer *et al.*, 1978), and it has been shown that a variety of radical recombination and hydrogen abstraction reactions are possible, so that the number of final products is quite large. Addition of the benzoyl radical to the *ortho* and *para* positions of the benzyl ether radical gives semiquinoid structures which may cause yellowing during cure (Pappas, 1981). The initial cleavage is thought to occur from the triplet state.

It has been found that tertiary aliphatic amines when used as synergists with DMPA provide a marked improvement in surface cure (Berner *et al.*, 1978). This has also been found with benzoin ethers (Belicar, 1977). However, Sandner *et al.* (1972) found that triethylamine caused a slight deterioration in the cure of methyl acrylate with benzoin methyl ether in bulk under nitrogen. The improvement in surface cure is thought to be due to the suppression of oxygen inhibition by a chain reaction which allows the amine to consume oxygen (Bartholomew and Davidson, 1971):

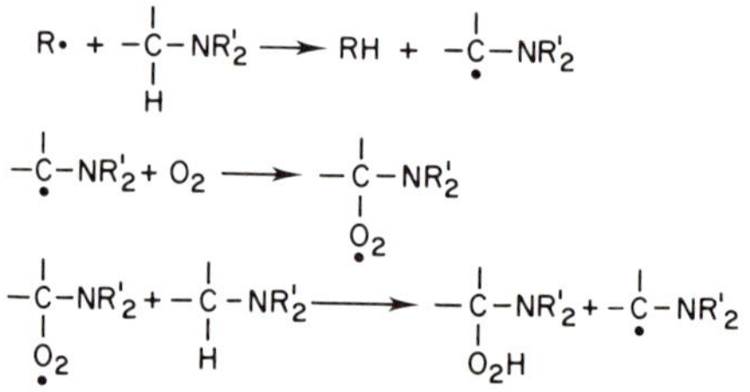

Amines act as chain transfer agents, and so although they enhance surface cure, they produce a softer film (Pappas, 1981).

3.3.3.3 α,α-Diethoxyacetophenone

α,α-Diethoxyacetophenone (DEAP) has been found to be a very effective photoinitiator in unpigmented coatings. Although there is an α hydrogen atom, this is non-benzylic and therefore does not affect can stability (Pappas, 1978). Product analysis (Osborn and Sander, 1974; Borer *et al.*, 1978) has shown that it is possible for radicals to be formed by intramolecular hydrogen transfer (Norrish Type II):

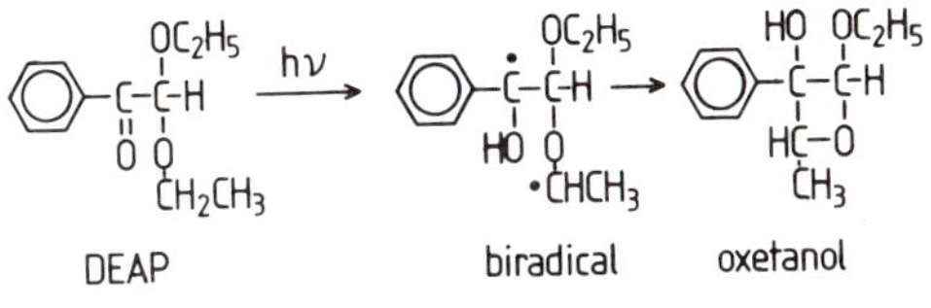

as well as by α cleavage, followed by secondary fragmentation:

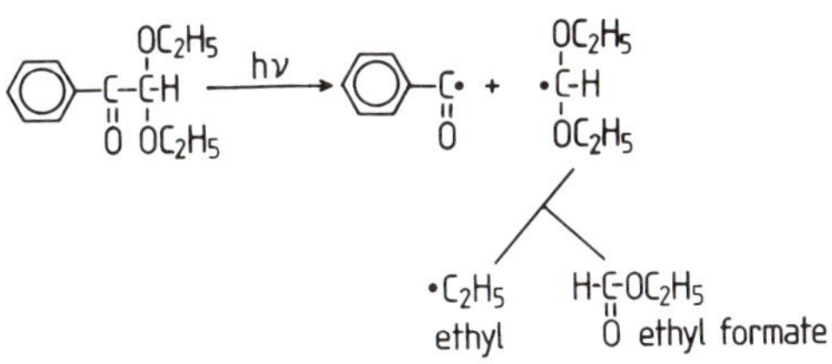

It is not clear whether the biradical is capable of initiating polymerization (Pappas, 1978).

As there is no benzylic radical produced during cure there is less yellowing.

3.3.3.4 Hydroxy alkylphenones

The two hydroxy alkylphenones shown in Fig. 3.7 have been commercialized under the trade name Darocur by E. Merck. They have been described by Herz and Eichler (1979). They are claimed to have excellent pot life, low toxicity, low viscosity and to be particularly suitable for clear varnishes as there are no yellow by-products, a problem with many photoinitiators. They fragment by α cleavage from the triplet state (Eichler *et al.*, 1980a, b).

In the same general class is the hydroxycycloalkylphenone Irgacure 184 (Ciba Geigy). Rates of polymerization and induction periods for trimethylolpropane triacrylate in isopropanol have been compared for Irgacure 651,

Irgacure 184 and Darocur 1173 in air and in nitrogen by Decker and Fizet (1980) using the scattering of a laser beam to measure polymerization.

3.3.3.5 O-acylated α-oximinoketones

O-Acylated α-oximinoketones, having the general formula

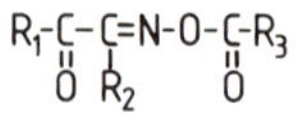

have been found to be effective photoinitiators (Delzenne *et al.*, 1970; Heine and Traenchner, 1975; McGinniss, 1975, 1976a). The primary cleavage is of the N—O bond, and is followed by secondary fragmentations, e.g.

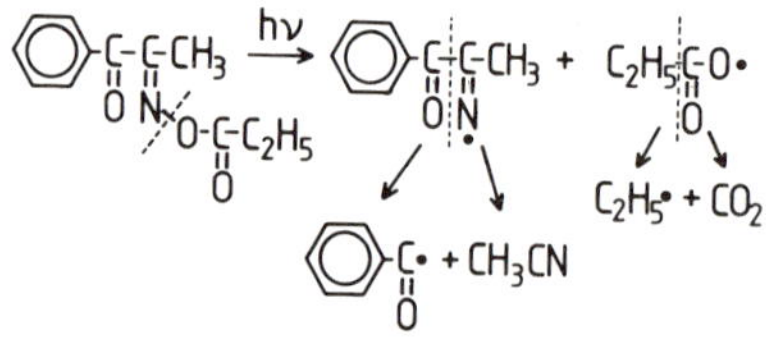

These photoinitiators work well in conjunction with benzophenone (Kurpiewski and McDowell, 1978). They are used in conjunction with dyes and amines for visible light curing of glass-reinforced polyesters (Alexander and Scott, 1980).

3.3.3.6 β cleavage

Although in α-chlorinated acetophenones α cleavage takes place to a small extent,

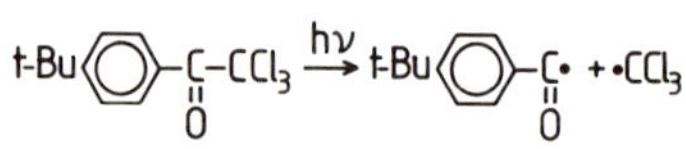

the main cleavage is the β cleavage of a C—Cl bond (Heine *et al.*, 1972):

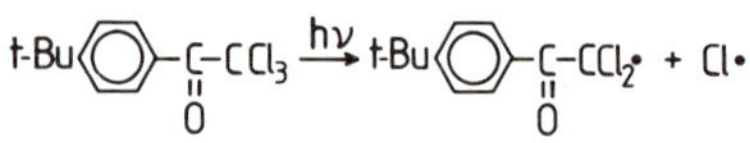

The Cl$^{\cdot}$ is an efficient initiator. It can also abstract hydrogen atoms and generate hydrogen chloride:

$$RH + Cl^{\cdot} \longrightarrow R^{\cdot} + HCl$$

It is claimed that because of the ability to generate a strong acid this photoinitiator can be used to cure acid curing resins such as urea formaldehyde and melamine types (Belicar, 1977).

The ease of photocleavage of carbon–halogen and sulphur–halogen bonds has lead to the design of a wide range of photoinitiators with the general structures RCH_2X and RSO_2X, where $X = Cl$, Br or I. Halogen radicals are generated by the following mechanisms (McGinniss, 1976b, 1979; Vrancken, 1977):

$$RCH_2X \xrightarrow{h\nu} RCH_2^{\cdot} + X^{\cdot}$$

$$RSO_2X \xrightarrow{h\nu} RSO_2^{\cdot} + X^{\cdot}$$

The ease of photocleavage of C—Cl and S—Cl bonds allows initiators with quite low triplet energies to fragment, and therefore fragmentation can result from long wavelength absorption. Sulphonyl chlorides have triplet energies which are about 10 kcal mol^{-1} lower than the chloromethyl derivatives (Pappas, 1978). A problem with halogenated photoinitiators is the possible corrosion which may be produced by their by-products.

3.3.4 Bimolecular radical production

3.3.4.1 Hydrogen abstraction reactions

Many aromatic carbonyl compounds do not photolyse when irradiated in isolation. However, they may be capable of generating radicals by a bimolecular reaction which involves hydrogen abstraction. One of the best documented reactions in photochemistry is the photo-oxidation of isopropanol by benzophenone (Pitts *et al.*, 1959; Beckett and Porter, 1963):

$$\phi_2C{=}O \xrightarrow{h\nu} \phi_2C{=}O^*$$

$$\phi_2C{=}O^* + Me_2CHOH \longrightarrow \phi_2\dot{C}{-}OH + Me_2\dot{C}{-}OH$$

$$Me_2\dot{C}{-}OH + \phi_2C{=}O \longrightarrow Me_2C{=}O + \phi_2\dot{C}{-}OH$$

$$2\phi_2\dot{C}{-}OH \longrightarrow \phi_2COH{-}COH\phi_2$$

On irradiation benzophenone first forms an excited singlet state which very quickly converts to the lowest triplet state by intersystem crossing, with almost 100% efficiency. Hydrogen abstraction from alcohols, and also ethers and alkanes, is dependent on this lowest triplet state being n,π^*. The electronic configuration of the n,π^* state provides the carbonyl group with a biradical character, i.e. $\phi_2\dot{C}{-}\dot{O}$.

Aromatic ketones with lowest triplet states which are n,π^* are able to abstract hydrogen atoms from thiols. This is the basis of the *thiol–ene* photopolymer system developed by W. R. Grace and Co. (Morgan and Ketley, 1973; Morgan *et al.*, 1977). A ketone photoinitiator abstracts hydrogen from a polyfunctional thiol, giving a radical which can then react with a polyene:

Excitation:

$$\phi_2C{=}O \xrightarrow{h\nu} \phi_2C{=}O^* \xrightarrow{IC} \phi_2C{=}O\ {}^1(n,\pi^*) \xrightarrow{ISC} \phi_2C{=}O\ {}^3(n,\pi^*)$$

Hydrogen abstraction:

$$\phi_2C{=}O\ {}^3(n,\pi^*) + \underset{\text{thiol}}{RSH} \longrightarrow \underset{\text{ketyl}}{\phi_2\dot{C}{-}OH} + \underset{\text{thiyl}}{RS^{\cdot}}$$

Chain reaction:

$$\begin{cases} RS^{\cdot} + CH_2{=}CH{-}CH_2R' \longrightarrow RSCH_2{-}\dot{C}H{-}CH_2R' \\ RSCH_2{-}\dot{C}H{-}CH_2R' + RSH \longrightarrow RSCH_2CH_2CH_2R' + RS^{\cdot} \end{cases}$$

The $RS^{\cdot}$ radical is electron poor because of the electronegativity of the heteroatom, and therefore it reacts most favourably with electron-rich monomers such as vinyl ethers rather than with electron-poor monomers such as acrylates. The thiol–ene system has been found to be resistant to oxygen inhibition and this is thought to be because thiols can react with hydroperoxide radicals to yield thiyl radicals which can initiate further polymerization (Morgan and Ketley, 1980). Note that the hydrogen atom abstracted from the thiol is attached directly to the sulphur atom (the S—H bond is weak), whereas with alcohols and ethers the hydrogen atom abstracted is in the α position, i.e. attached to the adjacent carbon atom (Cohen and Chao, 1968):

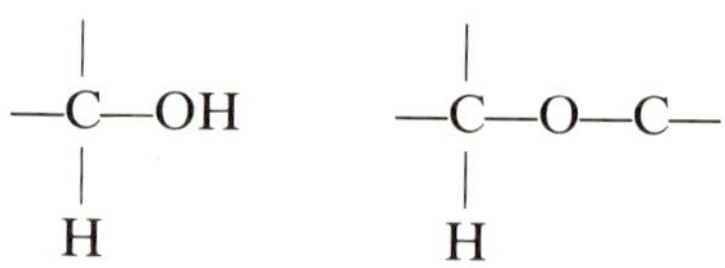

Ketone–alcohol and ketone–ether combinations are not effective photo-initiating systems because the hydrogen abstraction process is slow compared with quenching by oxygen (Pappas, 1978) or by monomer (Sandner *et al.*, 1972). However, although benzophenone causes almost no cure in methyl acrylate in solution in *t*-butyl alcohol (Sandner *et al.*, 1972), if the monomer contains an ether group, as in diethylene glycol diacrylate, some cure may be found (Heine *et al.*, 1972). Also cure may be found if tetrahydrofuran, an ether with abstractable hydrogen, is used as a solvent (Block *et al.*, 1971). Polyether groups in the oligomer have also been found to serve as hydrogen-donating sites (Schmidle, 1977).

The $\pi \rightarrow \pi^*$ transition has the effect of increasing the electron density of the oxygen atom and thereby strengthening the polar nature of the carbonyl group (Vranken, 1977). Aromatic ketones with electron-donating substituents frequently have lowest excited states which are π,π^* rather than n,π^*, and do not abstract hydrogen atoms from alcohols and ethers (Beckett and Porter, 1963). However, amines containing α hydrogen can also supply abstractable hydrogen atoms:

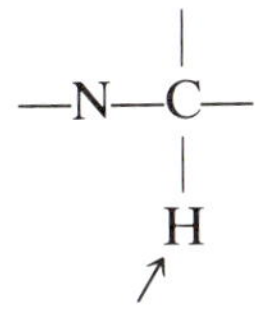

and both types of ketone are able to take part in bimolecular radical producing reactions with amines (Cohen *et al.*, 1973). The mechanism involves the formation of an excited complex which splits into radicals by proton transfer (Cohen and Cohen, 1967; Cohen and Chao, 1968; Davidson and Lambeth, 1968; Sander *et al.*, 1972):

$$\phi_2C{=}O^*\ (T_1) + RR'NCH_2R'' \longrightarrow [\phi_2\dot{C}{-}\bar{O}\ RR'\underset{+}{\dot{N}}CH_2R'']^*$$

$$\downarrow \text{exciplex}$$

$$\phi_2\dot{C}{-}OH \quad + \quad RR'N\dot{C}HR''$$

The polar nature of the π,π^* excited state actually encourages electron transfer (Pappas, 1974). The formation of the *exciplex*, which may be preceded by the formation of an "*encounter complex*" (Ledwith, 1976a, b), is rapid and competes effectively with other triplet quenching processes. Exciplex formation is possible because of the low ionization potentials of amines, and the lower the ionization potential the faster the rate of exciplex formation (Cohen and Stein, 1969). The exciplex may be more realistically represented as a "*charge transfer*" complex (Dack, 1973), i.e.

$$[\phi_2C{=}\overset{\delta-}{O}\ RR'\overset{\delta+}{N}CH_2R''],$$

with complete electron transfer being one, rather unlikely, decay process for the exciplex (Wagner and Kemppainen, 1969; Arimitsu and Masuhara, 1973; Pappas, 1978).

Ketone–amine systems are widely used for photoinitiation of free radical polymerization. Examples of suitable ketones are shown in Fig. 3.7. Some ketones in this category, such as benzil (Hulme *et al.*, 1976) and benzanthrone (McGinniss, 1976b), have absorption spectra which extend well into the visible. Derivatives of parent compounds are often chosen because of superior solubility or more favourable absorption characteristics. Tertiary amines are more suitable than primary or secondary amines, and a C—H group adjacent to the nitrogen atom is essential. Triphenylamine forms an exciplex with excited ketones (Turro and Engel, 1969) but subsequent proton transfer is not possible (Bartholomew *et al.*, 1972). The relative merits of a range of tertiary aliphatic amines have been measured by McGinniss and Kah (1977) and have been related to the basicity of the amine and steric factors.

In the example shown above the excited ketone accepts an electron from the ground state amine, which acts as a donor. Therefore, in more general terms,

$$\underset{\text{acceptor}}{A^*} + \underset{\text{donor}}{D} \longrightarrow \underset{\substack{\text{encounter}\\ \text{complex}}}{[A^*, D]} \longrightarrow \underset{\substack{\text{donor–acceptor}\\ \text{complex}}}{[\dot{A}^{-}\dot{D}^{+}]^*} \longrightarrow \text{radicals}$$

The intermediate is a *donor–acceptor complex*, and in principle either donor or acceptor may be excited (Ledwith, 1976b; Bosley *et al.*, 1978). This implies that the exciplex may be formed by the interaction of an *excited* amine with a *ground state* ketone. This is significant if highly absorbing amines such as 4,4′-bis(dimethylamino)benzophenone (Michler's ketone) or ethyl-4-dimethylaminobenzoate (Hulme *et al.*, 1976) are used.

When a photoinitiator like a benzoin ether fragments, each cleaving molecule yields two radicals, both of which are very efficient initiating species. When a ketone–amine system is used, each molecule involved can generate at the most only one radical, rather than two. Furthermore, although the amine-derived radical, being electron rich, is a very efficient initiator for the electron-deficient acrylate group (Pappas, 1974), the ketyl or semipinacol radical derived from the ketone is more likely to dimerize, or terminate a growing polymer chain, than to initiate by an addition reaction to monomer (Hutchison *et al.*, 1973; Kinstle and Watson, 1976). At temperatures significantly above ambient, however, semipinacol radicals may initiate by hydrogen abstraction (Braun and Becker, 1971). There is also the possibility of scavenging oxygen to regenerate the original ketone (Bartholomew and Davidson, 1971). The maximum possible effective radical production from a unimolecular process photoinitiator will always be quite a lot higher than from a ketone–amine system at the same concentration (Plews and Phillips, 1979). On the other hand, for C—C bond fragmentation to take place, a high energy lowest excited state is required and this inevitably means weak absorption above 350 nm. Triplet energies needed for exciplex formation are not generally as high as those needed for cleavage (Hulme *et al.*, 1976), and therefore longer wavelength absorption is possible, and if the lowest excited state is π,π^* an absorption peak above 350 nm with a high extinction coefficient may be obtained. This may be helpful in pigmented systems. Also, the amine helps to minimize

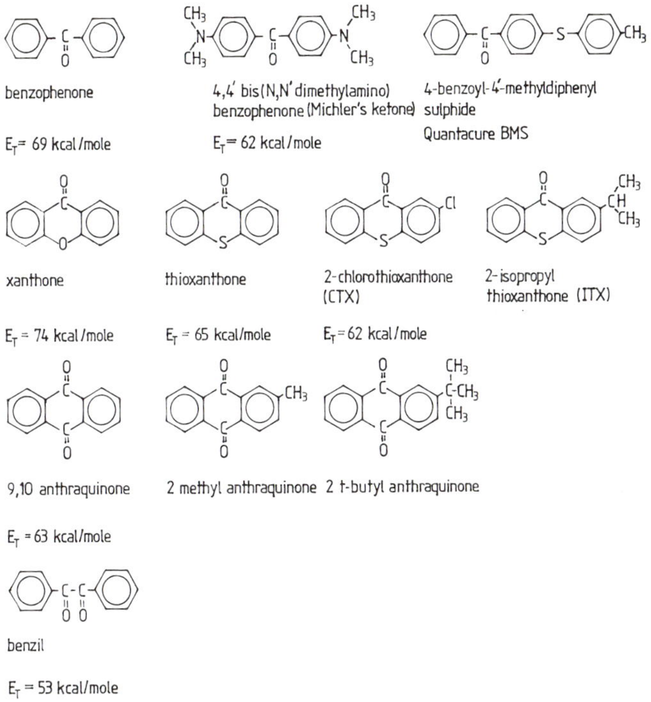

Fig. 3.7. Aromatic carbonyl compounds which act as photoiniators by a bimolecular process.

oxygen inhibition (Plews and Phillips, 1979) by a mechanism that has already been described (see Section 3.3.3.2).

Ketone–amine systems cannot of course be used in formulations which contain acidic components such as acrylic acid (Armbruster *et al.*, 1978).

3.3.4.2 Benzophenone/Michler's ketone synergism

Michler's ketone (MK) is of great interest because it contains, in one molecule, both ketone and amine functionality. It is therefore possible, in principle, for one molecule of Michler's ketone to abstract hydrogen from another molecule of Michler's ketone, to form two radicals (Koch and Jones, 1970). However, Michler's ketone on its own is a poor photoinitiator (Clarke and Shanks, 1980) because it functions poorly *as a ketone* because of the electron-rich amino groups (Pappas, 1974). Michler's ketone works best as a synergistic mixture with benzophenone (Wamser *et al.*, 1970). Intersystem crossing to the triplet state takes place with unit efficiency (Lamola and Hammond, 1965):

$$\underset{\text{A}}{\text{BP}} + \underset{\text{D}}{\text{MK}^*}\ (\text{T}_1) \longrightarrow \underset{\text{triplet exciplex}}{[\overset{\delta-}{\text{BP}} \cdots \overset{\delta+}{\text{MK}}]^*} \longrightarrow \begin{cases}\text{ketyl radical (from BP)} \\ \text{amine radical (from MK)}\end{cases}$$

Michler's ketone, like other *p*-amino benzophenones, has an absorption spectrum which is markedly dependent on the polarity of the solvent. In polar solvents (e.g. ethanol), MK has a very strong absorption at 365 nm which makes it very responsive to the 365 nm line of the mercury spectrum, whereas in non-polar solvents (e.g. benzene) the peak absorption is at about 335 nm. It is possible that the ketone and amine functionalities interact so that the lowest excited state is an *internal* charge transfer (CT) complex (Porter and Suppan, 1964; Cundall and Gilbert, 1970, pp. 64–66):

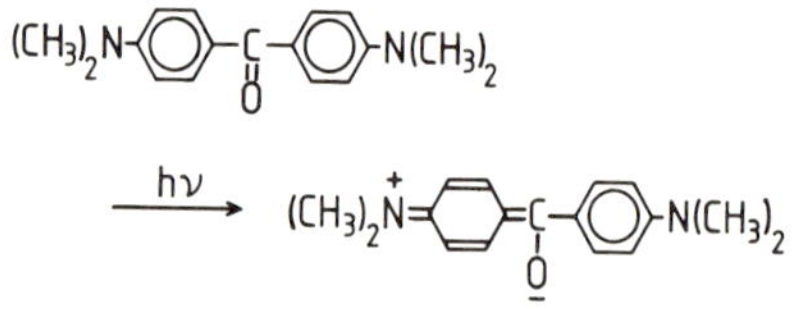

The mixture of benzophenone and Michler's ketone has been found to be very effective in pigmented inks (Ackerman *et al.*, 1972) and has been found to most effective when the two components are intimately mixed before incorporation in the ink vehicle (Hencken, 1975, 1978).

Despite the effectiveness of this synergistic mixture, it has fallen into disuse because of doubts about the toxicity of Michler's ketone, which is a suspect carcinogen (Pappas, 1976). However, the ethyl homologue is available commercially, and behaves in a similar manner (McGinniss, 1979).

3.3.4.3 Thioxanthones

Thioxanthones are derived notionally from benzophenone by the addition of a sulphur atom between the *ortho* positions on the two benzene rings. The presence of the sulphur atom in thioxanthone and its derivatives allows a broad moderately large absorption between about 360 nm and 400 nm with a peak at about 385 nm. Thioxanthone itself is not much used, although it has been suggested as a photosensitizer for triplet activation of quinoline sulphonyl chloride (McGinniss, 1979). Thioxanthones have found great success as photoinitiators for the curing of coatings containing TiO_2 (Hulme *et al.*, 1976; Hauser *et al.*, 1978; Noomen, 1980, 1981).

The thioxanthone derivatives which have been chosen for widespread use have been 2-chlorothioxanthone (CTX), 2-methylthioxanthone (MTX) and 2-isopropylthioxanthone (ITX). These are the derivatives which

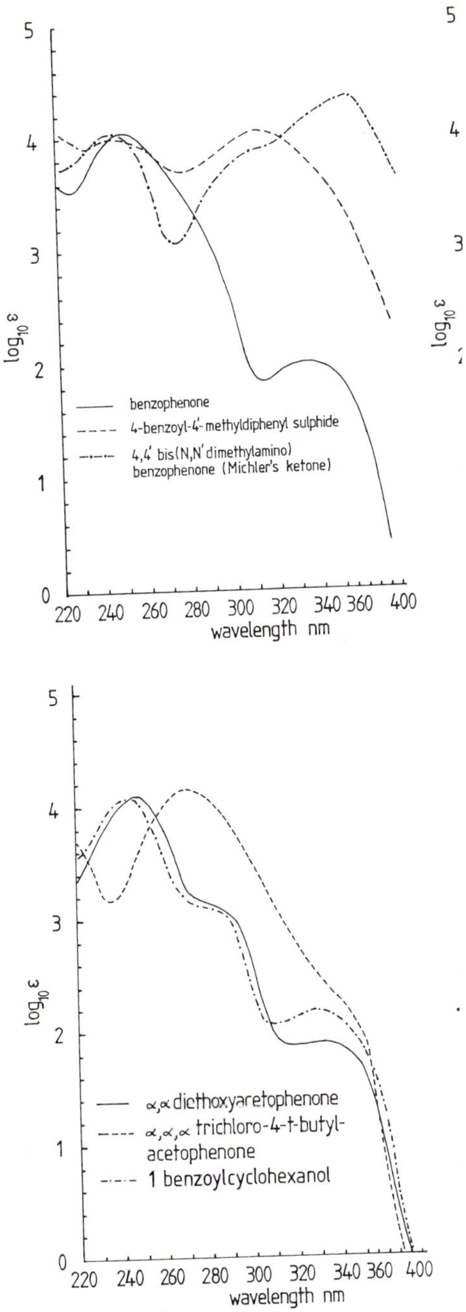

Fig. 3.9. Absorption spectra of photoi
mole-centimetre) calculated from abs

provide the best combination of reactivity, solubility and paleness of colour (Davis *et al.*, 1978). Thioxanthone derivatives require an amine synergist, and excellent results have been obtained with aminobenzoates such as ethyl 4-dimethylaminobenzoate (Hulme *et al.*, 1976; Parrish, 1977), ethyl 2-dimethylaminobenzoate (Davis *et al.*, 1978) and *n*-butoxyethyl 4-dimethylaminobenzoate (Green *et al.*, 1978). The last two aminobenzoates have the advantage of being liquids and solvents for the thioxanthone. The mercury discharge lamp is weak in output between 365 and 405 nm, and lamps dosed to give a high output in this region, in particular with iron iodide, have been found to give faster cure (Hulme *et al.*, 1976; Berner *et al.*, 1978).

Thioxanthone derivatives are effective initiators in formulations containing organic pigments (Plews and Phillips, 1979). Many common organic pigments have reasonable transmission somewhere in the 360–400 nm region, as can be seen from the spectra published by Hencken (1975, 1978) and Plews and Phillips (1979). Printing inks containing a combination of CTX, an aminobenzoate and DMPA have been found to cure very well (Berner *et al.*, 1978).

The amine radical derived from an aminobenzoate is more likely to be an effective initiator than the ketyl radical derived from the thioxanthone derivative. Figure 3.8 shows the changes in the UV absorption of a thin film of an urethane acrylate-based varnish, containing CTX (2% w/w) and ethyl 4-dimethylaminobenzoate (3% w/w) as a photoinitiator system, when

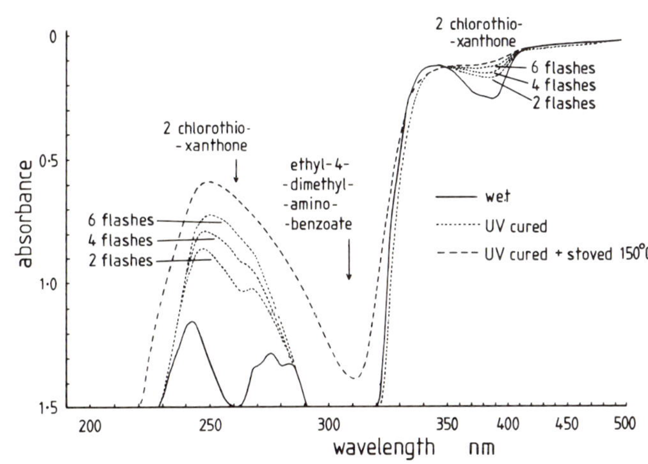

Fig. 3.8. Effects of UV cure and subsequent stoving on the UV absorption spectrum of an unpigmented coating of the following composition: aliphatic urethane acrylate, 68·7%; trimethylolpropane triacrylate, 27·3%; 2-chlorothioxanthone, 2%; ethyl 4-dimethylaminobenzoate, 3%.

cured with successive flashes
stoved for 10 min at 150 °C. It i
CTX chromophore is graduall
aminobenzoate absorption ren
reaction does not affect the ch
all traces of CTX chromopho
There remains a strong ami
benzoate chromophore is in
3.8 demonstrates that UV
technique for investigating t
system. Components which
spectrum of the complete fo
spectrum as a result of their

3.3.5 Absorption spectra of pl

Absorption spectra for some
3.9. There is no ideal solvent
of photoinitiators. Ethanol w
Fig. 3.9 because it combine
absorption in the region of n
that the polarity of the solve
measured. In general, n→π
absorptions a red shift as the
p. 108). These effects are
exception to this rule in the
the spectra of benzoin and it
significant differences when t
and Ledwith, 1973, 1974). In
hydrogen bonding in the c
molecular hydrogen bonding
ethanol, and results in a *red*

3.4 Cationic Photoiniti

Although the general classe
cationic polymerization are
there is much less certainty
The classes of most signifi

formed, the hydrogen ion being derived from the solvent. The anion remained unchanged.

Knapczyk and his coworkers concluded from their examination of the photolysis products, and the effects on these of radical scavengers, that the primary photochemical process is a homolytic cleavage, of a C—I bond in the case of iodonium salts, and a C—S bond in the case of sulphonium salts, i.e.

$$Ar_2I^+ \xrightarrow{h\nu} ArI^{+\cdot} + Ar^\cdot$$

$$Ar_3S^+ \xrightarrow{h\nu} Ar_2S^{+\cdot} + Ar^\cdot$$

When diphenyl iodonium tetrafluoroborate is irradiated at 254 nm in isopropanol, the major products are benzene, iodobenzene, H^+ and acetone, with very small amounts of biphenyl and phenyl isopropyl ether being formed (Knapczyk *et al.*, 1972). Crivello and Lam (1976, 1977) carried out similar photolyses with 4,4′-di-*t*-butyldiphenyliodonium tetrafluoroborate in acetonitrile, ethanol–water, or acetone solvents. They found 4-*t*-butyliodobenzene and 4-*t*-butylbenzene as major products, with minor products derived from radical combination reactions. On the basis of their own work, and of the results of Knapczyk and coworkers, Crivello and Lam suggested a scheme by which Brönsted acid can be formed. This scheme is illustrated here for iodonium salts:

Major

$$Ar_2I^+\ X^- \xrightarrow{h\nu} [Ar_2I^+\ X^-]^*$$

$$Ar_2I^+\ X^- \longrightarrow ArI^{+\cdot} + Ar^\cdot + X^-$$

$$ArI^{+\cdot} + SH \longrightarrow ArI^+H + S^\cdot$$

$$ArI^+H \longrightarrow ArI + H^+$$

Minor

$$[Ar_2I^+\ X^-] + SH \longrightarrow ArSH^+ + ArI + X^-$$

$$[ArSH]^+ \longrightarrow ArS + H^+$$

where SH represents a monomer or solvent.

There seems to be little doubt that a Brönsted acid is primarily responsible for the initiation of polymerization. However, Ledwith *et al.* (1981) have recently stated that ". . . extrapolation of photofragmentation patterns established for iodonium and sulphonium salts in polar media (e.g. acetone), where free ions predominate, to the much less polar photocurable epoxide coatings, where ion pairs and higher aggregates predominate, may not be justified".

It has been suggested that the decomposition of an onium salt may occur by a chain reaction induced by a radical derived from solvent, monomer or a free radical photoinitiator (Abdul-Rasoul *et al.*, 1978; Ledwith, 1978, 1979; Smets *et al.*, 1980). Quantum yields greatly in excess of unity have been measured for diazonium salts (Smets *et al.*, 1980) but only in the absence of oxygen. Oxygen strongly inhibits the chain decomposition mechanism, and therefore it may not be operational in surface coatings in air.

Crivello and Lam (1979a) found that the photolysis of triphenyl sulphonium hexafluorophosphate or arsenate was independent of the presence of free radical photoinitiators, oxygen, radical traps, triplet quenchers or temperature.

If the anion is non-nucleophilic it has very little effect on the absorption spectrum. Furthermore the anion has little effect on the rate of photolysis (Crivello and Lam, 1979a). Diaryl iodonium and triaryl sulphonium salts with common substituents on the aromatic rings absorb strongly in the region of 230–260 nm, with very little absorption above 300 nm. In contrast to the situation with diazonium salts, simple substituents have very little effect. Therefore, the simple diaryl iodonium and triaryl sulphonium salts are not particularly effective for the curing of thick or pigmented sections. Another problem is that the aryl iodide formed from the iodonium salt, and the diaryl sulphide formed from the sulphonium salt, have absorption spectra which are almost identical to those of the parent compounds, and therefore compete for radiation. There are two solutions to this problem. The first is to use a complex sulphonium salt with a larger chromophore. Two examples are shown below, both of which have significant absorption above 300 nm, and are more effective than triphenyl sulphonium salts (Smith and Olofson, 1979; Crivello and Lam, 1980; Chang, 1980). Apart from the larger absorption at longer wavelengths, these salts on photolysis give products which do not compete for radiation with the parent salt.

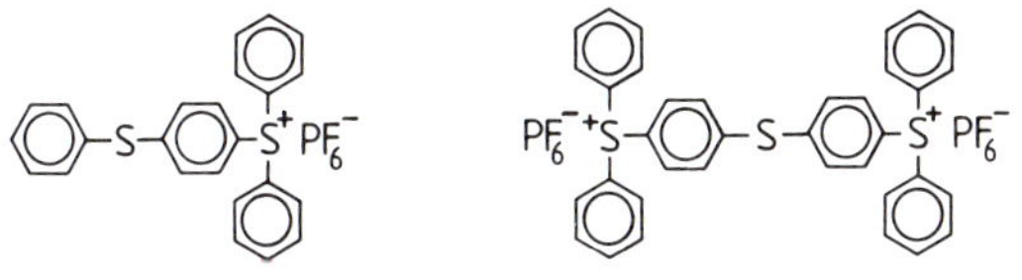

It is an interesting fact that much of the early laboratory work on "triphenyl sulphonium" salts was in fact carried out using mixtures of these complex salts without the workers involved realizing it, their samples being derived from impure commercially available "triphenyl sulphonium" chloride (Pappas, 1981). At Metal Box we developed a preference for using what we at first thought was triphenyl sulphonium hexafluorophosphate in formulations containing organic pigments, for this reason (see Fig. 3.12).

The second approach is to use a photosensitizer which will absorb at longer wavelengths. Certain dyes, e.g. acridine orange and benzoflavin, are able to activate systems containing iodonium salts to wavelengths in the near UV and visible (Crivello and Lam, 1978c). Perylene is able to activate systems containing sulphonium salts to wavelengths between 390 and 475 nm (Crivello and Lam, 1979b). The mechanism of photosensitization has been studied by Pappas and Jilek (1979), who concluded that electron transfer takes place from the photosensitizer to the onium salt. It is found that iodonium salts (with more favourable reduction potentials) can be photosensitized by a larger range of compounds than can sulphonium salts. Abdul-Rasoul *et al.* (1978), Ledwith (1978) and Pappas (1981) have found that free radical photoinitiators are able to act as photosensitizers for onium salts because the radicals formed on photolysis are powerful electron donors.

It is becoming clear that a variety of mechanisms are available by which a monomer cation can be produced, and which of these predominates may depend on the exact circumstances. In general one may expect the monomer to be involved in the mechanism by which the initiating cation is produced. However, it has been found possible to irradiate triphenyl sulphonium hexafluoroarsenate in chloroform, prior to the addition of monomer, and still induce polymerization (Crivello, 1978).

While the anion has little effect on either the absorption spectra or the rate of photolysis of iodonium and sulphonium salts, it does have a profound effect on the rate of polymerization. The order of merit is $SbF_6^- > AsF_6^- > PF_6^- \gg BF_4^-$, and is associated with the size of the anion. The larger the anion the less association there is with the propagating cation, and therefore the higher the rate of propagation.

3.4.4 Applications and problems

The potential benefits of cationic systems as described in Section 3.2.7 have been largely realized (Ludwigsen, 1974; Deviny, 1976; Strand, 1976; Kuhn, 1978). At Metal Box we have spent several years examining coatings of various types based on epoxides and other cationically polymerizable monomers and oligomers, using onium salts as photoinitiators. Varnishes and white pigmented coatings for metal decorating, based on diazonium salts, have been used commercially. Because of the poor thermal stability of systems containing diazonium salts they are supplied in two parts, to be mixed prior to use.

There has been very little use of iodonium and sulphonium salts. The reasons are partly commercial and economic, and partly technical. Iodonium and sulphonium salts have not been as readily available in large quantities as the common free radical photoinitiators, and the rather confused patent situation has had an inhibiting effect.

Although the range of monomers and resins which can be used is extensive, in practice, if rapid cure at ambient temperature is required, it is necessary to incorporate quite a large proportion of cycloaliphatic epoxide in any formulation. Vinyl ethers cure more rapidly, but vinyl ethers suitable for incorporation in surface coatings do not appear to be commercially available. Cycloaliphatic epoxides are quite a lot more expensive than the more common DGEBA or epoxy novolak types, and this fact, coupled with the high cost of iodonium and sulphonium salts, has meant that projected costs of formulations based on iodonium and sulphonium salts have often been found to be greater than those for comparable free radical curing formulations.

While direct comparison of the functional performance of cured coatings based on free radical and cationic mechanisms often shows the superiority of the cationic type (Aloye, 1976), experience shows that cationic systems have two quite serious problems associated with the curing mechanism. The first problem is that ring-opening polymerization is, at ambient temperature, quite a lot slower than the polymerization of acrylate double bonds, so that while the time taken to achieve an acceptable level of cure ("tack-free time") is usually less than a second with acrylate systems, it may be more than a minute with an onium salt–epoxide system (Watt, 1978). Figure 3.10 shows the result of an experiment in which an epoxide system containing an iodonium salt initiator is coated on a salt disc and irradiated, *in situ* in an infra-red spectrophotometer, with a single flash of light of 500 μs duration from a photographic flashgun. The aliphatic ether absorption at $1080\,cm^{-1}$ monitors the rate at which ring opening occurs. A comparison with Fig. 3.5 shows how much slower the curing process is. Note that the epoxide system contains 2-ethyl-9,10-dimethoxyanthracene as a photosensitizer. The iodonium salt by itself would not respond to the long wave UV and visible radiation from the flashgun. The tack-free time for ring-opening polymerization can be substantially reduced by raising the ambient temperature, particularly if the relative humidity is high (Watt, 1978).

The second, and perhaps more serious, problem is that even minute quantities of certain types of adventitious impurity can inhibit cure, partly or even totally. These impurities are nucleophiles which can interact either with the onium salt, or with a propagating cation. Amines are the most common example. Formulating unpigmented systems does not present any major difficulty, although inhibition may be observed if a coating machine is used which has previously been used with free radical curing coatings,

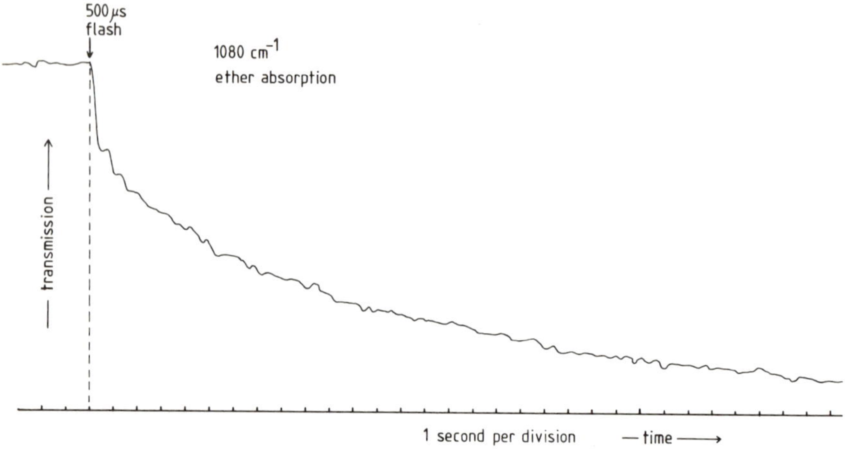

Fig. 3.10. Flash curing of epoxide-based coating *in situ* in an infra-red spectrophotometer (cf. Fig. 3.5). Coating composition: DEN 438 epoxy novolak (Dow), 48·5%; CY179 cycloaliphatic epoxide (Ciba), 48·5%; diphenyl iodonium hexafluoroarsenate, 2·0%; 2-ethyl-9,10-dimethoxyanthracene, 1·0%.

which often contain amines. A scrupulous cleaning with a solvent mix containing acetic acid is recommended. Likewise difficulties may be encountered if a cationic curing varnish is coated over a free radical curing ink (Aloye, 1976).

The formulation of printing inks presents a much greater problem, because many pigments contain basic components. At Metal Box our involvement with UV curing has been predominantly in the area of printing inks rather than other coatings, and for this reason we have made a detailed examination of the influence of pigments on the cure of epoxide mixes with iodonium and sulphonium salt photoinitiators. About 100 commercially available pigments for use in printing inks were investigated and it was found that about a third caused partial or total inhibition of cure for reasons which could not be related simply to the absorption spectrum of the pigment. It was found that the pattern of inhibition could not be related simply to the colour index of the pigment, which characterizes the chemical structure of the chromophore. Although certain classes of pigments, e.g. basic dye complexes, invariably caused inhibition, with other groups of pigments, such as lithol rubines which all have the same pigment number (C.I. Red 57.1), the degree of inhibition was found to vary greatly between different commercial grades. The situation is complicated firstly by the fact that pigments often have surface treatments to improve their properties, and this might introduce nucleophilic components, and secondly by the varying degrees of solubility of the pigment components in the resin system. When the pigment screening trials were carried out a range of iodonium and sulphonium salt initiators was used. The iodonium salts and some of the sulphonium salts were used as pure compounds. Other sulphonium salts were used in the form of "curatives" supplied by either 3M or GE. These curatives were sulphonium salts dissolved in solvents. We found that with quite a large number of pigments, the degree of inhibition varied markedly with the nature of the initiator. Least inhibition was found when pure salts were used, or with curatives using epoxides as solvents, and greatest inhibition was found with curatives containing propylene carbonate as solvent (Crivello, 1981). It seems that with some pigments propylene carbonate is able to dissolve the basic component and allow it to inhibit cure.

Water has a generally inhibiting but rather unpredictable effect on cure. Relative humidity influences cure, particularly at low temperatures (Watt, 1978). The influence of water is important because it has a bearing on whether it is possible to print cationically curing inks by the lithographic process.

3.5 Hybrid Systems

Sometimes there are advantages to be obtained by formulating coatings the cure of which involves two distinct mechanisms. The rationale is to use the advantages of each to compensate for the disadvantages of the other. It is possible to combine free radical and cationic propagation in the same formulation (Ketley and Tsao, 1978; Crivello and Lam, 1979c). Sulphonium salts are able to initiate both free radical and cationic polymerization, and therefore simultaneous polymerization of acrylates and epoxides.

Other types of hybrid system involve an initial rapid UV cure to provide a tack-free surface followed by a thermal cure either at ambient or elevated temperature. Cermak and Mleziva (1974) have advocated the use of a combination of a photoinitiator and a thermal initiator for the crosslinking of thick films of unsaturated polyesters. Noomen (1981) has described the preparation and use of resins containing both acrylate and isocyanate, or acrylate and epoxide functionality, particularly in pigmented coatings. Coates Bros (1980) have patented a formulation which consists of an acrylate monomer and photoinitiator combined with an epoxy resin and a polyamine. The system is supplied in two parts, the polyamine being added prior to use, and providing a slow ambient cure (24 h) of the epoxy resin. These UV–thermal hybrids may provide cheaper and more effective

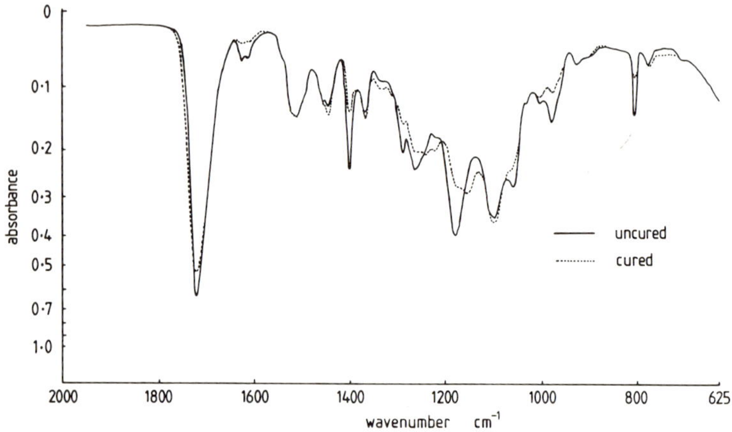

Fig. 3.11. Infra-red spectra of coating of composition aliphatic urethane acrylate, 68·7%; trimethylolpropane triacrylate, 27·3%; α,α-diethoxyacetophenone, 5·0%; before and after cure.

phthalocyanine blue. The inks were printed at a nominal 3 μm thickness on sodium chloride discs. The details of the resins and pigment used, and the exact formulations are shown in Table 3.3. The prints were cured with a medium pressure mercury lamp (input power 80 W cm^{-1}) mounted in a focusing semi-elliptical reflector angled across a conveyor moving at 1·5 m s^{-1}. Two passes were allowed. Infra-red spectra were measured before and after cure. The aliphatic ether absorption which appears on

Table 3.3. Details of resins, pigments, etc., for experiments described in Fig. 3.12. Resin mix: DEN439 Epoxy Novolak Resin (Dow), 42·5%; CY179 Cycloaliphatic Epoxy Resin (Ciba Geigy), 42·5%. Pigment: Irgalite GLA Phthalocyanine Blue (Ciba Geigy), 15%

Formulation	Photoinitiator (3%)	Photosensitizer†
a	$\phi_2I^+ PF_6^-$	—
b	$\phi_2I^+ AsF_6^-$	—
c	$(Me\phi)_2I^+ BF_4^-$	—
d	$(Me\phi)_2I^+ PF_6^-$	—
e	$(Me\phi)_2I^+ PF_6^-$	0·5%
f	$\phi_2I^+ PF_6^-$	0·5%
g	$\phi_3S^+ PF_6^-$‡	—
h	$\phi_3S^+ PF_6^-$‡	0·5%

† 2-Ethyl-9,10-dimethoxyanthracene.

‡ Impure, may contain 4-diphenylsulphoniodiphenylsulphide and 4,4′-bis(diphenylsulphonio)diphenylsulphide salts (Pappas, 1981).

cure has a very broad peak, and is superimposed on other peaks which do not change during cure. The ether absorption which develops is best measured by change in absorbance (ΔA) against wavelength. In order to allow for small variations in film thickness, ΔA is divided by the absorbance due to a group which does not change on cure. In this case the carbonyl absorption at $1720\,cm^{-1}$ due to the ester group in the cycloaliphatic epoxide was chosen.

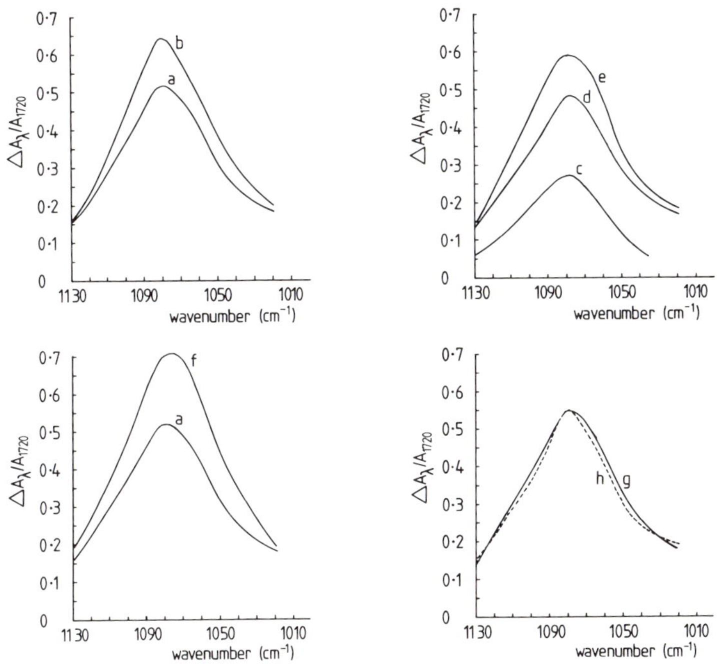

Fig. 3.12. Comparison of iodonium and sulphonium salts for the cure of a blue epoxide-based ink. Prints on NaCl discs. Cure measured by growth of aliphatic ether absorption at $1080\,cm^{-1}$.

The results shown in Fig. 3.12 demonstrate some of the effects described in Section 3.4.3. It can be seen that the degree of cure obtained depends on the anion in the ranking order $AsF_6^- > PF_6^- \gg BF_4^-$. The sulphonium salt gives better cure than the iodonium salts because being an impure mix of the complex sulphonium cations described in Section 3.4.3 it has greater long wavelength absorption. The photosensitizer has a greater effect on the cure with the iodonium salts than on the cure with the sulphonium salt.

References

Abdul-Rasoul, F. A. M., Ledwith, A. and Yagci, Y. (1978). *Polymer* **19,** 1219–1222.

Ackerman, J. F., Weisfeld, J., Savageau, R. G. and Beerli, G. (1972). US Pat. 3,673,140 (Inmont Corporation).

Alexander, I. J. and Scott, R. J. (1980). Eur. Pat. Appl. 14,293 (Scott Bader Ltd).

Aloye, J. A., Jr (1976). SME Tech. Pap. (Ser.) FC76-504.

Ando, W. (1978). *In* "Chemistry of Diazonium and Diazo Groups", Part 1 (Ed. S. Patai), pp. 341–487, John Wiley & Sons, Chichester.

Arimitsu, S. and Masuhara, H. (1973). *Chem. Phys. Lett.* **22,** 543–546.

Armbruster, D. C., Zeliznak, K. J., Rybny, C. B. and Vona, J. A. (1978). SME Tech. Pap. (Ser.) FC78-507. Also in *J. Radiat. Curing* **6** (4), 10–19 (1979).

Bailey, W. J. (1977). *ACS Div. Polym. Chem. Preprints* **18** (1), 17–22.

Balz, G. and Schiemann, G. (1927). *Berichte* **60,** 1186–1190.

Barker, P., Bottom, R. A., Guthrie, J. T., Godfrey, A. A., Green, P. N. and Young, J. R. A. (1981). *Res. Discl.* February, 93.

Barrett, J. L. (1979). *J. Radiat. Curing* **6** (4), 20–28.

Bartholomew, R. F. and Davidson, R. S. (1971). *J. Chem. Soc. C.* **1971,** 2342–2346.

Bartholomew, R. F., Davidson, R. S., Lambeth, R. F., McKellar, J. F. and Turner, P. H. (1972). *J. Chem. Soc., Perkin Trans. II* **1972,** 577–582.

Bassemir, R. W. and Bean, A. J. (1974). Paper presented at 26th Annual Meeting of TAGA, St Paul, Minn., 13–15 May 1974; *TAGA Proc.*, **1974,** 133–146.

Beckett, A. and Porter, G. (1963). *Trans. Faraday Soc.* **59,** 2038–2050.

Belicar, J. (1977). *Radiat. Curing* **4** (1), 15–19.

Berner, G., Kirchmayr, R. and Rist, G. (1978). *J. Oil Colour Chem. Assoc.* **61,** 105–113.

Billingham, N. C. and Ledwith, A. (1974). *In* "Reactivity, Mechanism and Structure in Polymer Chemistry" (Ed. A. D. Jenkins and A. Ledwith), pp. 1–30. John Wiley & Sons, London.

Block, H., Ledwith, A. and Taylor, A. R. (1971). *Polymer* **12,** 271–288.

Bolon, D. A. and Webb, K. K. (1978). *J. Appl. Polym. Sci.* **22,** 2543–2551.

Borer, A., Kirchmayr, R., and Rist, G. (1978). *Helv. Chim. Acta* **61,** 305–324.

Bosley, J. A., Ledwith, A. and Purbrick, M. D. (1978). *J. Oil Colour Chem. Assoc.,* **61,** 95–104.

Braun, D. and Becker, K. H. (1971). *Makromol. Chem.* **147,** 91–99.

Brown, K. H. (1977). *Radiat. Curing* **4** (4), 8, 9, 12–17.

Bush, R. W., Ketley, A. D., Morgan, C. R. and Whitt, D. G. (1980). *J. Radiat. Curing,* **7** (2), 20–25.

Butler, G. B. and Ferree, W. I., Jr (1975). *ACS Div. Org. Coat. Plast. Chem. Preprints* **35** (1), 84–89.

Calvert, J. G. and Pitts, J. N., Jr (1966). "Photochemistry". John Wiley & Sons, New York.

Carlblom, L. H. and Pappas, S. P. (1977). *J. Polym. Sci., Polym. Chem. Ed.* **15,** 1381–1391.

Cermak, V. and Mleziva, J. (1974). *Farbe Lack* **80,** 125–133.

Chang, K.-T. (1980). US Pat. 4,197,174 (American Can Co.).

Chang, Y. C. (1977). *Photogr. Sci. Eng.* **21,** 348–350.

Chinmayanandam, B. R. and Melville, H. W. (1954). *Trans. Faraday Soc.* **50,** 73–82.

Ciamician, G. and Silber, P. (1901). *Berichte* **34,** 1530–1543.

Clarke, S. R. and Shanks, R. A. (1980). *J. Macromol. Sci. A* **14,** 69–78.

Closs, G. L. and Paulson, D. R. (1970). *J. Amer. Chem. Soc.* **92,** 7229–7231.

Coates Bros (1980). Brit. Pat. (A) 2,029,433.

Cohen, S. G. and Chao, H. M. (1968). *J. Amer. Chem. Soc.* **90,** 165–173.

Cohen, S. G. and Cohen, J. I. (1967). *J. Amer. Chem. Soc.* **89,** 164–165.

Cohen, S. G. and Stein, N. (1969). *J. Amer. Chem. Soc.* **91,** 3690–3691.

Cohen, S. G., Parola, A. and Parsons, G. H., Jr (1973). *Chem. Rev.* **73,** 141–161.

Cohen, T. and Lipowitz, J. (1974). *Tetrahedron Lett.* **1974,** 3721–3725.

Collins, G. L., Young, D. A. and Costanza, J. R. (1976). *J. Coat. Technol.* **48,** (618), 48–51; SME Tech. Pap., (Ser.) FC76-484.

Colthup, N. B., Daly, L. H. and Wiberley, S. E. (1964). "Introduction to Infrared and Raman Spectroscopy". Academic Press, New York and London.

Cox, A., Kemp, T. J., Payne, D. R. and Pinot de Moira, P. (1977). *J. Photogr. Sci,* **25,** 208–215.

Crivello, J. V. (1978). US Pat. 4,081,276 (General Electric Co.).

Crivello, J. V. (1981). US Pat. 4,273,668 (General Electric Co.).

Crivello, J. V. and Lam, J. H. W. (1976). *J. Polym. Sci., Polym. Symp.* **56,** 383–395.

Crivello, J. V. and Lam, J. H. W. (1977). *Macromolecules* **10,** 1307–1315.

Crivello, J. V. and Lam, J. H. W. (1978a). *ACS Div. Org. Coat. Plast. Chem.* **39,** 31–35.

Crivello, J. V. and Lam, J. H. W. (1978b). *ACS Symp. Ser.* **114,** 1–16.

Crivello, J. V. and Lam, J. H. W. (1978c). *J. Polym. Sci., Polym. Chem. Ed.* **16,** 2441–2451.

Crivello, J. V. and Lam, J. H. W. (1979a). *J. Polym. Sci., Polym. Chem. Ed.* **17,** 977–999.

Crivello, J. V. and Lam. J. H. W. (1979b). *J. Polym. Sci., Polym. Chem. Ed.* **17,** 1059–1065.

Crivello, J. V. and Lam, J. H. W. (1979c). *J. Polym. Sci., Polym. Lett. Ed.* **17,** 759–764.

Crivello, J. V. and Lam, J. H. W. (1980). *J. Polym. Sci., Polym. Chem. Ed.* **18,** 2677–2695, 2697–2714.

Crivello, J. V., Lam, J. H. W. and Volante, C. N. (1977). *J. Radiat. Curing* **4** (3), 2–16.

Crivello, J. V., Lam, J. H. W., Moore, J. E. and Schroeter, S. H. (1978). *J. Radiat. Curing* **5** (1), 2–17.

Cundall, R. B. and Gilbert, A. (1970), "Photochemistry". Nelson, London.

Dack, M. R. J. (1973). *J. Chem. Educ.* **50,** 169–173.

Davidson, R. S. and Lambeth, P. F. (1968). *Chem. Commun.* **1968,** 511–512.

Davis, M. J., Doherty, J., Godfrey, A. A., Green, P. N., Young, J. R. A. and Parrish, M. A. (1978). *J. Oil Colour Chem. Assoc.* **61,** 256–263.

Decker, C. and Fizet, M. (1980). *Makromol. Chem. Rapid Commun.* **1,** 637–642.

Decker, C., Faure, J., Fizet, M. and Rychla, L. (1979). *Photogr. Sci. Eng.* **23,** 137–140.

Decker, C., Fizet, M. and Faure, J. (1980). *ACS Div. Org. Coat. Plast. Chem. Preprints* **42,** 710–715.

Delzenne, G. A. (1971). *J. Macromol. Sci. Revs Polym. Technol. D* **1,** 185–215.

Delzenne, G. A. (1976). *In* "Encyclopedia of Polymer Science and Technology", Suppl. 1, pp. 401–443. Wiley–Interscience, New York.
Delzenne, G. A. (1979a). *Makromol. Chem.* Suppl. 2, 169–188.
Delzenne, G. A. (1979b). *J. Radiat. Curing* **6** (4), 2–9.
Delzenne, G. A. (1979c). *Advanc. Photochem.* **11,** 1–103.
Delzenne, G. A., Laridon, U. and Peeters, H. (1970). *Eur. Polym. J.* **6,** 933–943.
DePoortere, M., Ducarme, A., Dufour, P. and Merck, Y. (1978). *J. Oil. Colour Chem. Assoc.* **61,** 195–203.
Deviny, E. J. (1976). SME Tech. Pap. (Ser.) FC76-503.
Dreyfuss, M. P. and Dreyfuss, P. (1965) *Polymer* **6,** 93–95.
Dreyfuss, M. P. and Dreyfuss, P. (1966). *J. Polym. Sci. A1* **4,** 2179–2200.
Eichler, J., Herz, C. P. and Schnabel, W. (1980a). *Angew. Makromol. Chem.* **91,** 39–54.
Eichler, J., Herz, C. P., Naito, I. and Schnabel, W. (1980b). *J. Photochem.* **12,** 225–234.
Ellinger, L. P. (1963). *Chem. Ind.* **1963,** 1982–1983.
Evans, A. J., Armstrong, C. and Tolman, R. J. (1978). *J. Oil Colour Chem. Assoc.* **61,** 251–255.
Feinberg, J. H. (1976). US Pat. 3,960,684 (American Can Co.).
Fields, J. F. (1978). SME Tech. Pap. (Ser.) FC78-504.
Fields, J. F. (1979). *Radiat. Curing* **6** (2), 14–19.
Fischer, E., Kuellmar, K. and Weissermel, K. (1965). US Pat. 3,210,297 (Hoechst).
Fuhr, K. (1977). *Polym. Paint Colour J.* **167,** 672–675.
Geary, J. T. (1977). *J. Coat. Technol.* **49** (626), 25–28.
Green, G. E. and Stark, B. P. (1981). *Chem. Brit,* **17,** 228, 230, 232, 234.
Green, P. N., Young, J. R. A. and Davis, M. J. (1978). *Paint. Manuf.* **48** (3), 32.
Hageman, H. J., Van Der Maeden, F. P. B. and Janssen, P. C. G. M. (1979). *Makromol. Chem.* **180,** 2531–2537.
Harm, W. (1980). "Biological Effects of Ultraviolet Radiation". Cambridge University Press, Cambridge.
Hauser, P., Osterloh, R. and Jacobi, M. (1978). *Proc. XIVth FATIPEC Congr., Budapest* 241–247.
Heine, H.-G. (1972). *Tetrahedron Lett.* **1972,** 4755–4758.
Heine, H.-G. and Traenchner, H. J. (1975). *Prog. Org. Coatings* **3,** 115–139.
Heine, H.-G., Rosenkranz, H. J. and Rudolph, H. (1972). *Angew. Chem. Internat. Edit.* **11,** 974–978.
Hencken, G. (1975). *Farbe Lack* **81,** 916–923.
Hencken, G. (1978). *Amer. Ink Maker* **56** (3), 57, 58, 60, 64, 66, 68, 70.
Hepher, M. (1964). *J. Photogr. Sci.* **12,** 181–190.
Herz, C. P. and Eichler, J. (1979). *Farbe Lack* **85,** 933.
Holman, R. J. and Rubin, H. (1978). *J. Oil Colour Chem. Assoc.* **61,** 189–194.
Hulme, B. E., Hird, M. J. and Marron, J. J. (1976). SME Tech. Pap. (Ser.) FC76-492.
Hutchison, J. and Ledwith, A. (1973). *Polymer* **14,** 405–408.
Hutchison, J. and Ledwith, A. (1974). *Advanc. Polym. Sci.* **14,** 49–86.
Hutchison, J., Lambert, M. C. and Ledwith, A. (1973). *Polymer* **14,** 250–254.
Irie, M., Tomimoto, S. and Hayashi, K. (1970). *J. Polym. Sci. Polym. Lett. Ed.* **8,** 585–588.

Irie, M., Tomimoto, S. and Hayashi, K. (1972). *J. Phys. Chem.* **76,** 1419–1424.
Irie, M., Masuhara, H., Hayashi, K. and Mataga, N. (1974). *J. Phys. Chem.* **78,** 341–347.
Irie, M., Yamamoto, Y. and Hayashi, K. (1977). *Pure Appl. Chem.* **49,** 455–461.
Jablonski, A. (1935). *Z. Phys.* **94,** 38–46.
Jirkovsky, J., Fojtik, A. and Becker, H. G. O. (1981). *Collect. Czech. Chem. Commun.* **46,** 1560–1565.
Kasha, M. (1950). *Disc. Faraday Soc.* **9,** 14–19.
Ketley, A. D. and Tsao, J.-H. (19768). *ACS Div. Polym. Chem. Preprints* **19,** 656–661.
Kinstle, J. F. and Watson, S. L., Jr (1976). *J. Radiat. Curing* **3** (1), 2–11.
Knapczyk, J. W. and McEwen, W. E. (1970). *J. Org. Chem.* **35,** 2539–2543.
Knapczyk, J. W., Lubinowski, J. J. and McEwen, W. E. (1972). *Tetrahedron Lett.* **1972,** 3739–3742.
Knappe, E. F., Ahlers, K. and Haring, E. (1978). SME Tech. Pap. (Ser.) FC78-500.
Koch, T. H. and Jones, A. H. (1970). *J. Amer. Chem. Soc.* **92,** 7503–7505.
Kornis, G. and de Mayo, P. (1964). *Canad. J. Chem.* **42,** 2822–2827.
Kosar, J. (1965). "Light Sensitive Systems". John Wiley & Sons, New York.
Krajewski, J. J., Packer, E. S. and Yamazaki, T. (1976). *J. Coat. Technol.* **48** (623), 43–49; *Met. Finish.* **74** (11), 58–63.
Kuhlmann, R. and Schnabel, W. (1977). *Polymer* **18,** 1163–1168.
Kuhn, W. M. (1978). SME Tech. Pap. (Ser.) FC78-549.
Kurpiewski, T. and McDowell, J. R. (1978). SME Tech. Pap. (Ser.) FC78-509.
Lamola, A. A. and Hammond, G. S. (1965). *J. Chem. Phys.* **43,** 2129–2135.
Ledwith, A. (1976a). *J. Oil. Colour Chem. Assoc.* **59,** 157–165.
Ledwith, A. (1976b). *Polymer* **17,** 957–964.
Ledwith, A. (1978). *Polymer* **19,** 1217–1219.
Ledwith, A. (1979). *Pure Appl. Chem.* **51,** 159–171.
Ledwith, A. and Sherrington, D. C. (1974). *In* "Reactivity, Mechanism and Structure in Polymer Chemistry" (Ed. A. D. Jenkins and A. Ledwith), pp. 244–309. John Wiley & Sons, London.
Ledwith, A., Russell, P. J. and Sutcliffe, L. H. (1972). *J. Chem. Soc. Perkin Trans. II* **1972,** 1925–1928.
Ledwith, A., Bosley, J. A. and Purbrick, M. D. (1978). *J. Oil Colour Chem. Assoc.* **61,** 95–104.
Ledwith, A., Al-Kass, S., Sherrington, D. C. and Bonner, P. (1981). *Polymer* **22,** 143–145.
Lee, G. A. and Doorakian, G. A. (1977). *J. Radiat. Curing* **4** (1), 2–7.
Lee, H. and Neville, K. (1967). "Handbook of Epoxy Resins". McGraw-Hill, New York.
Licari, J. J. and Crepeau, P. C. (1965). US Pat. 3,205,157 (North American Aviation Inc.).
Ludwigsen, R. J. (1974). SME Tech. Pap. (Ser.) FC74–533. See also *Radiat. Curing* **2** (1), 10–14 (1975).
Mao, T. J. and Eldred, R. J. (1967). *J. Polym. Sci. A1* **5,** 1741–1751.
McGinniss, V. D. (1975). *J. Radiat. Curing* **2** (1), 3, 4, 6–13.
McGinniss, V. D. (1976a). SME Tech. Pap. (Ser.) FC76-486.
McGinniss, V. D. (1976b). *ACS Symp. Ser.* **25,** 135–149.
McGinniss, V. D. (1979). *Photogr. Sci. Eng.* **23,** 124–136.

McGinniss, V. D. and Dusek, D. M. (1974). *ACS Div. Polym. Chem. Preprints* **15** (1), 480–484; *J. Paint Technol.* **46** (589), 23–30.
McGinniss, V. D. and Holsworth, R. M. (1975). *J. Appl. Polym. Sci.* **19,** 2243–2254.
McGinniss, V. D. and Kah, A. (1977). *Polym. Eng. Sci.* **17,** 478–483.
McGinniss, V. D. and Ting, V. W. (1975). *J. Radiat. Curing* **2** (1), 14–18.
Minsk, L. M., Smith, J. G., Van Deusen, W. P. and Wright, J. F. (1959). *J. Appl. Polym. Sci.* **2,** 302–307.
Mleziva, J. and Cermak, V. (1973). *Farbe Lack* **79,** 1153–1157.
Moore, J. E. (1976). *ACS Div. Org. Coat. Plast. Chem. Preprints* **36** (2), 747–753.
Moore, J. E. (1978). *In* "UV Curing: Science and Technology"' (Ed. S. P. Pappas), pp. 133–159. Technology Marketing Corporation, Stamford, Conn.
Moore, J. E., Schroeter, S. H., Shultz, A. R. and Stang, L. D. (1976). *ACS Symp. Ser.* **25,** 90–106.
Morgan, C. R. and Ketley, A. D. (1973). *ACS Div. Org. Coat. Plast. Chem.* **33** (1), 281–288.
Morgan, C. R. and Ketley, A. D. (1980). *J. Radiat. Curing* **7** (2), 10–13.
Morgan, C. R., Magnotta, F. and Ketley, A. D. (1977). *J. Polym. Sci. Polym. Chem. Ed.* **15,** 627–645.
Mustroph, H., Epperlein, J. and Pietsch, H. (1980). *Photogr. Sci. Eng.* **24,** 196–200.
Nishikubo, T., Imaura, M., Mizuko, T. and Takaoka, T. (1974). *J. Appl. Polym. Sci.* **18,** 3445–3454.
Noomen, A. (1980). *Proc. XVth FATIPEC Congr., Amsterdam* **1,** 346–359.
Noomen, A. (1981). *J. Oil Colour Chem. Assoc.* **64,** 347–354.
Norrish, R. G. W. and Smith, R. R. (1942). *Nature* **150,** 336–337.
Olaj, O. F., Breitenbach, J. W. and Kauffmann, H. F. (1971). *J. Polym. Sci. B* **9,** 877–882.
Osborn, C. L. and Sandner, M. R. (1974). *ACS Div. Org. Coat. Plast. Chem. Preprints* **34** (1), 660–663.
Osborn, C. L. (1976). *J. Radiat. Curing* **3** (3), 2–3, 5–11.
Oster, G. and Immergut, E. H. (1954). *J. Amer. Chem. Soc.* **76,** 1393–1396.
Oster, G. and Yang, N.-L. (1968). *Chem. Rev.* **68,** 125–151.
Pappas, S. P. (1974). *Prog. Org. Coat.* **2,** 333–347.
Pappas, S. P. (1976). SME Tech. Pap. (Ser.) FC76-490.
Pappas, S. P. (1978). *In* "UV Curing Science and Technology" (Ed. S. P. Pappas), pp. 1–22. Technology Marketing Corporation, Stamford, Conn.
Pappas, S. P. (1981). *Radiat. Curing* **8** (3), 28, 30, 32–35.
Pappas, S. P. and Chattopadhyay, A. K. (1973). *J. Amer. Chem. Soc.* **95,** 6484–6485.
Pappas, S. P. and Chattopadhyay, A. K. (1975). *J. Polym. Sci. Polym. Lett. Ed.* **13,** 483–486.
Pappas, S. P. and Jilek, J. H. (1979). *Photogr. Sci. Eng.* **23,** 140–143.
Parrish, M. A. (1977). *J. Oil Colour Chem. Assoc.* **60,** 474–478.
Petterson, R. C., DiMaggio, A., Herbert, A. L., Haley, T. J., Mykytka, J. P. and Sarkar, I. M. (1971). *J. Org. Chem.* **36,** 631–633.
Phillips, R. (1978). *J. Oil Colour Chem. Assoc.* **61,** 233–240.
Pitts, J. N., Jr, Letsinger, R. L., Taylor, R. P., Patterson, J. M., Recktenwald, G. and Martin, R. B. (1959). *J. Amer. Chem. Soc.* **81,** 1068–1077.
Plews, G. and Phillips, R. (1979). *J. Coat. Technol.* **51** (648), 69–77.

Porter, G. and Suppan, P. (1964). *Pure Appl. Chem.* **9,** 499–505.
Pryce, A. (1976). *J. Oil Colour Chem. Assoc.* **59,** 166–170.
Rabek, J. F. (1968). *Photochem. Photobiol.* **7,** 5–57.
Reiser, A. (1980). *J. Chim. Phys.* **77,** 469–481.
Rowe, W. (1976). SME Tech. Pap. (Ser.) FC76-495.
Rubin, H. (1974). *J. Paint Technol.* **46** (588), 74–81.
Rudolph, H., Rosenkranz, H. J. and Heine, H. G. (1975). *Appl. Polym. Symp.* **26,** 157–164.
Sandner, M. R. and Osborn, C. L. (1974). *Tetrahedron Lett.* **1974,** 415–418.
Sandner, M. R., Osborn, C. L. and Trecker, D. J. (1972). *J. Polym. Sci. Polym. Chem. Ed.* **10,** 3173–3181.
Schlesinger, S. I. (1974a). *Photogr. Sci. Eng.* **18,** 387–393.
Schlesinger, S. I. (1974b). *Polym. Eng. Sci.* **14,** 513–515.
Schmidle, C. J. (1977). *Amer. Ink Maker* **55** (3), 41–42, 44, 103.
Smets, G., Aerts, A. and Van Erum, J. (1980). *Polym. J. (Tokyo)* **12,** 539–547.
Smith, G. H. and Olofson, P. M. (1979). UK Pat. Appl. 2,014,144; Ger. Offen. 2,904,626 (Minnesota Mining and Manufacturing Company).
Strand, R. C. (1976). SME Tech. Pap. (Ser.) FC76-507.
Sukigara, M. and Kikuchi, S. (1967). *Bull. Chem. Soc. Jap.* **40,** 1077–1081.
Tazuke, S. (1977). *High Polym.* **29,** 681–723.
Tazuke, S., Asai, M., Ikeda, S. and Okamura, S. (1967). *J. Polym. Chem. B* **5,** 453–457.
Trommsdorff, E., Köhle, H. and Lagally, P. (1948). *Makromol. Chem.* **1,** 169–198.
Turro, N. J. (1978) "Modern Molecular Photochemistry". Benjamin/Cummings Publishing Co., Menlo Park, Calif.
Turro, N. J. and Engel, R. (1969). *Mol. Photochem.* **1,** 143–146, 235–238.
Van Neerbos, A. (1978). *J. Oil Colour Chem. Assoc.* **61,** 241–250.
Vrancken, A. (1977). *Farbe Lack* **83,** 171–179.
Wagner, P. J. and Kemppainen, A. E. (1969). *J. Amer. Chem. Soc.* **91,** 3085–3087.
Wagner, P. J. (1976). *In* "Topics in Current Chemistry 66: Triplet States III", pp. 1–52. Springer Verlag, Berlin.
Wamser, C. C., Hammond, G. S., Chang, C. T. and Baylor, C., Jr (1970). *J. Amer. Chem. Soc.* **92,** 6362–6363.
Watanabe, K. (1978). *Graphic Arts Japan* **20,** 14–18.
Watt, W. R. (1978). *ACS Div. Org. Coat. Plast. Chem.* **39,** 36–41.
Wayne, R. P. (1970). "Photochemistry". Butterworth, London.
Wight, F. R. (1978). *J. Polym. Sci. Polym. Lett. Ed.* **16,** 121–127.
Wight, F. R. and Hicks, G. W. (1977). *Soc. Plast. Eng. Tech. Pap.* **23,** 500–502; see also *Polym. Eng. Sci.* **18,** 378–381 (1978).
Yokota, K., Tomioka, H., Ono, T. and Kuno, F. (1972). *J. Polym. Sci. A1* **10,** 1335–1343.
Yokota, K., Tomioka, H. and Adachi, K. (1973). *Polymer* **14,** 561–564.

4 Radiometry

4.1 Definitions

Understanding the assessment of light sources is complicated by the bewildering number of concepts, definitions and units employed. The situation may be considerably clarified if the subject is divided into three areas:

(1) *Radiometry* is concerned with the total radiant energy at all wavelengths radiated by a source.
(2) *Spectral radiometry* is concerned with the radiation emitted at particular wavelengths, or in particular wavelength regions.
(3) *Photometry* is concerned with the response of the human eye to a light source, i.e. light sources are classed according to their ability to radiate the wavelengths to which the human eye is responsive. As most light sources are designed for illumination purposes, photometry has been the most important area of light measurement, and it is in photometry that most of the trivial units have proliferated.

Luckily, photometry is not relevant to the chemical applications of light, and therefore we can ignore the complexities of candelas, lumens, nits, phots, apostilbs, etc. We shall be concerned mainly with spectral radiometry.

There are six basic terms used in radiometry. These are shown in Table 4.1.

The terms given in Table 4.1 can be converted into spectroradiometric terms simply by specifying the wavelength or the wavelength interval. A unit wavelength interval term must be added to the units of the radiometric

Table 4.1. Terms used in radiometry

Term		Definition	Unit
Radiant energy	Q		joules (J)
Radiant power or radiant flux	$\frac{dQ}{dt} =$	Φ	watts (W)
Radiant emittance	$\frac{d\Phi}{dA}$	Flux per unit area emitted by an extended source	$W\,m^{-2}$
Irradiance	$\frac{d\Phi}{dA}$	Flux per unit area incident on a surface	$W\,m^{-2}$
Radiant intensity	$\frac{d\Phi}{d\omega}$	Flux per unit solid angle emitted by a point source	$W\,sr^{-1}$
Radiance	$\frac{d^2\Phi}{d\omega dA}$	Flux per unit solid angle per unit area emitted by an extended source	$W\,m^{-2}\,sr^{-1}$

term, e.g. spectral irradiance has units of watts per square metre per nanometre ($W\,m^{-2}\,nm^{-1}$).

The concepts which are most relevant to light sources for chemical applications are radiant power, radiant emittance and irradiance. In the lamps which are most commonly used (e.g. medium pressure gas discharges) the area of the source, i.e. the arc, is ill defined, and therefore, as lamps are almost always linear, it is common practice to modify the concept of radiant emittance slightly and refer to the power radiated per unit length of the arc.

4.2 Radiant Power and Radiant Emittance

It is really quite difficult to measure, in absolute terms, the spectral distribution of the total radiant output of a lamp. Firstly, it is difficult to actually collect all the radiation from a lamp and direct this to a suitable measuring device. Secondly, it is difficult to find a measuring device which will respond to all the wavelengths emitted.

Photodetectors measure only the radiation falling on them, and therefore essentially measure irradiance. Practical methods of determining radiant power, or radiant emittance, involve measurement(s) of irradiance (in watts per square metre) followed by a mathematical or experimental integration over a surface area which incorporates the detector surface and upon which essentially all the radiation from the lamp falls.

The classic method of measuring radiant power is to place the radiation source inside an "*integrating sphere*" (Walsh, 1958, p. 257; Grum and Becherer, 1979, p. 255). This is a hollow sphere, the interior of which is coated with a material of high diffuse reflectance. Sumpner (1892) showed that if a light source is placed inside such a sphere, any portion of the internal surface shielded from direct radiation will achieve a uniform radiant emittance which is proportional to the radiant power of the source, and is independent of its shape and dimensions if its size is small compared with the internal dimensions of the sphere. A detector set into the wall of the sphere will give a reading proportional to the radiant emittance of the sphere surface, and therefore to the radiant power of the lamp. A baffle is incorporated to prevent radiation from the lamp striking the detector directly. For an ideal integrating sphere, the irradiance of the detector is given by $(\Phi/4\pi R^2)[\rho/(1-\rho)]$, where R is the radius of the sphere, Φ the total radiant flux of the lamp, and ρ the reflectance of the interior of the sphere. The choice of material and coating method for the interior of the sphere is very critical. Magnesium oxide, formed as a smoke by the burning of magnesium ribbon in air, was popular for many years (Knowles-Middleton and Sanders, 1951; Tellex and Waldron, 1955), but has now been largely replaced by barium sulphate (Grum and Luckey, 1968), which shows better stability (Egan and Hilgeman, 1975). A recent innovation has been the use of fluorocarbon coatings (Eckerle *et al.*, 1976). It is essential, if spurious results are to be avoided, that the coating material does not fluoresce (Saunders and Ott, 1976).

The integrating sphere method is capable of comparing the radiant outputs of lamps of different shapes and radiation patterns. Unfortunately it may not be feasible to build an integrating sphere large enough to accommodate long linear lamps, although such spheres have been constructed by lamp makers and by national standards laboratories (Lambert and Preston, 1961).

I have used, with some success, two other methods to measure the radiant emittance (in watts per centimetre of arc length) in absolute terms for a linear medium pressure mercury lamp, in various spectral regions. Radiant emittance is a rather more useful concept than radiant power for long linear lamps, because the distance between the lamp and the material being processed is often small in comparison with the total length of the lamp.

The first method involved the measurement of the irradiance pattern around an isolated length of a bare lamp, i.e. the determination of "iso-irradiance" contours. By computing the total area of a contour through which all the radiation from the lamp passes, and multiplying this by the irradiance value for that contour, the total emittance from the isolated length of the lamp can be calculated. A photodiode detector with a well defined spectral response and an absolute calibration traceable to NBS standards was used. A linear fluorescent arc lamp radiates very much like an ideal diffuse cylindrical emitter which obeys Lambert's cosine law of emission, and it is possible to calculate to a reasonable approximation the relationship between radiant emittance and irradiance at any point around the lamp (Teele, 1965). Mercury and xenon lamps without fluorescent coatings appear to be rather more complicated. Figure 4.1 shows a *polar diagram* measured for a medium pressure mercury lamp using a narrow band 365 nm filter. The polar diagram shows how the irradiance, at a particular distance from a *small* section of the arc length, varies with the angle to the normal. A perfect Lambertian emitter would give a perfect circle, with the lamp as a tangent. Gavrilova *et al.* (1978) have measured polar diagrams for tubular xenon lamps, and have found that the shape of the curve depends on the optical density of the discharge at the wavelength being measured. As the optical density increases the curve approaches a circle, but at low optical densities, in the UV and visible, the curve is distorted to allow a higher irradiance at large angles to the normal than would be expected from a perfectly diffuse emitter.

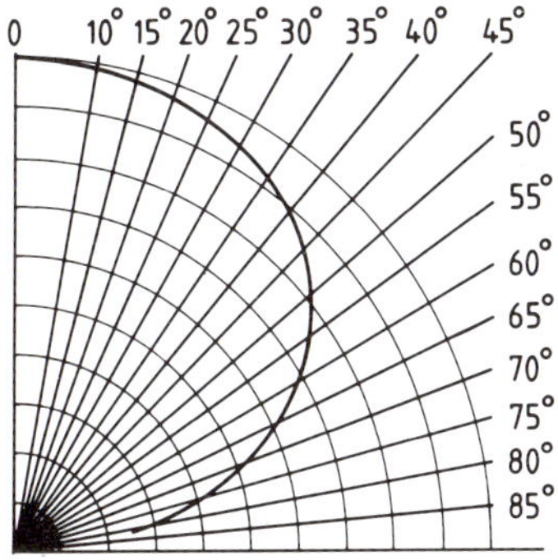

Fig. 4.1. Polar diagram for medium pressure mercury lamp.

The second method was particularly appropriate for a lamp housed in a reflector module in such a way that substantially all the radiation from the lamp falls on a plane. It is then possible in principle to integrate irradiance over the entire area of the plane, and, after making an appropriate correction for absorption of radiation by the reflector, derive the radiant power of the lamp. This can be done experimentally by using a detector which scans the plane at a constant speed and can integrate with respect to time, i.e. measure energy (in joules) per unit area. A knowledge of the scan speed then allows the radiant power of the lamp to be calculated.

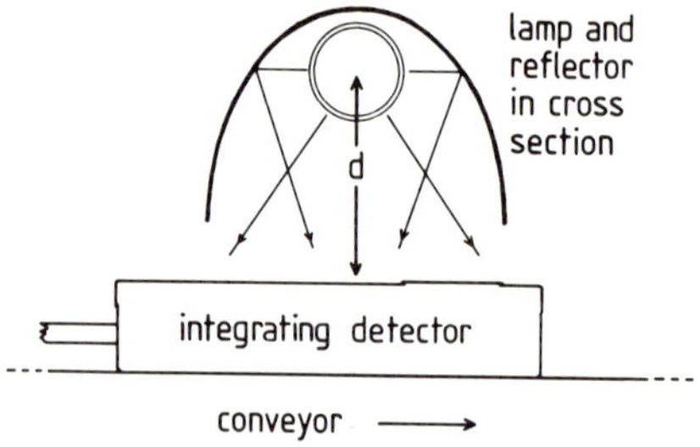

Fig. 4.2. Instrumental integration of irradiance.

It may be unnecessary to scan the entire irradiated area, particularly if it is radiant emittance which is required. The case of a linear medium pressure mercury discharge lamp housed in a reflector of uniform cross section (a typical arrangement for UV curing) is shown in Fig. 4.2. The detector is considered to have a response which is independent of the angle of incidence, and to have a sensitive element which is 1 cm square. Suppose that the distance *d* in Fig. 4.2 is small compared with the length of the lamp, and the detector, which integrates with respect to time, is passed at constant speed under the centre of the lamp and in a direction perpendicular to the length of the lamp. It is not difficult to show that the detector will *effectively* respond to the radiation emitted by a 1 cm length of the arc, part of which is of course lost at the reflector. If the detector gives a reading in joules per square centimetre, then multiplication by the speed in centimetres per second will give the radiant emittance of the lamp in watts per centimetre of arc, after a correction is made for the absorption by the reflector. The actual area of the detector is of no consequence provided that it has an absolute calibration in joules per square centimetre. The integrating detector I have used is the International Light "Light Bug", which will be described in Section 4.4.

I have found that the polar diagram method and the instrumental integration method give results which agree closely. The above discussion has presupposed the ability to measure spectral *irradiance* in absolute terms. Methods of doing this will be discussed in later sections of this chapter.

It is difficult to measure the radiant power of a lamp in absolute terms with any great accuracy, particularly in the ultraviolet part of the spectrum. In practice, what is often required is the *comparison* of lamps of similar shape, comparisons of the same lamp under differing operating conditions, or at different stages in its life, or comparison of the outputs of a particular lamp at different wavelengths. Measurement of the spectral distribution of the output of a lamp is made with a *spectral radiometer*, the operation of which will be described in the following section.

4.3 The Spectral Radiometer

A spectral radiometer is an instrument which measures irradiance as a function of wavelength. It consists essentially of two components, firstly a *selective device* which isolates a part of the spectrum for measurement, and secondly a *photosensitive detector.*

The selective device may be a filter which allows the wavelengths of interest to pass to the detector and rejects, e.g. by absorption, all other wavelengths. It may on the other hand be dispersive, giving a spatial separation of different wavelengths, as with a prism.

Photosensitive detectors fall into two main categories. *Instrumental* types rely on a *physical* response which is usually measured as a current or a voltage. Some instrumental types measure power directly from the thermal effects produced by its absorption, while others operate by some type of photoelectric effect, in which a photon induces a movement of charge. In the other main category, radiation induces a *chemical* change, usually irreversible. Examples are the photographic emulsion and the chemical actinometer.

4.3.1 Selective devices

4.3.1.1 Absorption filters

There are a variety of types of filter in common use. The less expensive varieties rely on the absorption/transmission characteristics of materials. These filters may be *solid*, of glass, gelatin or plastic, doped with other materials having appropriate absorption characteristics, or may consist of *liquid* solutions of appropriate chemical compounds in glass or quartz cells (Calvert and Pitts, 1966, p. 728). Filters may be of the "*edge*" or "*cut-off*" type for which transmission is very small up to a certain critical wavelength (long wave pass), or conversely with a high transmission below the critical wavelength, and a low transmission above it (short wave pass). "*Band-pass*" filters have a high transmission over a range of wavelengths, with (ideally) little transmission outside this range.

Liquid solution filters find little application in the measurement of radiation and are mainly used in photochemical experiments. Gelatin and plastic filters are only suitable for low intensities of irradiation, and

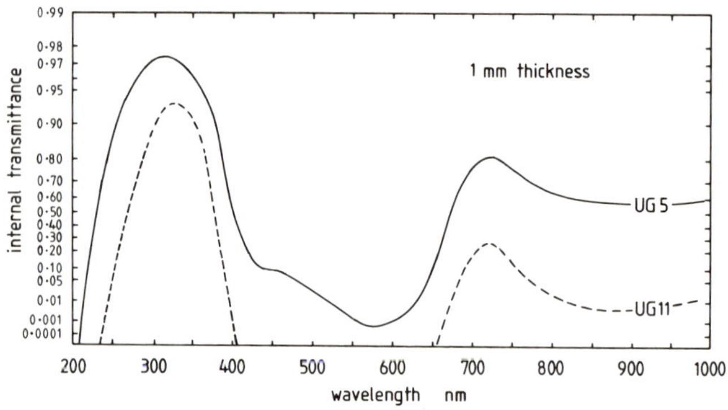

Fig. 4.3. Typical transmittance characteristics of glass bandpass filters (Schott Glass, with permission).

therefore coloured glass filters are the most important of the absorption types. Metal oxides are incorporated in the glass melt for bandpass filters, while cut-off filters use colloidal particles. The main attraction of glass absorption filters is their low cost. Stability is often poor if high intensities are used, due to solarization reactions. It is difficult to achieve both a high peak transmission and a narrow bandwidth in the bandpass type. Bandpass types are used for waveband selection in inexpensive instruments, while edge types are often used in more expensive instruments as blocking filters in combination with other selective devices.

Typical transmission curves for bandpass and edge filters are shown in Figs 4.3 and 4.4. Transmission outside the wavelength region of interest is a problem with the bandpass type, and most filters which transmit parts of the UV region also show some transmission in the red. Note that the

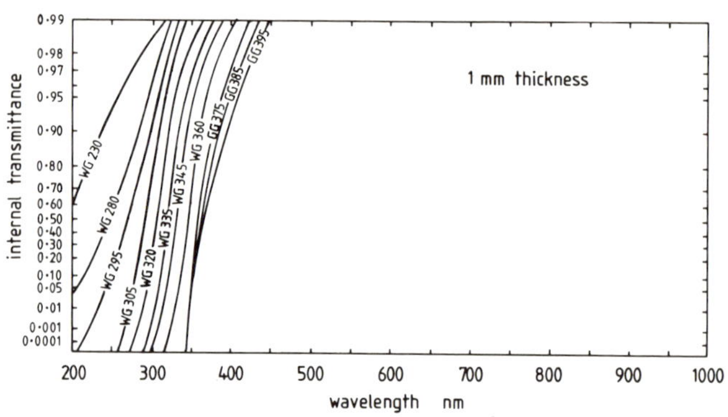

Fig. 4.4. Typical transmittance characteristics of glass edge filters (Schott Glass, with permission).

transmission curves shown in Figs 4.3 and 4.4 are for *internal* transmittance. There is always some reflection at air/glass interfaces, which depends on the refractive index of the glass and the angle of incidence.

4.3.1.2 Interference filters

For critical applications, and in particular for the isolation of very narrow wavebands, *interference* filters are often preferred, despite their complexity and high cost. The simplest is the metal–dielectric–metal (MDM) type, which operates on the principle of the Fabry–Perot interferometer. It consists of a thin layer of transparent dielectric material (e.g. magnesium fluoride) coated on both sides with a very thin, reflective, but semi-transparent metallic film, as shown in Fig. 4.5. The layers are formed by

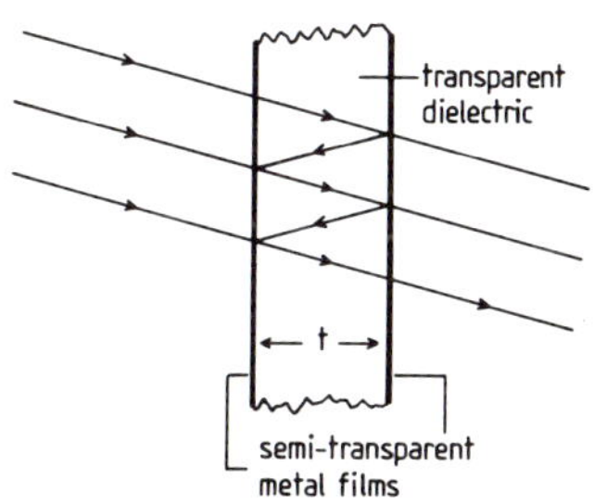

Fig. 4.5. The metal–dielectric–metal interference filter (single cavity).

vacuum deposition, and a glass or silica substrate is used. Vacuum deposition of the dielectric layer assures uniformity of thickness, even if the substrate is not a perfect optical flat (Baumeister, 1965). Part of the radiation falling on the face of the filter is reflected back and forth between the metal films, and if the optical path difference between adjacent rays is a multiple of one wavelength, constructive interference takes place. The condition for constructive interference at wavelength λ is given by

$$\lambda = 2nt/m$$

for normal incidence, where n is the refractive index and t the thickness of the dielectric film. m is the order number. Wavelengths which give constructive interference are largely transmitted by the filter, while others are rejected. It may be necessary to use blocking filters to eliminate transmission at undesirable values of m. When metal–dielectric–metal filters are designed it is necessary to allow for the phase change which occurs at the metal dielectric interface. This phase change acts as a positive optical thickness, and the thickness of the dielectric layer must be adjusted accordingly (Barr, 1975).

The bandpass type of interference filter has a bandshape which is symmetrical, as shown schematically in Fig. 4.6, which also indicates the terminology used to describe the performance characteristics of a filter. The bandwidth of a simple metal–dielectric–metal filter depends on the reflectance of the metal films (Barr, 1975). The bandwidth is reduced as the reflectance increases. Another way of reducing the bandwidth, and also improving the bandshape, is to use a "multicavity" filter, effectively a number of single cavity metal–dielectric–metal filters in series, having the construction MDMDM (double cavity), MDMDMDM (triple cavity) and so on.

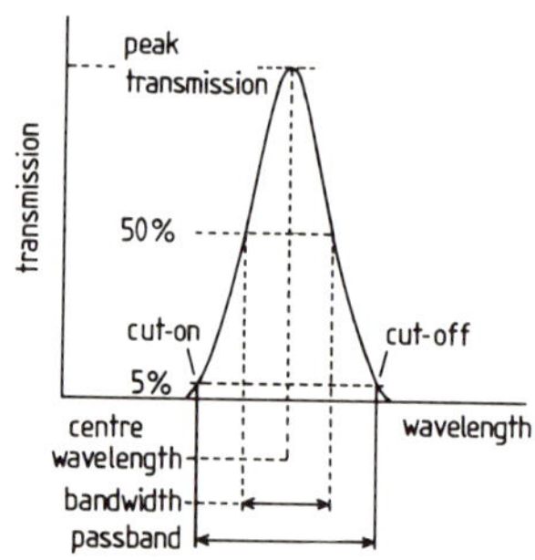

Fig. 4.6. Filter terminology.

Metal–dielectric–metal interference filters have the significant disadvantage that absorption by the metal films reduces the peak transmission of the filter. A solution to this problem lies in the replacement of each of the metal films by an all-dielectric "quarter wave stack". Quarter wave stacks consist of alternate layers of high (H) and low (L) index materials, each having an *optical* thickness of one-quarter of a wavelength. The construction of an all dielectric single cavity filter is shown in Fig. 4.7. A

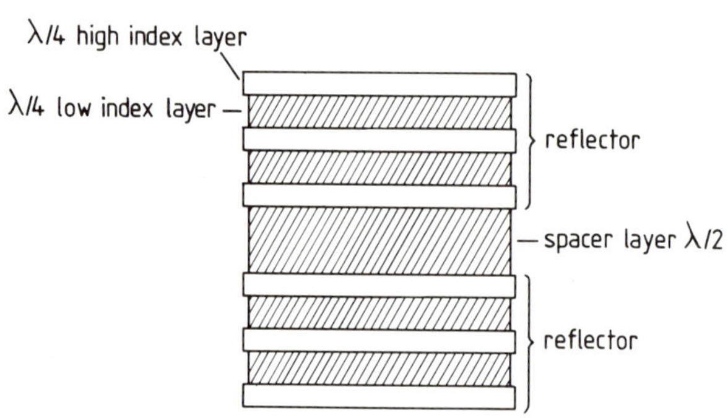

Fig. 4.7. The all-dielectric single cavity filter. (From Grum and Becher, 1979, p. 220, with permission.)

five-element quarter wave stack (HLHLH) is used for each reflector. As with the metal–dielectric–metal filter it is possible to improve the band-shape by increasing the number of cavities, as shown in Fig. 4.8.

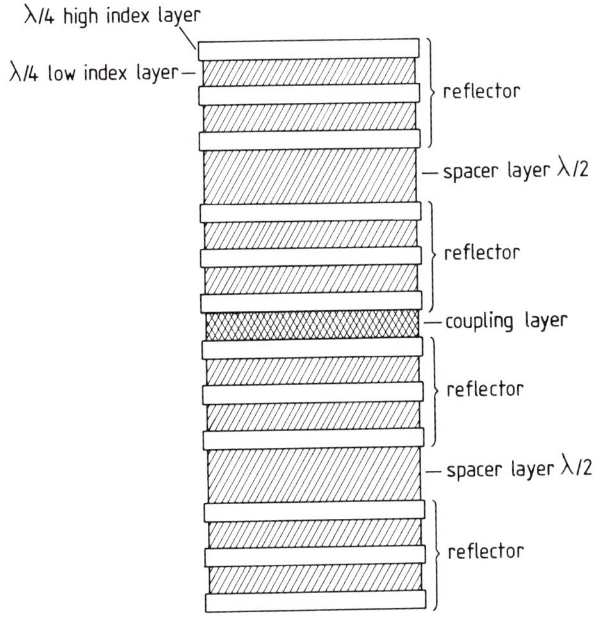

Fig. 4.8. The all-dielectric double cavity filter. (From Grum and Becher, 1979, p. 220, with permission.)

As the number of cavities increases the passband becomes more rectangular and the cut-off becomes sharper (Fig. 4.9). Dielectric stacks of alternating high and low index material are used not only for bandpass filters, but also for short wave and long wave pass filters (Baumeister, 1965). It is easier to design multilayer dielectric filters for the visible region than for the ultraviolet region, because most materials which have a high refractive index, such as zinc sulphide (2·3) and titanium dioxide (2·7), absorb strongly in the ultraviolet.

Although most interference filters are designed to give peak transmission at a single wavelength, it is possible, by using a wedge-shaped spacer layer of dielectric material between semitransparent metal films, to construct a "strip spectrum filter" so that the wavelength of peak transmission is continuously variable according to the position on the wedge. The filter strip may be linear, or may form an uncompleted circle. A slit system can be used to isolate the wavelength of interest, and filters of this type can

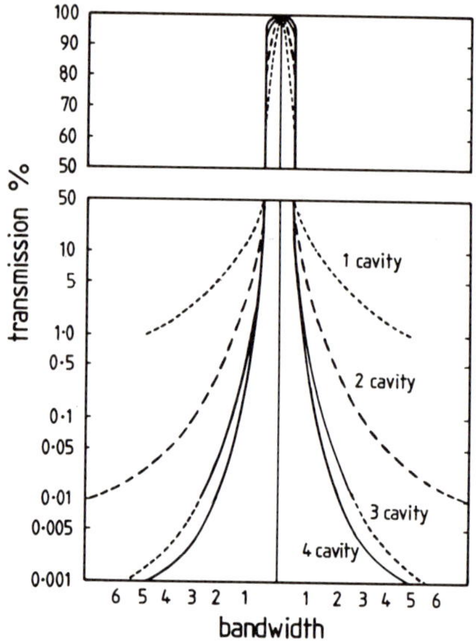

Fig. 4.9. Bandshape of all-dielectric filter as a function of the number of cavities. All curves have been normalized so that peak transmission is 100%.

form the basis of a simple, inexpensive and compact monochromator. The performance of a Barr and Stroud UMS type strip spectrum filter covering the range 250–400 nm is shown in Fig. 4.10. A slit width of 1 mm was used. The spectral output of a medium pressure mercury lamp measured with

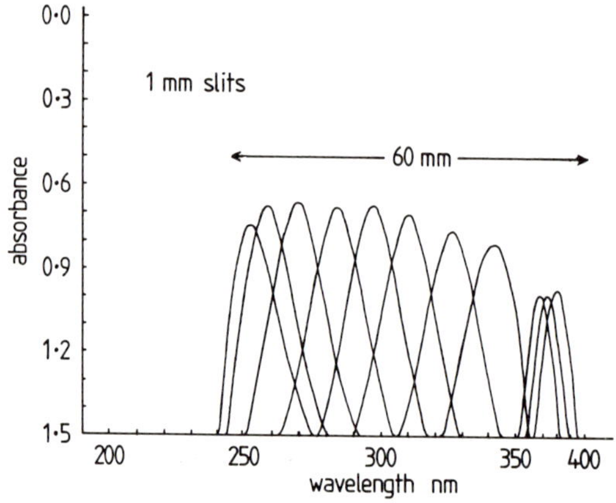

Fig. 4.10. Transmission characteristics of Barr and Stroud UMS strip spectrum filter.

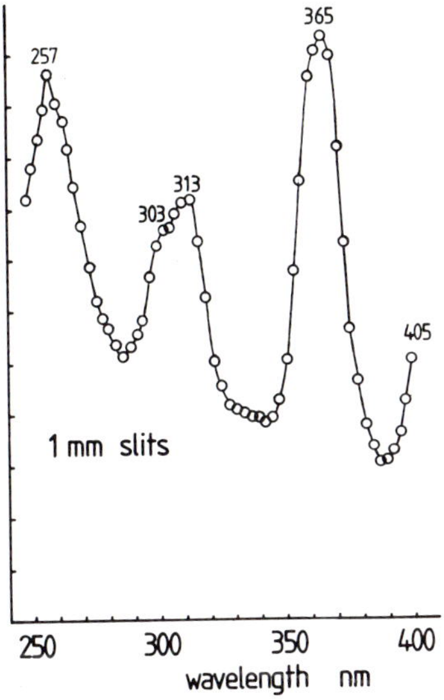

Fig. 4.11. Spectral distribution of medium pressure Hg lamp measured with strip spectrum filter.

this used as a monochromator is shown in Fig. 4.11. The resolution is poor compared with that obtainable with a prism or grating monochromator.

The edges of interference filters are normally sealed and escapsulated for protection and to prevent ingress of moisture which could cause delamination. Despite their precision construction they are fairly robust and, provided that they are not exposed to extremes of light intensity or temperature, will give consistent performance over long periods. A characteristic of interference filters not shared by absorption filters, which affects their use, is that the wavelength of peak transmission depends quite significantly on the angle of incidence, as this determines the path length in the spacer layer (Grum and Becherer, 1979, p. 221). Variation in temperature has a similar effect. Although these variations are normally a disadvantage, they can be used to fine tune a very narrow bandpass filter to an exact wavelength.

4.3.1.3 Monochromators

In a monochromator, light from the source falls on to an entrance slit, and then is converted into a parallel beam by a suitable lens or mirror and directed to a dispersing element which causes a deviation of the beam, the angle of deviation depending on the wavelength of the light. The deviated beam is then focused by another lens or mirror to an exit slit, at which a photosensitive detector may be placed. The position of the focus depends on the wavelength of the light, so that in principle the different wavelengths which comprise the original beam can be selected. In practice it is usually more convenient to keep the exit slit stationary and to rotate the dispersing element.

Monochromators separate different wavelengths by spreading them in space. There are several criteria by which monochromators of different types may be compared. The most fundamental is *angular dispersion*. If wavelengths λ and $\lambda + \delta\lambda$ leave the dispersing element with deviations of θ and $\theta + \delta\theta$, then the angular dispersion is defined as $\delta\theta/\delta\lambda$ or $d\theta/d\lambda$. The angular dispersion is dependent only on the dispersing element itself. Of more practical importance is the *linear dispersion* in the plane of the exit slit. This is given by $F d\theta/d\lambda$, where F is the focal length of the exit optics. Another concept related to but distinct from linear dispersion is *resolving power*, i.e. the ability to distinguish between two closely spaced spectral lines. If two lines at wavelengths λ and $\lambda + \delta\lambda$ can just be resolved, then the resolving power is defined as $\lambda/\delta\lambda$. Resolving power is proportional to linear dispersion, but also depends on the width of the entrance slit and the size of the dispersing element. *Stray light* can be a significant problem with monochromators. This occurs when small amounts of light of other wavelengths are superimposed on the principal monochromatic beam at the exit slit. In some cases the *transmission* characteristics of the monochromator are of importance, particularly when it is used as part of a source of monochromatic radiation.

Rothwell and Keeffe (1980) have described a multi-exit slit grating spectrometer which they term a "polychromator". Each slit has its own detector. This instrument allows several different spectral lines to be examined simultaneously. Another approach is to dispense with the slit and use a photodiode array or a photographic film to allow instantaneous recording of polychromatic spectra.

4.3.1.4 Prisms

A prism is a dispersive device which relies on the variation of refractive index with wavelength. A prism monochromator is shown schematically in Fig. 4.12. Prisms are normally used at minimum deviation, in order to minimize problems associated with slight divergence or convergence of the incident light (Jenkins and White, 1976, p. 32). For minimum deviation, i.e. with the angle of incidence ϕ_1 chosen to give a minimum value of θ, the refracted ray inside the prism makes equal angles with the two prism faces (Jenkins and White, 1976, p. 30). There is a general symmetry, and $\phi_1 = \phi_2$, $\phi_1' = \phi_2'$. Also it can be shown that the prism angle $\alpha = 2\phi_1'$. For

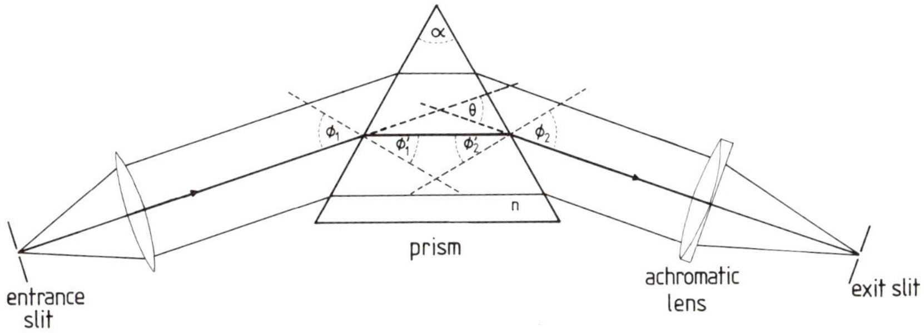

Fig. 4.12. The prism monochromator.

minimum deviation the angular dispersion is given by the following expression:

$$\frac{d\theta}{d\lambda} = \frac{2 \sin \alpha/2}{\sqrt{(1 - n^2 \sin^2 \alpha/2)}} \frac{dn}{d\lambda},$$

where n is the refractive index of the prism material. A common value of α is 60°, in which case it can be shown that for values of n between 1·4 and 1·6, to a close approximation,

$$\frac{d\theta}{d\lambda} = \frac{n dn}{d\lambda}$$

(Thorne, 1974, p. 135).

For spectral radiometry, prism monochromators have a number of serious shortcomings. The main problem is that $dn/d\lambda$ varies greatly with wavelength and is greatest near the absorption limit of the prism material. The complex variation of $d\theta/d\lambda$ with wavelength gives a non-linear wavelength scale.

4.3.1.5 Diffraction gratings

Diffraction gratings may be of the transmission or reflection type, and may be either plane or concave. The plane reflection grating is the most commonly used dispersive device for monochromators used in spectral radiometry.

Master gratings are made by first evaporating aluminium on to optically flat glass, and then cutting regularly and closely spaced grooves in the aluminium coating (Richardson, 1969). Gratings are characterized by the *groove spacing d* and the *blaze angle ϕ*. A reflection grating is shown in Fig. 4.13.

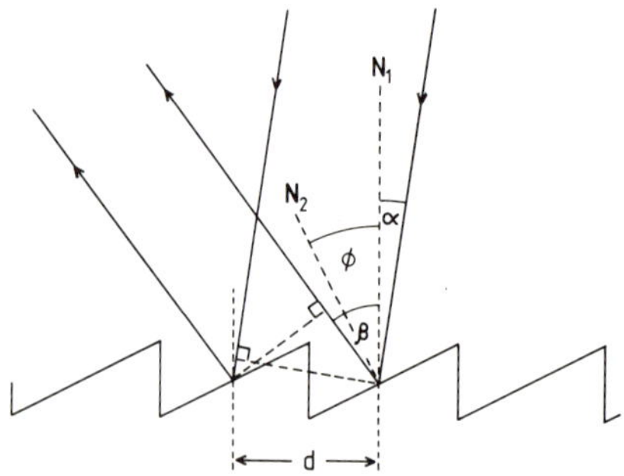

Fig. 4.13. The reflection grating. N_1, normal to grating; N_2, normal to groove face; d, groove spacing; α, angle of incidence; β, angle of diffraction; ϕ, blaze angle.

If the rays are incident at an angle α to the normal grating surface, and are diffracted at an angle β to the normal, then it is easy to show that rays diffracted from adjacent grooves have a path difference of $d(\sin\alpha \pm \sin\beta)$. The plus sign is used when both the incident rays and the diffracted rays are on the same side of the grating normal, the minus sign when they are on different sides. If the path difference is an integral multiple of the wavelength, constructive interference will take place, i.e. when

$$m\lambda = d(\sin\alpha \pm \sin\beta),$$

which is the *grating equation*. m is the order number.

The angular dispersion is given by $\mathrm{d}\beta/\mathrm{d}\lambda$, and it can be found by differentiating the grating equation, treating α as a constant:

$$\mathrm{d}\beta/\mathrm{d}\lambda = m/d\cos\beta.$$

It can be seen that the angular dispersion is proportional to the number of grooves per millimetre, and increases with the order selected. The value of d cannot be reduced much below the wavelength of the light being examined, otherwise the grating acts as a plane mirror, with $\alpha = \beta$, and $m = 0$, and no dispersion will take place.

A diffraction grating is "blazed" for a particular wavelength. This wavelength is such that the angle of the *diffracted* ray to the groove face normal is equal to the angle of the *incident* ray to the groove face normal. A diffraction grating operates most efficiently at the blaze wavelength.

It is easy to obtain a linear wavelength scale on a grating monochromator, by driving the grating rotation with a sine bar. It is generally possible to obtain better dispersion with a grating monochromator than with a prism monochromator. The resolving power of a grating is largely a function of the total number of grooves ruled on it, i.e. the length of the grating multiplied by the number of grooves per unit length.

The most significant problem associated with diffraction gratings is that of overlapping orders. For instance if the wavelength scale is calibrated for first-order spectra, the low pressure mercury lamp will give a main peak at 253·7 nm, but a peak will also be observed at 507·4 nm on the wavelength scale, which is due to the second-order spectrum. If the blaze angle is chosen correctly for the wavelength region of interest, second and higher orders of spectra will tend to be weak compared with the first-order spectrum. However, it is possible to eliminate undesirable orders by using order sorting filters, which absorb wavelengths below the region of interest. For instance, to examine the range between 500 and 600 nm it may be desirable to use an order sorting filter with a cut-off at 300 nm or above.

When stray light is a problem it is usual to use a double monochromator, with the diffracted light from the first grating being focused at the entrance slit of the second, the wavelength scales being precisely aligned.

4.3.2 Photosensitive detectors

4.3.2.1 General performance criteria

There are many types of device which will respond to electromagnetic radiation by producing a signal (current or voltage) which can be measured quantitatively. The characteristics of the ideal detector depend very much on the application. The ideal detector for *photometric* measurements would have a spectral response similar to that of the human eye. The ideal detector for radiometry on the other hand would respond uniformly to radiation of all wavelengths. For *spectral radiometry*, what is needed is a uniform response over the wavelength region of interest, but zero response at other wavelengths, in order to minimize the effects of stray light and overlapping orders from interference filters and diffraction gratings. Sometimes a chemical or biological effect has an "action spectrum" which indicates the relative efficiencies of different wavelengths in producing that effect. It may be convenient to assess a radiation source by using a photosensitive detector having a spectral response curve which matches the action spectrum of the effect (Duval and Ryer, 1977). Photometry is one example of this general concept. Another is the measurement of the physiological hazard (see Chapter 15).

The *responsivity* R_λ of a detector is defined as the output signal per unit input, i.e. amps or volts per watt. The responsivity must be appropriate for the irradiance being measured. The concept of responsivity, and its definition, imply that photosensitive detectors have a *linear* response. The commonly used detectors give a linear response over several orders of magnitude of input, but tend to "saturate" at very high light levels. However, the range of linearity can be extended by the use of neutral density filters. The responsivity should ideally remain stable over a long operational lifetime, and should be insensitive to ambient conditions such as room temperature.

The smallest signal which can be detected from a source of radiation is limited by the *noise* which arises from random electrical fluctuations in the detector and its associated circuitry, fluctuations in the input radiation, and radiation from the background surrounding the source. The minimum detectable irradiance is normally considered to be that which produces a signal equal to the noise, i.e. a signal to noise ratio of unity. This minimum irradiance is known as the *noise equivalent power*, and its reciprocal as the *detectivity*.

Photosensitive detectors vary in their speed of response to sudden changes of input. Speed of response is characterized by the *response time* τ (Dennis, 1979). If the input is a rectangular wave, the output signal will take a finite time to reach a maximum value at the beginning of the wave, and will take a finite time to decay to zero at the end of the wave. Growth and decay curves are exponential, as shown in Fig. 4.14. Note that the

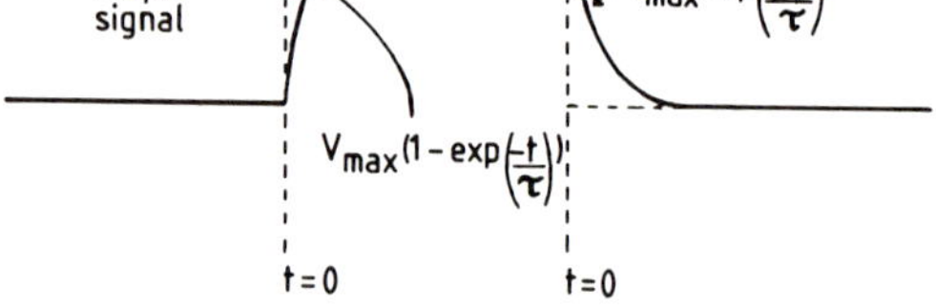

Fig. 4.14. Detector response time.

output wave shape is distorted. To minimize this when measuring transient phenomena, e.g. pulses of light from flash tubes, it is necessary to use a detector with a value of τ which is very small compared with the width of the pulse. This is a rather oversimplified treatment of the subject of response time. In a real detector there may be a number of processes occurring simultaneously or successively, and in reality there are a number of different response times which interact to produce pulse delay and pulse distortion.

An important practical aspect of detectors is their field of view and angular response. When a beam of radiation strikes a surface the irradiance produced depends on the angle θ between the beam and the normal to the surface. As the angle is increased the beam is spread over a larger area of the surface, and the irradiance decreases, being proportional to $\cos\theta$. A "cosine response" is normally a desirable feature in a detector, because it implies that all photons striking the detector produce the same response. The construction of most types of detector prevents a true cosine response over a 180° field of view, and therefore the input optics must be modified to give this response. This may be done by using a quartz diffuser or an integrating sphere (Mielenz *et al.*, 1975; Eckerle *et al.*, 1976; Parrish *et al.*, 1978, p. 52).

4.3.2.2 Thermal detectors

If a uniform response over a wide spectral range is the prime consideration it may be desirable to use some form of *thermal detector.* A detector surface is painted black so that it will absorb all wavelengths falling on it. This produces a heating effect which leads to a physical response. Typical is the *radiation thermocouple*, in which a small piece of blackened gold foil is welded to the junction of two dissimilar metals chosen to give a large thermoelectric e.m.f. (the Seebeck effect). An array of radiation thermocouples is known as a *thermopile.* Another device is the *bolometer*, which is a miniature resistance thermometer which relies on the change of resistance of either a metal or a semiconductor. The *Golay cell* depends on the thermal expansion of gas. More sophisticated is the *pyroelectric detector* (Putley, 1970, 1973). Certain materials, e.g. triglycine sulphate, which have a permanent internal electrical polarization, exhibit the *pyroelectric effect.* A change in temperature causes a change in the surface charge and, if electrodes are attached, a current will flow in an external circuit proportional to the *rate of change* of temperature. The pyroelectric detector is used with a chopper which modulates the input radiation.

With the exception of the pyroelectric detector, thermal detectors have response times usually significantly longer than those of photon detectors (Grum and Becherer, 1979, p. 201), and they require very careful experimental procedures. They have the attraction that they can be calibrated by using an electric current to produce the same response as the radiation (Gillham, 1962).

Vacuum photodiodes respond very quickly to changes in light level. The time delay between the incident light and the photoemission is less than 10^{-12} s (Vine, 1965). The speed of response is determined by the time of transit of electrons from cathode to anode. However, the ability to respond to high frequencies and accurately record pulse shapes is more a function of the *variation* in transit time (Williams and Becklund, 1972, p. 291). Photocurrents are small, of the order of microamps, and therefore current amplification is necessary. It is important to realize that while the response time of the photocurrent may be very small, the amplification and measurement circuit may impose a time constant. The photoemission is independent of the temperature of the cathode, but there is a very small amount of thermionic emission which is temperature dependent, and gives rise to a dark current in the region 10^{-12}–10^{-16} A cm^{-2} at 25 °C. This is usually of little consequence for vacuum photodiodes.

When the irradiance to be measured is very small, a version of the photoemissive tube known as the *photomultiplier* may be used. This contains a built-in electron current amplifier which utilizes the phenomenon of secondary emission. Between the cathode and anode is a series of between nine and 16 "dynodes". Each dynode is at a different d.c. voltage, with a potential drop of about 100 V between adjacent dynodes provided by a voltage dividing resistor chain. A schematic diagram of the photomultiplier is shown in Fig. 4.16. The photomultipier shown in Fig. 4.16 is a

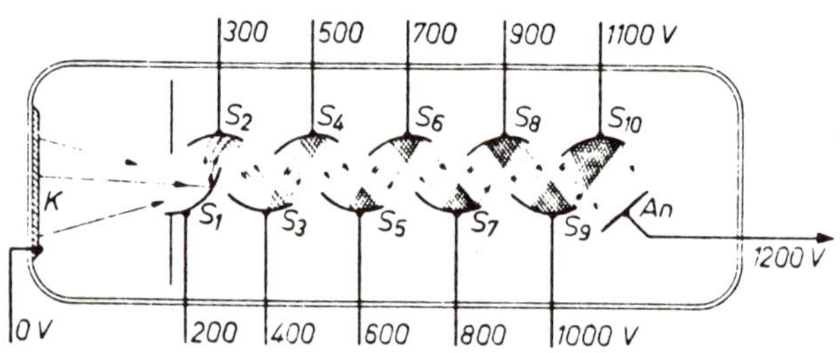

Fig. 4.16. Principle of the photomultiplier. K is the photocathode; S_1–S_{10} are dynodes; An is the anode. (From Piétri and Nussli, 1968.)

"head on" type with a semi-transparent cathode. An alternative is the "side on" type with an opaque cathode. There are four main types of dynode configuration in use, each with its own merits. These are shown in Fig. 4.17. The dynodes have a surface coating, e.g. Cs_3Sb, so that when an electron from the cathode strikes the first dynode, several (from two to five) secondary electrons are emitted, which flow to the second dynode, where each electron causes the emission of several more secondary

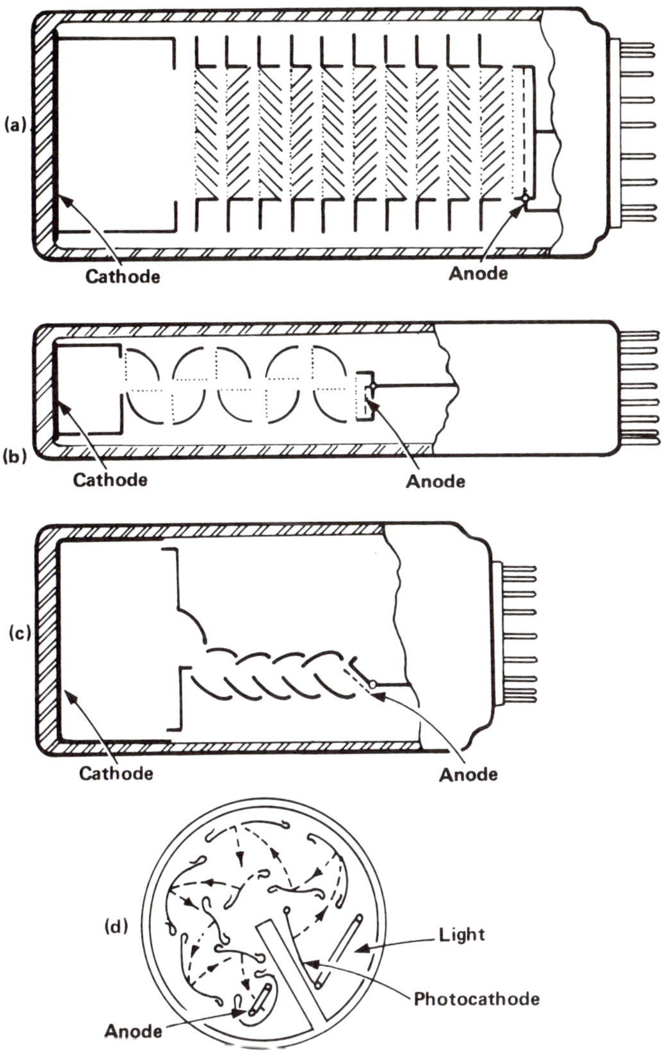

Fig. 4.17. Photomultiplier dynode configurations: (a) venetian blind; (b) box and grid; (c) linear focused; (d) circular focused. (From "EMI Photomultipliers", copyright EMI Industrial Electronics Ltd, 1979, with permission.)

electrons and so on. It is easy to see that each electron emitted from the cathode will lead to a large number being collected at the anode. A typical amplification is in the range 10^6–10^8. The level of amplification can be varied at will by altering the potential drop between adjacent dynodes, so changing the coefficient of secondary emission (Piétri and Nussli, 1968).

However, the sensitivity of the amplification of the photomultiplier to the dynode voltage does necessitate the use of an efficiently stabilized high voltage d.c. source to supply the voltage divider.

The response time of the photomultiplier is a little longer than that of the vacuum photodiode. The time delay (or transit time) between the photon striking the cathode and the shower of electrons reaching the anode is 10–100 ns. Of more consequence is the *distribution* of transit times which is caused by the distribution of kinetic energies of electrons emitted from the cathode and dynodes.

Photomultipliers are expensive, bulky and relatively fragile detectors, and are generally only used when their unique sensitivity is essential. Thermionic emission from the cathode is amplified by the dynode chain, but its effect can be reduced either by cooling the photomultiplier tube to below ambient temperature, or by chopping the incident light and using an a.c. amplifier and phase sensitive detection. Photoemissive tubes are prone to fatigue (reversible) and ageing (irreversible), both of which reduce the sensitivity. It is desirable to avoid high intensities of irradiation and to store tubes in the dark (Tassell, 1980).

4.3.2.4 Solid state photodetectors

Solid state detectors rely on the *internal photoelectric effect*, and tend to be more sensitive to longer wavelengths than photoemissive tubes. There are two basic types. *Photoconductive cells* make use of the photoconductive effect in intrinsic semiconductors. Photons promote electrons from the valence band to the conduction band, producing electron–hole pairs which act as current carriers. The material is sandwiched between two electrodes, and the electrical resistance between the electrodes is measured. Photoconductive cells are seldom used in the UV, but find applications in the visible and infra-red regions. Their response times are relatively long compared with those of photoemissive tubes, and are normally in the microsecond region. They are temperature dependent, and for critical work are often cooled to liquid nitrogen temperature to decrease the thermal noise and increase the sensitivity, particularly to longer wavelengths.

The second type of solid state detector is the *solid state photodiode*. This is most commonly a silicon p–n junction, an interface between p-type and n-type material. p-type material is doped with an *acceptor* impurity, i.e. atoms with three valence electrons, while n-type material is doped with a *donor* impurity, i.e. atoms with five valence electrons. When the junction is made there is controlled diffusion of donor impurity into the p-type material, and acceptor impurity into the n-type material, to form a *depletion layer* at the interface.

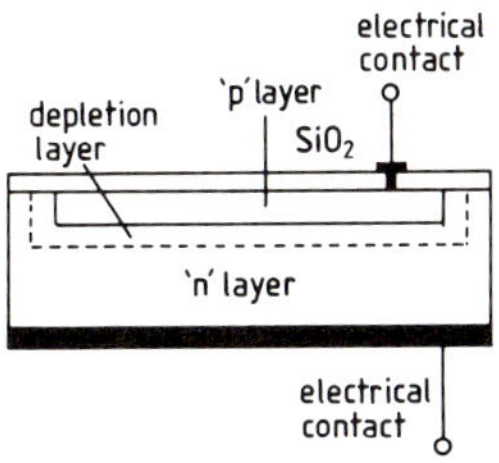

Fig. 4.18. The silicon photodiode.

Figure 4.18 shows a schematic diagram of a simple photodiode. In the absence of radiation, and with thermal equilibrium, the Fermi levels are equal in the p and n layers, and the energy level diagram is as shown in Fig. 4.19. When the device is irradiated, electron–hole pairs are generated in

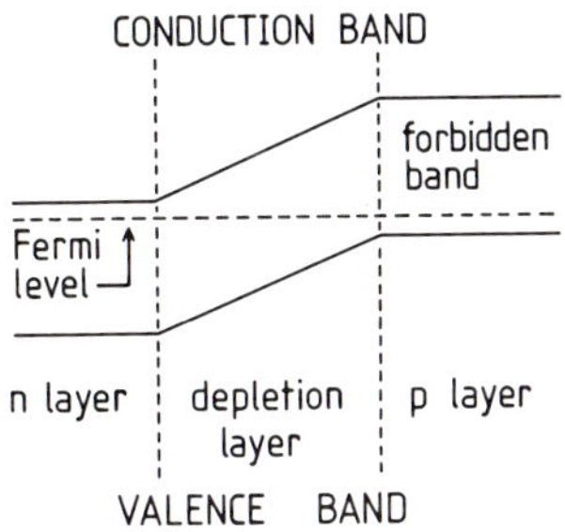

Fig. 4.19. Energy level diagram (dark).

the various layers. Holes tend to diffuse to the p layer and electrons to the n layer, with the result that a voltage is generated across a load resistor placed across the device. The electrical characteristics are shown in Fig. 4.20. Each curve is for a different irradiance, the uppermost curve showing the relationship between voltage and current when there is no incident radiation. When the device is short circuited, the current flowing is shown by the intercepts on the vertical axis, while when the load resistor is an open circuit, the open circuit voltage is given by the intercepts on the horizontal axis. "Load lines" are shown for various intermediate load resistor values. It can be seen that when the device is short circuited there is excellent linearity between the short circuit current and the irradiance. The relationship between the open circuit voltage and the irradiance is logarithmic. For satisfactory linearity therefore, it is desirable to operate with a small load resistor. There are two ways in which photodiodes may be used. Use in the *photovoltaic* mode involves operation in the lower right quadrant of Fig. 4.20, near to the vertical axis. The circuit is shown in Fig.

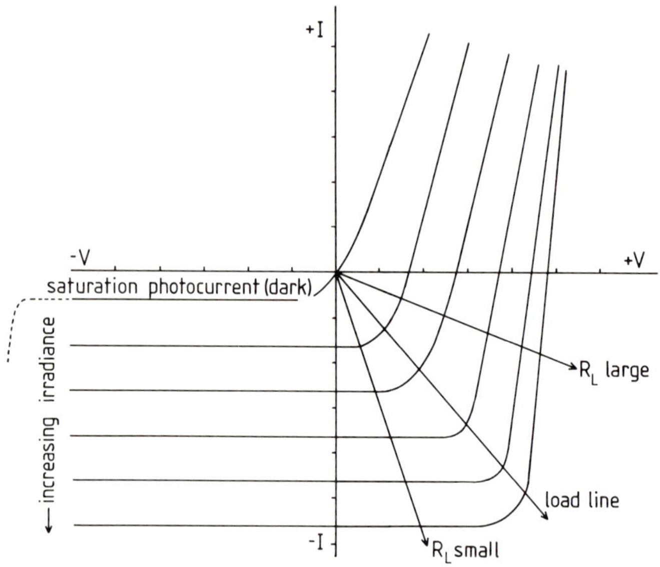

Fig. 4.20. Electrical characteristics of silicon photodiode. Current–voltage relationships at increasing levels of irradiance.

4.21(a). In the *photoconductive* mode the p side of the junction is reverse biased, and the photodiode acts as a photoconductive circuit element. An example of a suitable circuit is shown in Fig. 4.21(b).

Operation in the photoconductive mode gives better linearity, and improved response speed, but at the expense of increased dark current and noise levels. The general concepts of silicon photodiodes have been described. In reality there are many different types (e.g. PIN, PNN$^+$, planar diffusion, avalanche) and full details can be found in manufacturers' catalogues. Response times vary greatly and depend on the device itself

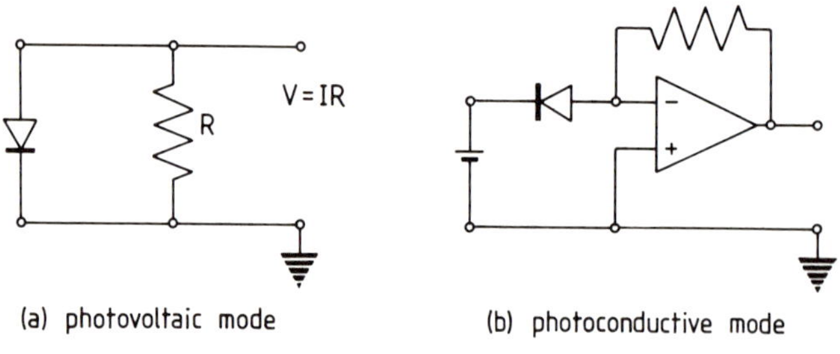

Fig. 4.21. Modes of operation of solid state photodiodes: (a) photovoltaic mode; (b) photoconductive mode.

and the associated circuitry, and typical values lie between nanoseconds and microseconds. Spectral response peaks in the near infra-red, at about 800 nm. Significant response down to 200 nm can be obtained by using an anti-reflection coating of SiO_2, and making the top layer as thin as possible to allow as much light as possible to reach the junction. Typical spectral response curves are shown in Fig. 4.22. Radiant sensitivities and quantum efficiencies tend to be higher than with photoemissive tubes.

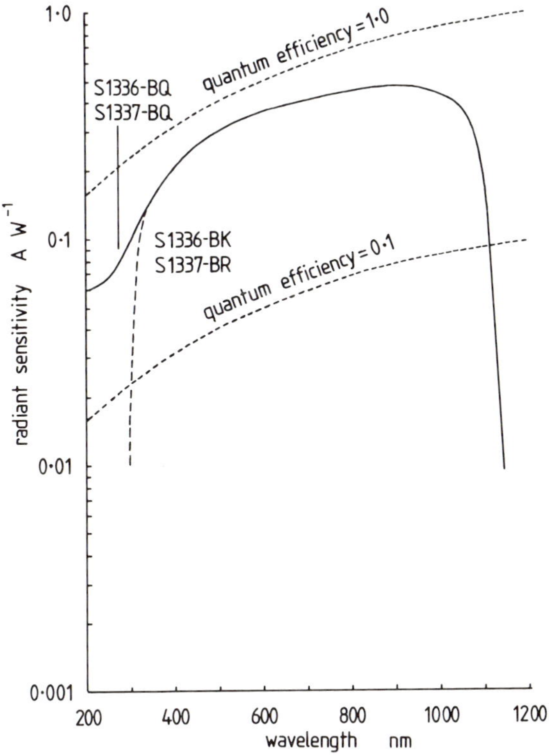

Fig. 4.22. Spectral response curves for silicon photodiodes. BQ, SiO_2 windows; BK, borosilicate glass window; BR, resin window. (From 1981 Hamamatsu Silicon Photocells catalogue, with permission.)

Solid state photodiodes are very compact and the associated circuitry is simple. They are often inexpensive, and they are robust and have better long term stability than photoemissive tubes. The photosensitive area can be made very small, which is an advantage for some applications. A detailed comparison of silicon photodiodes and photomultipliers has been given by Swindell (1980).

operate automatically (Wharmby, 1980). A very simple, but effective, manual technique is to use a voltage divider, consisting of a large fixed resistor, and a small variable resistor (in fact a decade resistance box), in series across the radiometer terminals. The chart recorder is connected to the terminals of the decade box. As the spectrum is run, the calibration factor, as a resistance, is dialled into the decade box, and the signal to the recorder is continuously corrected. This technique works well if the scan speed is reasonably slow. The decade box correction technique was used to derive the spectral distributions shown in Fig. 8.25.

4.3.3.3 Use of computers

A more sophisticated technique is to use a small computer which can store the calibration factors in its memory and automatically apply them to the detector signal as the wavelength range is scanned. The corrected spectrum can be displayed on a chart recorder in real time, or the corrected data can be stored and then retrieved at a later time.

The experimental arrangement for the examination of linear lamps with the computer manipulation of the detector signal is shown in Fig. 4.23. It was found desirable where possible to blacken the reflector so that only radiation directly from the lamp would be measured. This blackening was achieved with a specially prepared aerosol containing only carbon black and propellant. Most commercially available matt black spray paints were found to turn grey or white when exposed for lengthy periods to the UV, heat and ozone generated from a medium pressure lamp. The lamp was

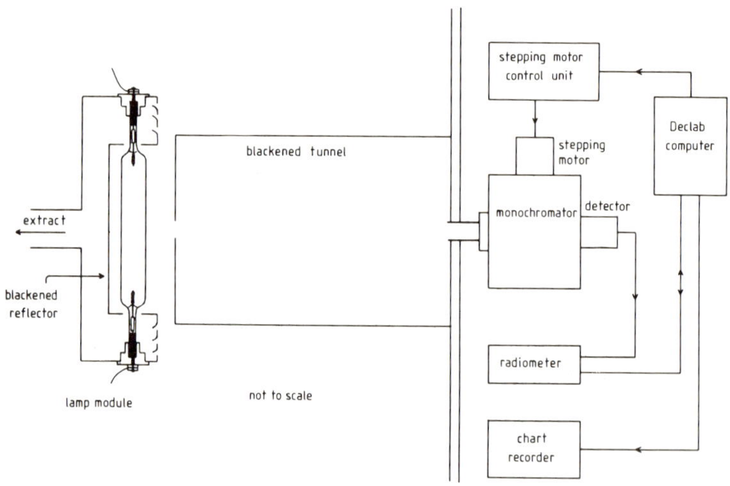

Fig. 4.23. Measurement of spectral distribution using a Declab computer (not to scale).

viewed through a blackened tunnel, and only a 2·8 cm length of lamp was examined. There were two reasons for this. Firstly it allowed the direct comparison of lamps of different lengths, and secondly it allowed the examination of the uniformity of a lamp's radiant emittance, particularly after deterioration processes had affected the output of the lamp.

4.3.3.4 Display of spectral distributions

One attraction of using a computer is that it is possible to display the spectral distribution data in a variety of ways. The most common and most informative (in that it contains the most information) is the graph of watts per square centimetre per nanometre versus wavelength, either on a absolute or on an arbitrary scale (see for instance Fig. 9.2). Other methods of display simplify the data to make it more readily understandable in particular circumstances. Lamp makers often give lists of outputs in the more significant spectral lines, and this information is useful, e.g. as a guide to plasma temperature (see Chapter 8). It must be realized that polychromatic sources such as the medium pressure mercury lamp have significant amounts of continuum radiation, particularly at the higher filling pressures, so that the sum of the outputs in the more significant lines does not represent the total output of the lamp. Histogram representation is a third method of display and is probably the most useful for lamps for practical purposes, as the interval used can be chosen to suit the selectivity of the operation. Finally one can plot a cumulative graph, where the percentage of the total output *between* two wavelengths which lies *below* any particular wavelength is plotted against wavelength. The Digital Declab computer (programmed in FORTRAN) produced histograms in graphical form (See Figs 8.26, 8.30, 9.5, 10.5, 10.6, 10.7, 10.8 and 10.10) and in tabular form (see Table 10.2), and also cumulative graphs (see Fig. 8.27).

4.3.3.5 Light waveforms and transients

If a lamp is operated from a stabilized d.c. supply, the light output remains constant in time. If an a.c. or transient supply is used, the light output will respond to the fluctuations in the power input. With an incandescent lamp, the light output fluctuation when operated on a normal mains a.c. supply is hardly perceptible because of the heat capacity of the filament. With a gas discharge the situation is very different, and it is often instructive to examine the way in which the light output at a particular wavelength varies during the mains cycle. This can be done by using a photomultiplier as a detector, and passing the anode current through a suitable resistor to earth. The voltage developed across the resistor is monitored with an oscilloscope, which displays the waveform of the light. This technique was used for Figs 8.34 and 10.9. The same technique was used to record pulse shapes from xenon flashlamps, and there are several examples shown in Chapter 11. The display of transients requires a storage oscilloscope. The photomultipliers used were an RCA 1P28 (side window, circular focused dynode configuration, CsSb cathode, UV-transmitting glass envelope) and a Hamamatsu R928 (similar, but with a multialkali cathode).

4.4 Instrumental Dosimeters

The thermal and photon detectors described in the previous sections measure the photon flux falling on the detector, i.e. the response is a function of the *power* per unit area. If actinic radiation is used to bring about a chemical or biological change, the irradiance of the system will control the rate of the primary photochemical process, and the total conversion will be a function of the irradiance integrated with respect to time, i.e. the dose, with units of *energy* per unit area. The function may be rather complicated if secondary non-photochemical processes are involved, and the total conversion may depend on both irradiance and integrated irradiance, as in photoinitiated free radical polymerization. However, dose measurements, using a *dosimeter*, have considerable practical value. If irradiance remains constant during processing, then the dose is simply calculated by multiplying the irradiance value by the time of exposure. In many practical cases, however, the irradiance varies during processing, from zero at the start, through a peak, and then to zero again. Examples are irradiation with pulses of light from a flash tube, and UV curing using a focusing semi-elliptical reflector. It is possible to measure dose by recording irradiance with respect to time using a storage oscilloscope or a chart recorder, and calculating the area under the curve obtained. I have obtained excellent results by passing a silicon photodiode fitted with a 365 nm filter under the mercury lamp of a UV conveyor drier, using trailing leads to connect the voltage output of the diode circuit to an oscilloscope. Irradiance versus time curves are given in Chapter 11 to illustrate the shapes of pulses of light from xenon flash tubes. However, manual integration is tedious and time consuming, and electronic methods of integration are readily available (Schultze and Vogel, 1973; Amrein *et al.*, 1974; Zimmerman *et al.*, 1977).

At Metal Box we have made extensive use of a radiometer designed specifically for application to the UV curing process. It is the International Light IL745 UV Curing Radiometer, manufactured by International Light Corporation, of Newburyport, Mass., USA. It is illustrated in Fig. 4.24. It

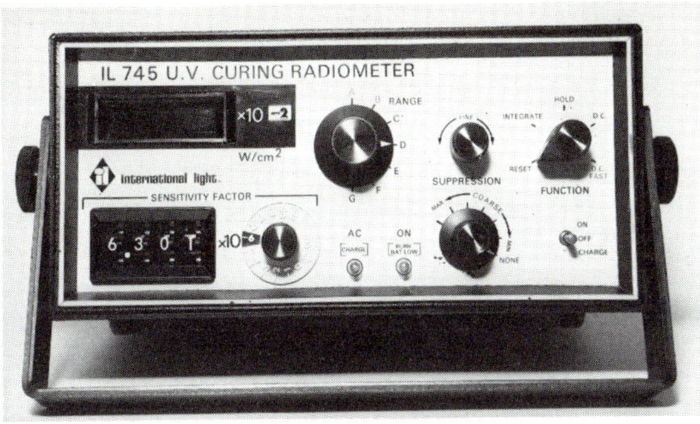

Fig. 4.24. The International Light IL745 radiometer.

can be used with a variety of detector–filter combinations and each is supplied with an absolute calibration traceable to NBS standards, and a sensitivity factor which can be dialled into the instrument to allow automatic correction of the digital display. The function control has a setting which allows integration of the measured irradiance for as long as the instrument is in the integrate mode. This can be used in two ways. The instrument can integrate the signal from a detector which is continuously connected. This method of operation is of limited value for assessing UV driers, because what is really required is a detector which can follow the same path as the coated material being processed. An accessory available for use with the IL745 is the "Light Bug" which is shown in Fig. 4.25. It consists of a disc 11.4 cm in diameter and 2.4 cm thick to which is attached an armoured cable and a multipin connector. Prior to use the Light Bug is plugged into the host instrument, with the control switch in the "reset" position, until a "full charge" indicator lamp glows. The Light Bug is then disconnected, the control switch moved to the "expose" position, and the

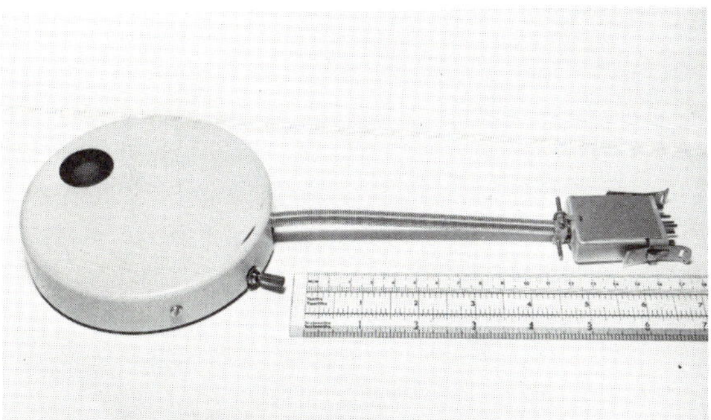

Fig. 4.25. The Light Bug.

unit passed on a conveyor under the lamp(s). The unit is then reconnected to the host instrument, the switch moved to the "readout" position, and the integrated irradiance is displayed in joules per square centimetre.

The detector used is a vacuum photodiode fitted with a wideband interference filter and a cosine correction diffuser. The region of sensitivity is between about 250 and 400 nm, and the spectral sensitivity curve is shown in Fig. 4.26. The Light Bug is provided with an absolute calibration for monochromatic radiation at 350 nm (the wavelength of peak sensitivity), so with a polychromatic source the value of integrated irradiance measured will always be an underestimate. The true value can be calculated from a knowledge of the spectral output (in relative terms) of the source.

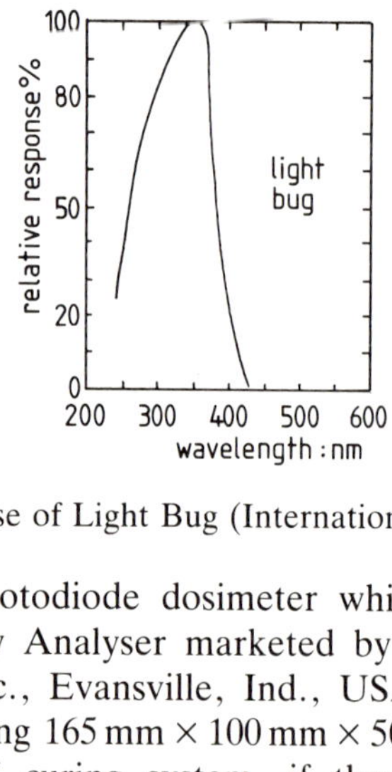

Fig. 4.26. Spectral response of Light Bug (International Light data).

Another type of photodiode dosimeter which we have used is the Ashdee Power Density Analyser marketed by the Ashdee Division of George Koch Sons Inc., Evansville, Ind., USA. This is a totally self-contained unit measuring 165 mm × 100 mm × 50 mm thick, which can be passed through a UV curing system, if there is sufficient clearance available. This instrument can be used not only to measure the integrated irradiance, but also to record the peak irradiance, which is useful for optimizing the focusing of high intensity radiation. Unlike the Light Bug this instrument responds only to 365 nm radiation. Both types of photodiode dosimeter have been described by Fadner and Schaeffer (1978).

An instrumental UV dosimeter suitable for long term monitoring of solar UV during weathering trials has been described by Harris (1968). A very compact lightweight waterproof dosimeter sensitive to erythemal solar radiation, and suitable for use as a personnel dosimeter, has been developed by Davidson *et al.* (1981). The photosensitive element is a $MgWO_4$ fluor which fluoresces in the region 420–560 nm. The green fluorescence is measured by a GaAsP diode detector and the cumulative exposure stored in digital form using specially designed hybrid micro-circuits.

Instrumental dosimeters can be used not only to measure radiation dose, but also to control the dose. Integrators are commonly used with light sources for photoresist exposure to control the exposure time.

4.5 Chemical Dosimeters and Actinometers

Instrumental detectors respond to photon flux, and when the flux ceases there is no residual effect. Measurement of dose requires a method of integration. Photosensitive detectors which rely on a *chemical* change respond to the *total* number of photons falling on the detector during the period of measurement, and therefore to *energy*. They are therefore primarily *dosimeters*. Power can be derived by the use of a shutter or other device to define the interval during which the energy is received.

In principle any photochemical reaction can be used as a photochemical detector if it is brought about by radiation in the right region of the spectrum, and it has a well defined quantum yield, and its progress can be easily measured. Chemical detectors can be divided into two main categories. Firstly there are films and papers which change in optical density or colour on irradiation in a more or less irreversible manner (as opposed to photochromic materials in which the colour change is more or less reversible). The second category involves chemical change in bulk fluids, either liquids or gases, and this is the area of *chemical actinometry*. Both categories of chemical detector have particular advantages, and complement rather than compete with instrumental types.

4.5.1 Photosensitive papers and films

The photographic silver halide type of film or plate has been widely used as a detector in spectral radiometry, although less so now than in the past. If the film or plate is used with a prism or grating monochromator there is no need for an exit slit, and the whole spectrum can be measured *at the same instant*, with no need for scanning. This is of great value when examining sources of transient radiation such as flashlamps, and sources with a continuously changing spectral output. However, silver halide emulsions have very considerable disadvantages. They need development. The relationship between the optical density (d) of the developed emulsion and the integrated irradiance (E) is complicated. The optical density d is defined as $\log_{10} 1/T$, where T is the transmission of the emulsion, as measured by a microdensitometer. A graph of d against log E (Fig. 4.27) reveals a threshold value of E which must be exceeded before any change in d is observed. There is then a linear portion, the slope of which defines the "contrast" of the emulsion, followed by saturation. The linear portion defines also the rather limited range over which the emulsion can be used to give quantitative results.

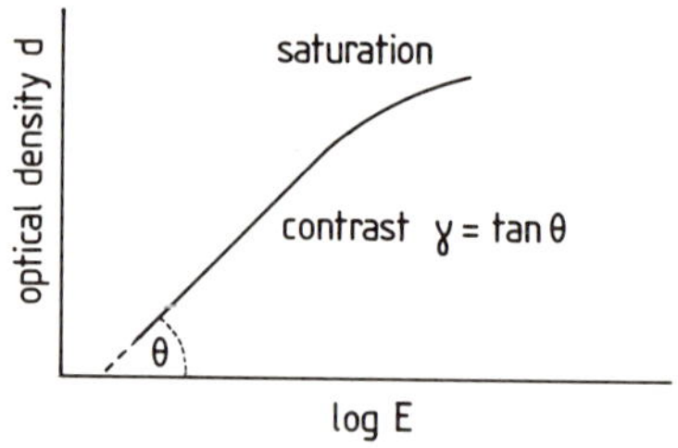

Fig. 4.27. Response curve for silver halide emulsion.

There has been much effort expended in devising chemical systems which give a permanent colour change or other measurable change in optical properties on irradiation with UV, and incorporating these systems into a paper, plastic or other matrix for use as a practical dosimeter for monitoring lamp output or solar radiation or as a personal dosimeter in the control of phototherapy. At Metal Box we have made regular use of a material sold as "Green Detex" by W. Sessions Ltd of York, UK. This is based on dosimeter papers developed by the UKAEA in the 1960s for ionizing radiation (Farrell and Vale, 1963). It is a green coloured paper with an adhesive backing. On irradiation with short wavelength UV (less than 300 nm) it changes colour to purple-red through various shades of brown. The dose required for the colour changes is well matched to the dose required for the UV curing of coatings. Visual assessment of the colour change is normally sufficient, but it can be measured quantitatively by a reflection colourimeter. Detex operates by the photodegradation of a chlorinated polymer, which releases HCl and causes an acid sensitive dye to change colour. Its main shortcoming is that the colour change after irradiation is not stable but tends to continue very slowly in the dark. It is obviously not as precise as an instrumental method, but is considerably more versatile. In particular it is well suited to the assessment of patterns of irradiation, e.g. to ensure that the entire width of a sheet or web, or the entire area of print on a can or pot, is uniformly irradiated.

Several chemical dosimeter systems have been devised which rely on a photoelimination reaction which introduces a double bond which then links two pre-existing chromophores to generate a larger highly coloured chromophore. A widely studied system has been that based on triphenyl methane dye cyanides (Chalkley, 1952; McLaughlin *et al.*, 1977; Kantz and Humpherys, 1978, Shaw and Hatzakis, 1981). The colourless cyanide on irradiation with wavelengths shorter than 330 nm yields a highly coloured carbonium ion (Fig. 4.28). The carbonium ion is stable if a small

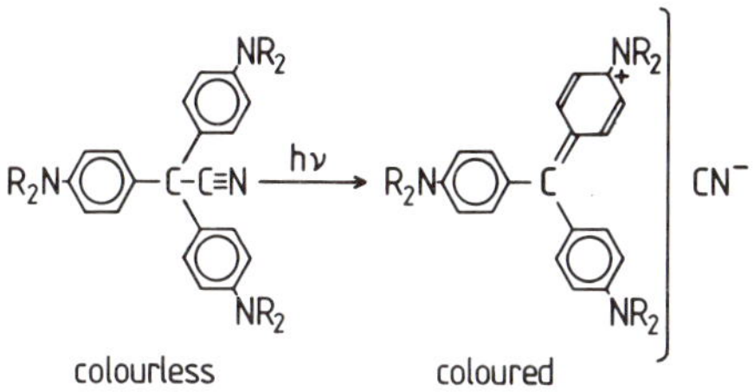

Fig. 4.28. Formation of colour by irradiation of triphenylmethane dye cyanides.

amount of acid is present. Various dye precursors can be used, including malachite green, pararosaniline and crystal violet. In the pure form the dye precursors are insensitive to radiation, and an activator is needed. A wide range of compounds is suitable for this (Chalkley, 1952). If the dye precursor is incorporated in a gelatin matrix with a trace of acetic or citric acid to stabilize the carbonium ion, the matrix provides the activation required (Kantz and Humpherys, 1978). Another system, sensitive to the mid UV and designed for use as a sunburn dosimeter, has been described by Zweig and Henderson (1976). It relies on the irreversible photochemical conversion of 3′-(*p*-(dimethylamino)phenyl)spiro(fluorene-9,4′-oxazolidine)-2′,5′-dione to *p*-dimethylamino-*N*-fluoren-9-ylidine aniline, with the elimination of carbon monoxide and carbon dioxide (Fig. 4.29). A

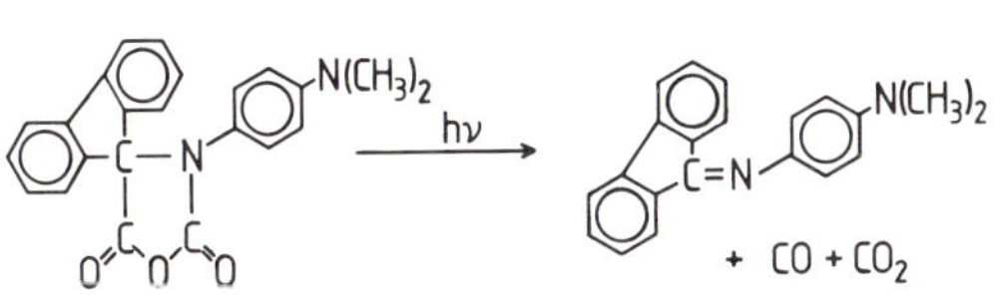

Fig. 4.29. Mechanism of the Zweig and Henderson sunburn dosimeter.

polycarbonate matrix is used. When a transparent matrix is used, the colour change can be easily measured quantitatively by absorption spectrophotometry. A system which does not produce a visible change, but one which can be easily measured by infra-red spectroscopy, is the photoisomerization of *o*-nitrobenzaldehyde to *o*-nitrosobenzoic acid (Fig. 4.30). This reaction has a quantum yield of 0·5 (Leighton and Lucy, 1934). The

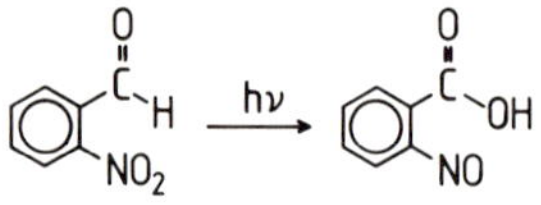

Fig. 4.30. Photoisomerization of *o*-nitrobenzaldehyde.

o-nitrobenzaldehyde can be incorporated in a KBr disc (Pitts *et al.*, 1964), or in a poly(methyl methacrylate) film (Cowell and Pitts, 1968; Pitts *et al.*, 1968). The conversion can be measured by monitoring the nitro group absorption at 1530 cm^{-1}. This type of dosimeter is suitable for the wavelength range 280–400 nm.

There has been some use of diazo systems. Normally, when a diazonium salt is irradiated, *decomposition* takes place, and the salt *loses* its ability to take part in colour-forming coupling reactions. However, Thielemann (1971, 1975) has devised an ingenious reaction sequence which involves the photoinduced *synthesis* of a diazonium salt, followed by a coupling reaction with α-naphthylamine to give a rose-red colour (Fig. 4.31).

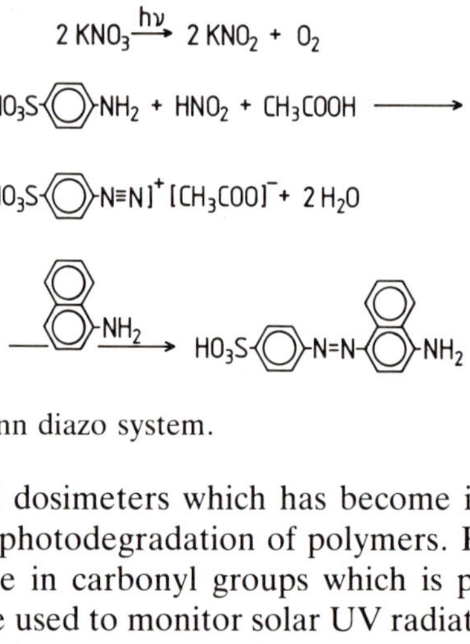

Fig. 4.31. The Thielemann diazo system.

A class of chemical dosimeters which has become important in recent years is based on the photodegradation of polymers. Poly(vinyl chloride) films show an increase in carbonyl groups which is proportional to UV irradiation, and can be used to monitor solar UV radiation (Martin, 1973). Polysulphone and poly(phenylene oxide) films have also been used for this purpose (Davis *et al.*, 1976b, 1979, 1981; Davis and Gardiner, 1982). These films show an increase in UVA absorption when irradiated with wavelengths lower than 400 nm in the case of poly(phenylene oxide) and 340 nm in the case of polysulphone (Challoner *et al*, 1978). Polysulphone has been used as a personal dosimeter, because it responds to the region of solar UV radiation which causes erythema (Davis *et al.*, 1976a). The increase in absorbance at 330 nm is measured. The dependence of the increase of absorbance on the UV dose is not linear because there is a protective *inner filtering* effect of the new absorption centres formed. When patients are being treated with photosensitizing drugs, it is appropriate to incorporate these or similar compounds into the polymer films to modify their spectral response. Hence films containing 8-methoxypsoralen (Diffey and Davis, 1978), phenothiazine (Diffey *et al.*, 1977), and nalidixic acid (Tate *et al.*, 1980) have been developed.

4.5.2 Chemical actinometers

Fluid actinometers are widely used by photochemists for measuring radiation fluxes in photochemical reactors. They possess many advantages in comparison with instrumental systems for this particular purpose (Parker, 1968, p. 208; Taylor, 1971). They are basically simple and require no specialized equipment. They are less prone to systematic error, and give more reproducible results. The actinometer fluid may be used in the actual reaction cell, so eliminating any problems associated with reflectances at interfaces, uneven light distribution, beam geometry, etc. Another potential attraction of the fluid actinometer is that it can be used in a cell which totally surrounds a light source, and therefore can collect almost all the radiant energy (Demas *et al.*, 1976).

Fluid actinometers may be gaseous or liquid. The quantum yield for the actinometer reaction should be independent of reactant concentration, light intensity, temperature, and wavelength over a reasonable range. Quantum yields must be determined by the use of thermopiles or other detectors with an absolute calibration.

The early history of actinometry has been discussed by Wightman (1959). Early chemical actinometers were devised for the measurement of solar radiation, and include the evolution of oxygen from chlorine water, and the reaction of hydrogen with chlorine. A large number of actinometer systems have been described in the literature, but only a few have been widely used.

4.5.2.1 The malachite green leucocyanide actinometer

The triphenylmethane dye cyanide system already described for the preparation of papers and films which change colour on irradiation has also been the basis of a solution actinometer. The reaction was first discovered by Lifschitz (1919). Weyde and Frankenburger (1931) developed an actinometer from crystal violet leucocyanide for the measurement of erythema-producing wavelengths. Later workers (Harris *et al.*, 1935; Harris and Kaminsky, 1935; Calvert and Rechen, 1952; Fisher *et al.*,

1967) carried out further development of the system, and found that malachite green was to be preferred to crystal violet. Irradiation of malachite green leucocyanide in ethanol gives a green colour with a broad absorption at 622 nm which can be measured with a spectrophotometer. The system is very sensitive, and is capable of measuring low intensities of light. The quantum efficiencies which have been measured are close to unity, in the region of 225–334 nm, and are fairly insensitive to changes in temperature and intensity. There is a small discrepancy between the results of Fisher *et al.* (1967) and those of previous authors (Fig. 4.32) which has been attributed to the difference in the value of the extinction coefficient of MG^+ used in the calculations. The main disadvantages of the actinometer are the restricted wavelength region over which it can be used, the inner filter effect caused by the coloured photoproduct absorbing UV radiation and therefore effectively reducing the quantum efficiency, and the fading of the colour in the dark.

4.5.2.2 The uranyl oxalate actinometer

Until the mid 1950s the most widely used chemical actinometer was that based on the decomposition of oxalic acid photosensitized by uranyl ions. Irradiation of a dilute aqueous solution containing oxalic acid and a uranyl salt, usually uranyl sulphate or uranyl oxalate, causes the oxalic acid to decompose:

$$H_2C_2O_4 \longrightarrow H_2O + CO_2 + CO$$

This is an oversimplified representation of the reaction, as the stoicheiometry varies with acidity (Carter and Weiss, 1940). The reaction has been known for many years (Fay, 1896). Early workers who measured the quantum efficiency for the loss of oxalic acid (Boll, 1913; Büchi, 1924; Anderson and Robinson, 1925; Bowen and Watts, 1926) found widely varying results. The first reproducible values were given by Leighton and Forbes (1930), for wavelengths between 253·7 and 435.0 nm. Brackett and Forbes (1933) extended the wavelength range down to 208 nm. A careful reevaluation by Discher *et al.* (1963) gave values almost identical to those of Leighton and Forbes. Quantum efficiencies for the uranyl oxalate actinometer are shown as a function of wavelength in Fig. 4.32, along with those for other significant actinometers. The uranyl oxalate actinometer has been widely accepted because the quantum efficiency is insensitive to dilution (Forbes and Heidt, 1934) and varies little with temperature (Leighton and Forbes, 1930; Norton, 1934, 1966). The quantum efficiency is insensitive to intensity and the actinometer is suitable for high intensity flash sources (Christie and Porter, 1952). There is no inner filter effect, and negligible dark reaction. The loss of oxalic acid is determined by measuring

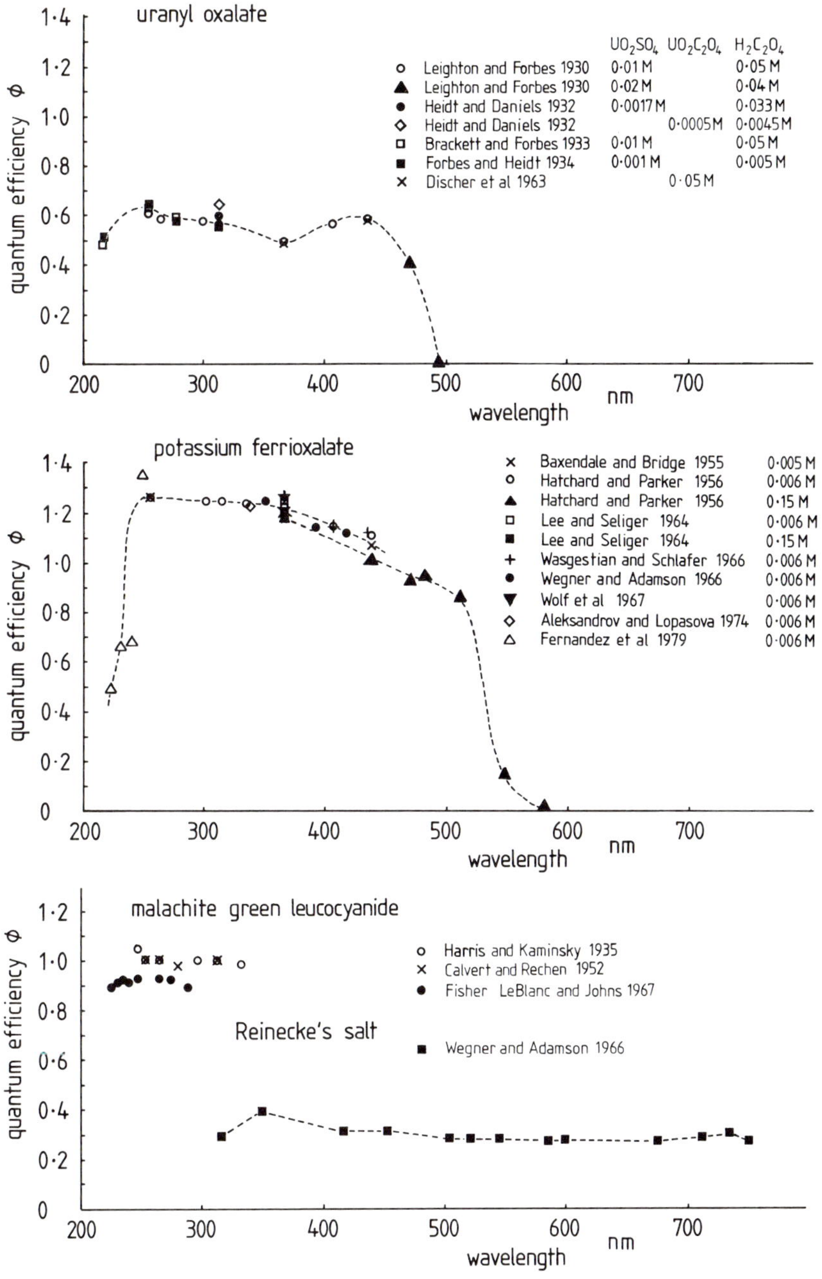

Fig. 4.32. Quantum efficiencies for some solution actinometers.

its concentration before and after irradiation by titration with an oxidant, usually potassium permanganate or ceric sulphate.

The main shortcoming of this actinometer in its classic version is its lack of sensitivity. Low intensities imply the measurement of a small difference between two large titration figures unless lengthy irradiation times are used. Various modifications have been employed to improve the sensitivity. Pitts *et al.* (1955) used ceric sulphate as a standard oxidant and differential absorption spectrophotometry as a more sensitive technique for estimating the oxalic acid decomposed. Various workers have chosen to measure the CO and/or the CO_2 produced by the reaction either manometrically (Zill, 1955) or by gas chromatography (Porter and Volman, 1962; Volman and Seed, 1964).

4.5.2.3 The potassium ferrioxalate actinometer

During the past 25 years the most widely used solution actinometer has been the potassium ferrioxalate type as developed by Parker (Parker, 1953; Hatchard and Parker, 1956). The overall reaction involved is

$$2\,Fe^{III}(C_2O_4)_3^{3-} \xrightarrow{h\nu} 2\,Fe^{II}(C_2O_4) + 3\,(C_2O_4)^{2-} + 2\,CO_2$$

The ferrous ions produced are measured by the addition, after irradiation, of *o*-phenanthroline, which forms a highly coloured complex (molar absorptivity $1{\cdot}11 \times 10^4$ at 510 nm) with the ferrous ions. This actinometer has all the attractions of the uranyl oxalate type, i.e. adherence to reciprocity, insensitivity to temperature changes and dilution, lack of inner filter effect and dark reaction, but with the additional advantages of much greater sensitivity and a wavelength range which stretches further into the visible (Fig. 4.32). It is normally used as Parker originally described it, although Kurien (1971) has suggested adding the *o*-phenanthroline prior to irradiation. Experimental procedures can be found in the original papers, and have also been given by Calvert and Pitts (1966, p. 783) and by Taylor (1971). It is simple to use, gives highly reproducible results and has proved to be quite versatile. For instance, under the right conditions the irradiated ferrous solution is quite stable (Nicodem *et al.*, 1977) and can be stored for a considerable period before analysis, making it suitable for use on long duration space missions (Wrighton and Witz, 1972). Its use at wavelengths below 254 nm has been described by Fernandez *et al.* (1979). It has been found suitable for measuring the intensity of short pulse laser radiation (Aleksandrov and Lopasova, 1974; Grüter, 1980).

4.5.2.4 Miscellaneous chemical actinometers

The malachite green leucocyanide, uranyl oxalate and ferrioxalate actinometers are those which have been most widely accepted and used, with the ferrioxalate type currently the most important. However, many other types of solution actinometer have been suggested and used, and some have particularly merits for particular applications. The characteristics of selected solution actinometers are shown in Table 4.2. Gas-phase actinometers are generally more difficult to use than solution actinometers, and give less reliable results, but several can be used in the vacuum UV. The characteristics of selected gas phase actinometers are shown in Table 4.3.

4.6 Thermoluminescent Dosimeters

Personnel dosimetry is essential for workers who may be exposed to ionizing radiations, and one well established type of dosimeter is based on the phenomenon of *thermoluminescence*, which is displayed by certain inorganic materials. These materials contain electron traps which may be populated when the material is irradiated. If the traps are deep, and the temperature low, the electrons will remain for long periods in these metastable states. If the temperature is gradually raised the electrons will receive enough thermal energy to escape the traps. If the freed electrons can move to luminescent centres, visible light will be emitted, and the total light energy emitted is a measure of the original irradiation dose. The glow curve may have several peaks corresponding to different trap depths. With some materials the photon energy of UV radiation is sufficient to induce thermoluminescence, and thermoluminescent dosimeters have been used for monitoring personnel exposure to UV, and also for monitoring solar UV radiation (McCullough *et al.*, 1972; Marwaha *et al.*, 1979, 1980).

There are two types of thermoluminescent dosimeter used for UV. The *intrinsic* type is heated prior to use to make sure that all the traps are empty. Traps are then populated by exposure to UV and, when the dose is to be measured, the dosimeter material is heated up and the glow curve measured with a photomultiplier. Suitable materials are thulium-activated calcium sulphate (Lakshmanan and Bhatt, 1979), bismuth-activated calcium sulphide, dysprosium-activated calcium fluoride, single crystal magnesium oxide and terbium-activated magnesium silicate (Marwaha *et al.*, 1979). The second type of thermoluminescent dosimeter involves *tranferred* thermoluminescence. Prior to use the material is exposed to gamma radiation, and then is heated up to say 400 °C to empty all but the deepest traps. Subsequent exposure to UV then transfers electrons from the deep traps to shallower traps, and then the glow curve below 400 °C is measured to reveal the populations of the shallower traps, and therefore the UV dose. Materials used are usually undoped, e.g. lithium fluoride, beryllium oxide, calcium fluoride, calcium sulphate.

Table 4.2. Miscellaneous solution actinometers

Actinometer	Reaction involved	Method of analysis	Wavelength region	Quantum yield	References
Monochloroacetic acid, aqueous	Hydrolysis: $CH_2ClCOOH + H_2O \xrightarrow{h\nu} CH_2OHCOOH + HCl$	Polarographic determination of chloride ion, or potentiometric titration with silver nitrate	$\leqslant 254$ nm	0·3–0·4 at 25 °C; very dependent on temperature	Leighton *et al.* (1938); Smith *et al.* (1939); Thomas (1940); Kemula and Grabowska (1955)
Anthracene or 9-methyl anthracene in benzene, or ethanol	Dimerization	Dimer precipitates from benzene. Reduction of spectral absorbance above 300 nm in ethanol	240–390 nm		Dore (1958); Zinina and Vember (1972)
Azobenzene in methanol	Isomerization	Change in UV absorbance	268–365 nm		Gauglitz (1976)
2,2′,4,4′-tetra-isopropyl azobenzene in methanol	Isomerization	Change in UV absorbance	350–390 nm	0·112	Frank and Gauglitz (1977)
Benzophenone/benzhydrol in benzene	Oxidation/reduction: $\phi{-}\underset{\underset{O}{\Vert}}{C}{-}\phi + \phi{-}\underset{OH}{\overset{H}{C}}{-}\phi \xrightarrow{h\nu} \phi_2\underset{OH}{C}{-}\underset{OH}{C}\phi_2$	Change in UV absorbance	300–400 nm	0·81	Wagner (1969); Rosenthal and Bercovici (1976); Horspool (1976, p. 47)
Aqueous hydroxyethyl-cellulose	Degradation	Change in viscosity of solution	Depends on dye sensitization		Dixon *et al.* (1975)

1,3-Cyclohexadiene in benzene	Dimerization sensitized by benzophenone	Chromatographic analysis of products	313, 365 nm	0·97	Lam *et al*. (1967); Vesley and Hammond (1973)
Aqueous hydrazoic acid	Hydrolysis: $HN_3 + H_2O \xrightarrow{h\nu} N_2 + NH_2OH$	Measurement of N_2 evolved	200–260 nm	1·00	Shapira and Treinin (1973)
Benzoic acid	Decomposition to fluorescent products	Measurement of fluorescence	$\leqslant$ 300 nm		Moroson and Gregoriades (1964)
Reinecke's salt in aqueous solution	Aquation: $[Cr(NH_3)_2(NCS)_4]^- + H_2O \xrightarrow{h\nu} Cr(NH_3)_2(NCS)_3(H_2O) + NCS^-$	NCS^- complexed with Fe^{3+} and absorbance measured at 450 nm	316–750 nm	~0·27 (see Fig. 4.32)	Wegner and Adamson (1966); Adamson and Fleischauer (1975); Cowan and Drisko (1976)
Tetramethylethylene (TME)	Photooxygenation sensitized by tris(2,2′-bipyridine)ruthenium: $[Ru(bpy)_3]^{2+} \xrightarrow{h\nu} {}^*[Ru(bpy)_3]^{2+}$; ${}^*[Ru(bpy)_3]^{2+} + O_2 \longrightarrow [Ru(bpy)_3]^{2+} + {}^1O_2$; ${}^1O_2 + TME \longrightarrow (CH_3)_2COOHC(CH_3){=}CH_2$	O_2 uptake measured by gas burette	280–560 nm	0·86	Demas *et al*. (1976)
Valerophenone in benzene	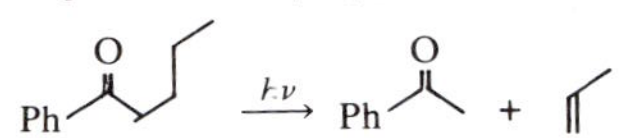	Acetophenone measured by gas–liquid chromatography	313 nm	0·33	Horspool (1976, p. 49)
Decafluoro-benzophenone/isopropanol	Photoreduction of DFBP	Chromatography or change of UV absorbance	290–370 nm	0·60	Minn *et al*. (1970); Horspool (1976, p. 49)
Aqueous ethanol	Photolysis; main products are H_2, CH_4, CO_2	Measurement of H_2 evolved	180–200 nm	Various values quoted between 0·4 and 0·8	Farkas and Hirschberg (1937); Bernas *et al*. (1965); Fox (1970); Davies *et al*. (1976)

Table 4.3. Miscellaneous gas phase actinometers

Actinometer	Reaction involved	Method of analysis	Wavelength region	Quantum yield	References
Oxygen	$3O_2 \xrightarrow{h\nu} 2O_3$	Absorption of Hg 254 nm radiation, or passage through acidified KI, to release I_2, followed by titration with thiosulphate, or spectrophotometric estimation of I_2 (353 nm)	130–190 nm	$\phi_{O_3} = 2$	Driscoll and Warneck (1971); Davies *et al.* (1976); Volman (1963)
Carbon dioxide	$2CO_2 \xrightarrow{h\nu} 2CO + O_2$	Measurement of CO	< 170 nm	$\phi_{CO} = 1{\cdot}06$	Mahan (1960); Warneck (1966)
Nitrous oxide	$2N_2O \xrightarrow{h\nu} 2N_2 + O_2$	Measurement of N_2	147–185 nm	$\phi_{N_2} = 1{\cdot}44$	Calvert and Pitts (1966, p. 783)
Hydrogen bromide	$2HBr \xrightarrow{h\nu} H_2 + Br_2$	Br_2 condensed by liquid N_2 followed by measurement of H_2	180–250 nm	$\phi_{H_2} = 1{\cdot}0$	Calvert and Pitts (1966, p. 782)
Acetone or diethyl ketone	e.g. $CH_3COCH_3 \xrightarrow{h\nu} 2C_2H_6 + CO$	Measurement of CO	250–320 nm	$\phi_{CO} = 1{\cdot}0$	Calvert and Pitts (1966, p. 782)
Hexafluoroacetone	$CF_3COCF_3 \xrightarrow{h\nu} C_2F_6 + CO$	Measurement of CO	147 nm	$\phi_{CO} = 1{\cdot}0$	Magenheimer and Timmons (1970)
Perfluoroacetic anhydride	$(CF_3CO)_2O \xrightarrow{h\nu} CO + CO_2 + C_2F_6$	Measurement of CO	254, 265 nm	$\phi_{CO} = 0{\cdot}29$ at 25 °C 0·34 at 200 °C	Chamberlain and Whittle (1975)
Perfluoropropionic anhydride	$(C_2F_5CO)_2O \xrightarrow{h\nu} CO + CO_2 + n\text{-}C_4F_{10}$	Measurement of CO	254, 265 nm	$\phi_{CO} = 0{\cdot}24$ at 26 °C 0·29 at 200 °C	Chamberlain and Whittle (1975)

References

Adamson, A. W. and Fleischauer, P. D. (1975). "Concepts of Inorganic Photochemistry". John Wiley, New York.

Aleksandrov, E. I. and Lopasova, T. A. (1974). *Sov. J. Quantum Electron.* **4,** 813.

Allison, R., Burns, J. and Tuzzolino, A. J. (1964). *J. Opt. Soc. Amer.* **54,** 747–751.

Amrein, W., Gloor, J. and Schaffner, K. (1974). *Chimia* **28,** 185–188.

Anderson, W. T., Jr and Robinson, F. W. (1925). *J. Amer. Chem. Soc.* **47,** 718–725.

Barr, E. E. (1975). *Proc. Soc. Photo-Opt. Instr. Eng.* **50,** 87–118.

Baumeister, P. (1965). *In* "Applied Optics and Optical Engineering", Vol. I (Ed. R. Kingslake), pp. 285–323. Academic Press, New York.

Baxendale, J. H. and Bridge, N. K. (1955). *J. Phys. Chem.* **59,** 783–788.

Bernas, A., Bodard, M. and Saghattchian, D. (1965). *J. Chim. Phys.* **62,** 1418–1422.

Boll, M. (1913). *Compt. Rend.* **156,** 1891–1896.

Bowen, E. J. and Watts, H. G. (1926). *J. Chem. Soc.* **1926,** 1607–1612.

Brackett, F. P., Jr and Forbes, G. S. (1933). *J. Amer. Chem. Soc.* **55,** 4459–4466.

Büchi, P. F. (1924). *Z. Phys. Chem.* **111,** 269–314.

Calvert, J. G. and Pitts, J. N., Jr (1966). "Photochemistry". John Wiley & Sons, New York.

Calvert, J. G. and Rechen, H. J. L. (1952). *J. Amer. Chem. Soc.* **74,** 2101–2103.

Carter, A. H. and Weiss, J. (1940). *Proc. Roy. Soc. A* **174,** 351–370.

Chalkley, L. (1952). *J. Opt. Soc. Amer.* **42,** 387–392.

Challoner, A. V. J., Corbett, M. F., Davis, A., Diffey, B. L., Leach, J. F. and Magnus I. A. (1978). *Natl. Cancer Inst. Monogr.* no 50, 97–100.

Chamberlain, G. A. and Whittle, E. (1975). *J. Chem. Soc. Faraday Trans. 1* **71,** 1978–1990.

Christie, M. I. and Porter, G. (1952). *Proc. Roy. Soc. A* **212,** 380–397.

Clarke, F. J. J. (1979). *Proc. Soc. Photo-Opt. Instr. Eng.* **146,** 76–80.

Cowan, D. O. and Drisko, R. L. (1976). "Elements of Organic Photochemistry". Plenum, New York.

Cowell, G. W. and Pitts, J. N., Jr (1968). *J. Amer. Chem. Soc.* **90,** 1106–1110.

Dance, J. B. (1969). "Photoelectronic Devices". Iliffe Books, London.

Davidson, G., Sluder, R. B., Kofsky, I. L. and Goldsmith, G. (1981). *Photochem. Photobiol.* **34,** 287–296.

Davies, A. K., Khan, K. A., McKellar, J. F. and Phillips, G. O. (1976). *Mol. Photochem.* **7,** 343–347, 389–398.

Davis, A. and Gardiner, D. (1982). *Polym. Degrad. Stab.* **4,** 145–157.

Davis, A., Deane, G. H. W. and Diffey, B. L. (1976a). *Nature* **261,** 169–170.

Davis, A., Deane, G. H. W., Gordon, D., Howell, G. V. and Ledbury, K. J. (1976b). *J. Appl. Polym. Sci.* **20,** 1165–1174.

Davis, A., Howes, B. V., Ledbury, K. J. and Pearce, P. J. (1979). *Polym. Degrad. Stab.* **1,** 121–132.

Davis A., Diffey, B. L. and Tate, T. K. (1981). *Photochem. Photobiol.* **34,** 283–286.

Demas, J. N., McBride, R. P. and Harris, E. W. (1976). *J. Phys. Chem.* **80,** 2248–2253.

Dennis, P. N. J. (1979). *Sci. Prog. (Oxford)* **66,** 267–294.

Diffey, B. L. and Davis, A. (1978). *Phys. Med. Biol.* **23,** 318–323.
Diffey, B. L., Davis, A., Johnson, M., and Harrington, T. R. (1977). *Brit. J. Dermatol.* **97,** 127–130.
Discher, C. A., Smith, P. F., Lippman, I. and Turse, R. (1963). *J. Phys. Chem.* **67,** 2501.
Dixon, G. D., Davies, D. H. and Voytko, J. D. (1975). *Environ. Sci. Technol.* **9,** 234–237.
Dore, W. G. (1958). *Ecology* **39,** 151–152.
Driscoll, J. N. and Warneck, P. (1971). *Photochem. Photobiol.* **13,** 283–287.
Duval, K. and Ryer, D. V. (1977). *Opt. Spectra* **1977** (June), 26–28.
Eckerle, K. L., Venable, W. H., Jr and Weidner, V. R. (1976). *Appl. Opt.* **15,** 703–707.
Egan, W. G. and Hilgeman, T. (1975). *Appl. Opt.* **14,** 1137–1142.
Fadner, T. A. and Schaeffer, W. D. (1978). *Brit. Ink Maker* **20,** 105–110.
Farkas, L. and Hirschberg, Y. (1937). *J. Amer. Chem. Soc.* **59,** 2450–2453.
Farrell, J. J. and Vale, R. L. (1963). *Nucleonics* **21** (11), 78–80, 82, 84–85.
Fay, H. (1896). *Amer. Chem. J.* **18,** 269–289.
Fernandez, E., Figuera, J. M., and Tobar, A. (1979). *J. Photochem.* **11,** 69–71.
Fisher, G. J., LeBlanc, J. C. and Johns, H. E. (1967). *Photochem. Photobiol.* **6,** 757–767.
Forbes, G S. and Heidt, L. J. (1934). *J. Amer. Chem. Soc.* **56,** 2363–2365.
Fox, M. F. (1970). *Quart. Rev. (Lond.)* **24,** 565–584.
Frank R. and Gauglitz, G. (1977). *J. Photochem.* **7,** 355–357.
Gauglitz, G. (1976). *J. Photochem.* **5,** 41–47
Gauglitz, G. and Kölle, U. (1975). *J. Photochem.* **4,** 309–312.
Gavrilova, L. I., Doinikov, A. S., Eremin, E. A., and Pakhomov, V. K. (1978). *Svetotekhnika* (1), 11–14.
Gillham, E. J. (1962). *Proc. Roy. Soc. A* **269,** 249–276.
Grum, F. and Becherer, R. J. (1979). "Optical Radiation Measurements", Vol. 1, "Radiometry". Academic Press, New York.
Grum, F. and Luckey, G. W. (1968). *Appl. Opt.* **7,** 2289–2294.
Grüter, H. (1980). *J. Appl. Phys.* **51,** 5204–5206.
Harris, P. B. (1968). *J. Sci. Instr. (J. Phys. E)* **1,** 1007–1010.
Harris, L. and Kaminsky, J. (1935). *J. Amer. Chem. Soc.* **57,** 1154–1159.
Harris, L., Kaminsky, J., and Simard, R. G. (1935). *J. Amer. Chem. Soc.* **57,** 1151–1154.
Hatchard, C. G. and Parker, C. A. (1956). *Proc. Roy. Soc. A* **235,** 518–536.
Heidt, L. J. and Daniels, F. (1932). *J. Amer. Chem. Soc.* **54,** 2384–2391.
Horspool, W. M. (1976). "Aspects of Organic Photochemistry". Academic Press, London.
Jenkins, F. A. and White, H. E. (1976). "Fundamentals of Optics", 4th edn. McGraw-Hill, New York.
Johnson, F. S., Watanabe, K. and Tousey, R. (1951). *J. Opt. Soc. Amer.* **41,** 702–708.
Kantz, A. and Humpherys, K. (1978). *Radiat. Curing* **5** (2), 4–10.
Kemula, W. and Grabowska, A. (1955). *Roczniki Chem.* **29,** 834–838.
Key, P. J. and Preston, R. C. (1980). *J. Phys. E* **13,** 866–870.
Knowles-Middleton, W. E. and Sanders, C. L. (1951). *J. Opt. Soc. Amer.* **41,** 419–424.
Koller, L. R. (1965). "Ultraviolet Radiation", 2nd edn. John Wiley, New York.

Kurien, K. C. (1971) *J. Chem. Soc. B* **1971,** 2081–2082.
Lam, E. Y. Y., Valentine, D. and Hammond, G. S. (1967). *J. Amer. Chem. Soc.* **89,** 3482–3487.
Lambert, G. E. V. and Preston, J. S. (1961). *Light Lighting* **54,** 327–330.
Lakshmanan, A. R. and Bhatt, R. C. (1979). *AMPI Med. Phys. Bull.* **4,** 221–224.
Lee, J. and Seliger, H. H. (1964). *J. Chem. Phys.* **40,** 519–523.
Leighton, P. A. and Lucy, F. A. (1934). *J. Chem. Phys.* **2,** 756–759, 760–766.
Leighton, W. G. and Forbes, G. S. (1930). *J. Amer. Chem. Soc.* **52,** 3139–3152.
Leighton, W. G., Smith, R. N. and Leighton, P. A. (1938). *J. Amer. Chem. Soc.* **60,** 2566.
Levi, L. (1980). "Applied Optics", Vol. 2, "A Guide to Optical Systems Design". John Wiley and Sons, New York.
Lifschitz, J. (1919). *Berichte* **52,** 1919–1926.
McCullough, E. C., Fullerton, G. D. and Cameron, J. R. (1972). *J. Appl. Phys.* **43,** 77–82.
McLaughlin, W. L., Levine, H. and Rosenstein, M. (1977). US Pat. 4,006,023 (USDHEW).
Magenheimer, J. J. and Timmons, R. B. (1970). *J. Chem. Phys.* **52,** 2790–2791.
Mahan, B. H. (1960). *J. Chem. Phys.* **33,** 959–965.
Marette, G. (1976). *Opt. Commun.* **19,** 112–115.
Martin, K. G. (1973). *Brit. Polym. J.* **5,** 443–450.
Marwaha, G. L., Singh, N., Vij, D. R. and Mathur, V. K. (1979). *Mater. Res. Bull.* **14,** 1489–1495.
Marwaha, G. L., Singh, N., Vij, D. R. and Mathur, V. K. (1980). *J. Phys. C* **13,** 1559–1565.
Melhuish, W. H. (1955). *New Zealand J. Sci. Technol. B* **37,** 142–149.
Melhuish, W. H. (1962). *J. Opt. Soc. Amer.* **52,** 1256–1258.
Mielenz, K. D., Mavrodineanu, R. and Cehelnik, E. D. (1975). *Appl. Opt.* **14,** 1940–1947.
Minn, F. L., Filipescu, N. and Pinion, J. P. (1970). *J. Chem. Soc. D* **1970,** 1413.
Moroson, H. and Gregoriades, A. (1964). *Nature* **204,** 676–678.
Nettleton, D. H. and Preston, R. C. (1981). *Appl. Opt.* **20,** 1274–1275.
Nicodem, D. E., Cabral, M. L. P. F. and Ferreira, J. C. N. (1977). *Mol. Photochem.* **8,** 213–238.
Norton, B. M. (1934). *J. Amer. Chem. Soc.* **56,** 2294–2297.
Norton, B. M. (1966). *Ohio J. Sci.* **66,** 421–425.
Parker, C. A. (1953). *Proc. Roy. Soc. A* **200,** 104–116.
Parker, C. A. (1968). "Photoluminescence of Solutions, with Applications to Photochemistry and Analytical Chemistry". Elsevier, Amsterdam.
Parrish, J. A., Anderson, R. R., Urbach, F. and Pitts, D. (1978). "UV-A: Biological Effects of Ultraviolet Radiation with Emphasis on Human Responses to Longwave Ultraviolet". John Wiley & Sons, Chichester.
Piétri, G. and Nussli, J. (1968). *Philips Tech. Rev.* **29,** 276–287.
Pitts, J. N., Jr, Margerum, J. D., Taylor, R. P., and Brim W. (1955). *J. Amer. Chem. Soc.* **77,** 5499–5501.
Pitts, J. N., Jr, Wan, J. K. S. and Schuck, E. A. (1964). *J. Amer. Chem. Soc.* **86,** 3606–3610.
Pitts, J. N., Jr, Cowell, G. W. and Burley, D. R. (1968). *Environ. Sci. Technol.* **2,** 435–437.
Porter, K. and Volman, D. H. (1962). *J. Amer. Chem. Soc.* **84,** 2011–2012.

Putley, E. H. (1970). *In* "Semiconductors and Semimetals", Vol. 5 (Eds R. K. Willardson and A. C. Beer). Academic Press, New York.
Putley, E. H. (1973). *Phys. Technol.* **4,** 202–222.
Richardson, D. (1969). *In* "Applied Optics and Optical Engineering," Vol. V (Ed. R. Kingslake), pp. 17–46. Academic Press, New York.
Rosenthal, I. and Bercovici, T. (1976). *Atoms. Environ.* **10,** 1139–1140.
Rothwell, H. L. and Keeffe, W. M. (1980). *J. Illum, Eng. Soc.* **10,** 40–46.
Saunders, R. D. and Ott, W. R. (1976). *Appl. Opt.* **15,** 827–828.
Schaarschmidt, B. (1970). *Z. Naturforsch. B* **25,** 330–331.
Schultze, H. and Vogel, H.-R. (1973). *Mol. Photochem.* **5,** 223–225.
Shapira, D. and Treinin, A. (1973). *J. Phys. Chem.* **77,** 1195–1198.
Shaw, J. M. and Hatzakis, M. (1981). *J. Vac. Sci. Technol.* **19,** 1343–1347.
Smith, R. N., Leighton, P. A. and Leighton, W. G. (1939). *J. Amer. Chem. Soc.* **61,** 2299–2301.
Sommer, A. H. (1968). "Photoemissive Materials: Preparation, Properties and Uses". John Wiley and Sons, New York.
Stair, R. (1970). *Advanc. Geophys.* **14,** 83–109.
Stair, R., Schneider, W. E. and Jackson, J. K. (1963). *Appl. Opt.* **2,** 1151–1154.
Sumpner, W. E. (1892). *Proc. Phys. Soc.* **12,** 10.
Swindell, W. (1980). *In* "Applied Optics and Optical Engineering", Vol. VIII (Eds. R. R. Shannon and J. C. Wyant), pp. 317–347. Academic Press, New York and London.
Tassell, C. (1980). *Opt. Laser Technol.* **12,** 271–273.
Tate, T. J., Diffey, B. L., and Davis, A. (1980). *Photochem. Photobiol.* **31,** 27–30.
Taylor, H. A. (1971). *In* "Analytical Photochemistry and Photochemical Analysis" (Ed. J. M. Fitzgerald), pp. 91–115. Dekker, New York.
Teele, R. P. (1965). *In* "Applied Optics and Optical Engineering", Vol. I (Ed. R. Kingslake), pp. 1–42. Academic Press, New York.
Tellex, P. A. and Waldron J. R. (1955). *J. Opt. Soc. Amer.* **45,** 19–22.
Thielemann, H. (1971). *Sci. Pharm.* **39,** 297–298.
Thielemann, H. (1975). *Z. Chem.* **15,** 451.
Thomas, L. B. (1940). *J. Amer. Chem. Soc.* **62,** 1879–1880.
Thorne, A. P. (1974). "Spectrophysics". Chapman and Hall, London.
Van Laar, J. and Scheer, J. J. (1968). *Philips Tech. Rev.* **29,** 54–66.
Vesley, G. F. and Hammond, G. S. (1973). *Mol. Photochem.* **5,** 367–369.
Vine, B. H. (1965). *In* "Applied Optics and Optical Engineering", Vol. II (Ed. R. Kingslake), pp. 229–278. Academic Press, New York.
Volman, D. H. (1963). *Advanc. Photochem.* **1,** 43–82.
Volman, D. H. and Seed, J. R. (1964). *J. Amer. Chem. Soc.* **86,** 5095–5098.
Wagner, P. J. (1969). *Mol. Photochem.* **1,** 71–87.
Walsh, J. W. T. (1958). "Photometry", 3rd edn. Constable, London.
Wareham, E. M. (1979). *Proc. Soc. Photo-Opt. Instr. Eng.* **146,** 22–28.
Warneck, P. (1966). *J. Opt. Soc. Amer.* **56,** 408–409.
Wasgestian, F. and Schläfer, H. F. (1966). *Z. Phys. Chem. Frankfurt* **51,** 208–223.
Watanabe, K. and Inn, E. C. Y. (1953). *J. Opt. Soc. Amer.* **43,** 32–35.
Wegner, E. E., and Adamson, A. W. (1966). *J. Amer. Chem. Soc.* **88,** 394–404.
Weyde, E. and Frankenburger, W. (1931). *Trans. Faraday Soc.* **27,** 561–572.
Wharmby, D. O. (1980). *Light. Res. Technol.* **12,** 221–225.
Wightman, E. P. (1959). *Photogr. Sci. Eng.* **3,** 64–86.

Williams, C. S. and Becklund, O. A. (1972). "Optics: A Short Course for Engineers and Scientists". Wiley–Interscience, New York.
Wolf, H. P., Bohning, J. J., Schnieper, B. A. and Weiss, K. (1967). *Photochem. Photobiol.* **6,** 321–329.
Wrighton, M. and Witz, S. (1972). *Mol. Photochem.* **3,** 387–394.
Yguerabide, J. (1968). *Rev. Sci. Instr.* **39,** 1048–1052.
Zill, L. P. (1955). *Arch. Biochem. Biophys.* **59,** 473–483.
Zimmerman, H. E., Cutler, T. P., Fitzgerald, V. R. and Weigt, T. J. (1977). *Mol. Photochem.* **8,** 379–385.
Zinina, E. M. and Vember, T. M. (1972). *J. Appl. Spectrosc.* **16,** 503–507.
Zweig, A. and Henderson, W. A., Jr (1976). *Photochem. Photobiol.* **24,** 543–549.

5 Incandescent Sources

5.1 Black Body Radiators

The radiation emitted by hot bodies was the subject of much experimentation and theoretical work during the latter half of the 19th century, and there was the development of the concept of the "black body radiator". This is an idealized source of radiant energy which absorbs totally all radiation at all wavelengths falling on it and neither reflects nor transmits any. It is a perfectly diffuse radiator which radiates at all wavelengths, and at any particular wavelength and temperature its radiant emittance is the maximum possible for any real source in thermal equilibrium at the same temperature and wavelength. The black body is a theoretical concept in so far as no perfect black body exists in nature, but a close approximation consists of a small opening in a hollow body of which the internal walls are absorbent. The concept of the black body plays an important part in the understanding of all radiation sources in which the component parts of the radiating body are in thermal equilibrium, and in particular incandescent solids and medium and high pressure gas discharges.

The amount and spectral distribution of the radiation emitted by a black body is dependent only on the temperature. The total radiant power emitted by a black body is given by the *Stefan–Boltzmann law*:

$$M(T) = \sigma T^4,$$

where $M(T)$ is the total power (in watts) radiated into one hemisphere (i.e. 2π solid angle) by 1 cm^2 of surface at T kelvins. The law was first derived by Stefan in 1879 from experimental data, and theoretically in 1884 by Boltzmann. The Stefan–Boltzmann constant σ is $5{\cdot}6703 \times 10^{-12}$ W cm^{-2} K^{-4}.

Real radiators always radiate less than a black body at the same temperature, and the Stefan–Boltzmann law becomes

$$M(T) = \varepsilon\sigma T^4,$$

where ε is the *emissivity factor.* For a black body ε is unity, while for real radiators ε is always less than unity. ε is zero for a perfect transmitter or a perfect reflector.

The wavelength at which the radiant emission of a black body has its highest value decreases as its temperature is raised. This is expressed by the *Wien displacement law* (1894):

$$\lambda_{max} = 2{\cdot}8978 \times 10^3\ T^{-1} \qquad \mu\text{m}.$$

The first satisfactory theoretical treatment of the spectral distribution of radiation emitted by a black body was a result of Planck's quantum theory (1900) which led to *Planck's radiation law.* The spectral radiant emittance of a black body radiator at temperature T kelvins into one hemisphere, per unit wavelength interval at wavelength λ, is given by

$$M_\lambda(T) = \frac{2\pi hc^2\lambda^{-5}}{e^{hc/k\lambda T} - 1}$$

where h is Planck's constant, c is the velocity of light and k is Boltzmann's constant. Planck's radiation law is often simplified to

$$M_\lambda(T) = \frac{c_1\lambda^{-5}}{e^{c_2/\lambda T} - 1} \qquad \text{W cm}^{-2}$$

per centimetre wavelength interval, where c_1 is Planck's first radiation constant $(2\pi hc^2) = 3{\cdot}7418 \times 10^{-12}$ W cm^2 and c_2 is Planck's second radiation constant $(hc/k) = 1{\cdot}4388$ cm K.

If the wavelength interval considered is 1 μm,

$$M_\lambda(T) = \frac{37\,418\ \lambda^{-5}}{e^{14\,388/\lambda T} - 1} \qquad \text{W cm}^{-2}\ \mu\text{m}^{-1},$$

and for a real radiator

$$M_\lambda(T) = \varepsilon_{\lambda,T}\frac{37\,418\ \lambda^{-5}}{e^{14\,388/\lambda T} - 1} \qquad \text{W cm}^{-2}\ \mu\text{m}^{-1},$$

where $\varepsilon_{\lambda,T}$ is the *spectral emissivity.* The spectral emissivity of most materials varies with both wavelength and temperature.

If a black body radiator is to produce a high proportion of its output in the UV (say between 200 and 400 nm), the form of Planck's radiation law requires that it must be at a very high temperature. Figure 5.1 shows what

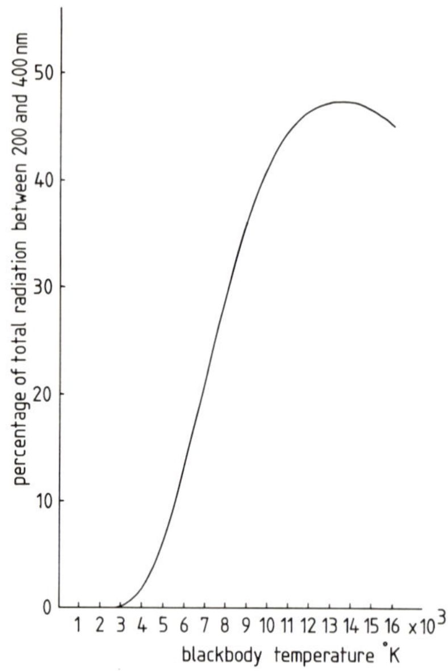

Fig. 5.1. UV (200–400 nm) emission from black body radiators.

percentage of the total radiant emission lies within this wavelength region, as a function of temperature. It can be seen that at the maximum temperature at which incandescent solids can retain their integrity (not much above 3000 K), very little of the total output appears in the UV. However, while Planck's radiation law shows that the *efficiency* with which a black body at around 3000 K emits UV radiation is very low (about 0·2%), the actual *quantity* of radiation emitted in watts per square centimetre between 200 and 400 nm is really quite respectable (0·9 W cm^{-2} at 3000 K). A black body radiator with an area of a few square centimetres could produce at a substantial amount of UV radiation at 3000 K. Unfortunately the power which would be required to maintain this temperature would be very large. On the other hand, in principle, if an incandescent solid could be found with a spectral emissivity approaching unity in the UV part of the spectrum, and very low elsewhere, it could be used as the basis of an efficient UV source.

The spectral emissivity of real radiators always varies with wavelength, and they are sometimes referred to as *selective emitters*. If the variation is very small, as with carbon, the radiator may be called a *grey body*. A selective emitter may have a radiation curve which, at least over a limited wavelength range, has the same shape as that of a black body at a different temperature. The black body temperature giving the best match is called the "*colour temperature*" of the real radiator.

5.2 Filament Lamps

Electrical filament lamps were developed from the early work of Swan (in the UK) and Edison (in the USA). Many different materials have been tried, including carbon, tantalum and osmium, but virtually all lamps now have tungsten filaments. The normal tungsten filament lamp consists of a coiled filament of fine tungsten wire enclosed in a glass envelope which is either evacuated or filled with an inert gas. The filament is coiled to minimize convection loss to the filling gas. Tungsten has the highest melting point of any metallic element, 3655 K. However, it is the evaporation rate of the tungsten which limits the temperature at which the filament can be operated. Evaporation from the filament causes dark deposits to be formed on the inside wall of the envelope, so reducing the transmission, and after about a 5% weight loss from the filament, breakage of the filament may occur.

The tungsten filament is capable of operating in the region of 3000 K. Figure 5.2 shows the hemispherical radiant emittance of a black body

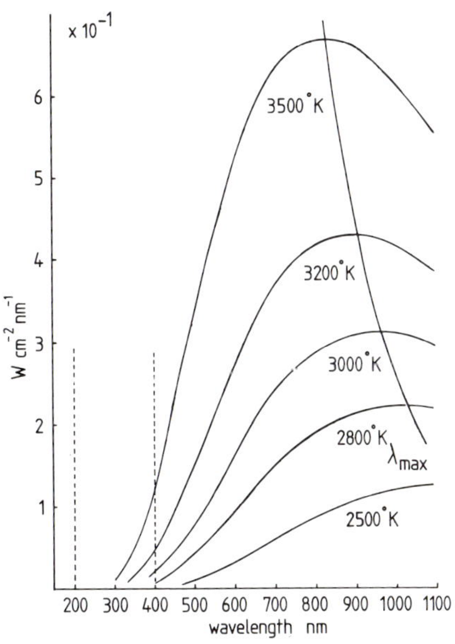

Fig. 5.2. Black body curves for temperatures between 2500 and 3500 K.

radiator for a variety of temperatures relevant to incandescent solids. Because tungsten is a selective radiator, the total radiant emittance of a tungsten surface is much less than that of a black body at the same temperature. Figure 5.3 shows how the spectral emissivity of tungsten varies with temperature and wavelength. There is a maximum in the UV and visible, and it decreases markedly as the wavelength is increased into the infra-red. This means that the ratio of visible light to infra-red is greater than that for a black body, and therefore tungsten is a more efficient source of visible light than a black body. Tantalum shows an even more beneficial selectivity in spectral emissivity. The colour temperature of incandescent tungsten is higher than its actual temperature.

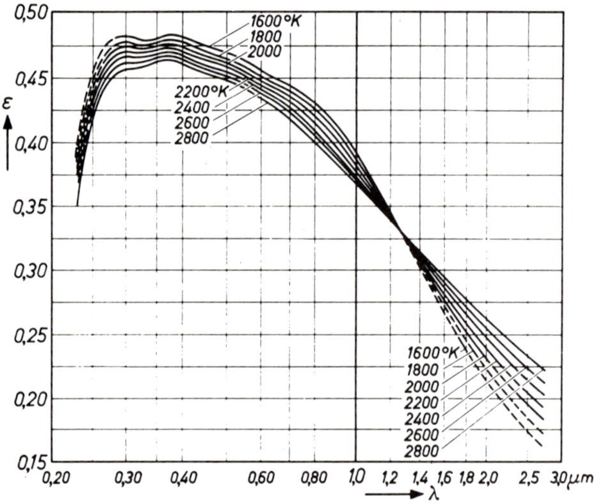

Fig. 5.3. Spectral emissivity of tungsten. (From de Vos, 1954, with permission of North Holland Publishing Company.)

The spectral output of a tungsten filament lamp will differ somewhat from that of a flat tungsten surface at the same temperature. Firstly the envelope will absorb some of the infra-red and, unless it is silica, some of the UV also. Secondly, the spectral emissivity of a material depends to some extent on its shape. The apertures between the coils of the filament act rather like black bodies because of inter-reflections, and therefore tungsten in a coiled filament form does not show quite the selectivity that is shown by a flat tungsten surface.

The Stefan–Boltzmann law shows that the radiant output of a tungsten filament lamp increases rapidly with temperature. Also, the visible to infra-red ratio increases, and therefore it is beneficial to operate at the highest possible temperature. The filling gas decreases the rate of evaporation of the tungsten and therefore allows a higher filament temperature, but at the expense of a greater convective and conductive loss. Filament lamps with a lifetime of about 1000 h have a filament temperature of about 2700 K. For certain specialized purposes (e.g. photographic lighting) the filament may be run at a temperature as high as 3300 K, but lifetime is then reduced to only a few hours.

The lifetime of the tungsten filament lamp is determined by a gradual deterioration process, blackening of the inside wall of the lamp envelope, followed by catastrophic failure when the filament breaks. Lifetime before catastrophic failure can be increased by operating the lamp at a low voltage and a high current. This implies a lower resistance filament, and therefore one which is shorter and fatter and can tolerate a higher percentage evaporation before breakage occurs.

5.3 Tungsten Halogen Lamps

The *tungsten halogen* lamp represents a very ingenious method of improving the performance of the tungsten filament lamp. In the tungsten halogen lamp a small amount of a halogen, iodine or bromine, is added to the gas filling. The halogen has the function of combining with tungsten evaporating from the filament, so preventing it from forming an opaque deposit on the inside wall of the envelope. The tungsten halide diffuses back to the vicinity of the hot filament where it decomposes and the tungsten is redeposited on the filament. A "tungsten halogen" cycle is established which minimizes net evaporation of the tungsten, prevents wall blackening and allows a higher filament temperature with increased lamp life.

Iodine can be used in its elemental form. Bromine is too reactive and will cause erosion of the cooler parts of the filament and its supports, and therefore it is added as HBr, an organic compound such as CH_3Br or CH_2Br_2, or as a phosphonitrile. There is evidence that when iodine is used a trace of oxygen present in the gas fill is beneficial to the regenerative cycle.

A simple tungsten filament lamp requires a large envelope so that tungsten evaporating from the filament can be spread thinly over a large surface area. The envelope of a tungsten halogen lamp can be made very much smaller, and indeed this is necessary as the wall temperature must be exceed 250 °C for the tungsten halogen cycle to operate. In practice wall temperatures are similar to those for medium pressure gas discharge lamps, and therefore the envelope must be made from silica. Silica is necessary for a second reason. The impurities in glass would interact with the halogen and affect the cycle.

Although the tungsten halogen cycle causes tungsten which evaporates from the filament to be redeposited on the filament, this, *per se*, allows only a limited improvement in filament life. This is because tungsten evaporates from hot spots on the filament and is redeposited on cooler parts of the filament and its supports. In other words the cycle does not provide a healing process for sites of potential failure. However, the small, strong envelope allows higher gas filling pressures and the use of expensive krypton or xenon, and this reduces the rate of evaporation. It is of interest that if fluorine could be used instead of iodine or bromine, the tungsten would redeposit on the hot spots. This is because tungsten fluoride (WF_6) is stable up to about 3400 K, and would dissociate only at the filament surface, and preferentially at the hot spots (Coaton and Fitzpatrick, 1980). Unfortunately fluorine and fluorides would react with the silica of the envelope, and a viable tungsten fluorine lamp would require a transparent fluorine-resistant coating on the inside wall of the envelope.

5.4 The Practical Importance of Incandescent Sources

Filament lamps are very inefficient sources of UV and blue light. However, they are cheap and easy to operate and do not require complex and expensive control gear, and there are occasions when the tungsten halogen lamp in particular is an appropriate source of actinic radiation. In some cases the large amount of non-actinic visible and infra-red radiation may cause problems. Dichroic reflectors may help. Although filament lamps are inefficient sources of actinic radiation, even high powered lamps have filaments of very small surface area, so that the emittance is very high. This allows easy optical manipulation of the radiation.

"Black light" filament lamps are available. These have envelopes which transmit long wavelength UV but are opaque to visible radiation. They are useful for exciting fluorescence.

Tungsten filament lamps, when designed for the purpose, and operated from stabilized power supplies, have predictable and reproducible spectral outputs, and are useful as secondary standards for the calibration of radiometry equipment (see Section 4.3.3.1).

Perhaps the most significant aspect of incandescent sources in the present context is that most lamp types which are practical and efficient sources of actinic radiation have components in their construction which are incandescent when the lamp is operating. Electrodes in discharge lamps usually operate at incandescent temperature in order to maintain current flow by thermionic emission, and high powered lamps have envelopes which may need to be at 600–900 °C. There is therefore an incandescent component in the spectral distribution, which adds considerably to the infra-red. This may make a useful contribution to the effectiveness of the lamp, or may alternatively be a nuisance, particularly when heat-sensitive materials are being irradiated. Whereas the radiation emitted by the arc of a discharge lamp ceases immediately the lamp is extinguished, the infra-red radiated by the electrodes and envelope continues for quite a number of seconds as the lamp itself cools down. This can cause considerable practical problems, particularly in the design of UV driers for the photopolymerization of surface coatings.

References

Coaton, J. R. and Fitzpatrick, J. R. (1980). *IEE Proc. A* **127,** 142–148.
De Vos, J. C. (1954) *Physica* **20,** 690–714.

6 Gas Discharges

6.1 Introduction

When a tungsten filament conducts electricity the current is carried entirely by the free electrons. The "ions" are firmly located in the lattice and can only vibrate about their equilibrium positions. In a gas, however, both electrons and ions are free to move and carry current, although the higher charge : mass ratio of the electron gives it a greater mobility in the electric field and in general the electrons carry almost all the current.

Gas discharges are extremely complex. There are many types, and some form the basis of very useful light sources.

Discharges can be classified as *open*—the atmosphere between the electrodes is air—or *enclosed*. Enclosed discharges may be *wall confined*, with an electrode separation considerably greater than the diameter of a tube containing the discharge, or *electrode confined*, with a small electrode separation and a spherical bulb.

Steady state discharges may be distinguished from *transient* discharges. The former may be subdivided into those which are *self-sustaining* and those which are not self-sustaining. In a self-sustaining discharge the passage of current is sufficient to maintain the required concentration of current carriers.

Self-sustaining discharges may be further classified as *arc discharges* or *glow discharges*. The cathode of an arc is hot while that of a glow is cold (Elenbaas, 1972, p. 67). Arcs are formed when there is a high current and/or a cathode of low heat capacity, so that cathode heating causes significant thermionic emission of electrons. Glow discharges operate with lower currents and higher voltages than arcs. Positive ion bombardment and field emission cause the cathode to eject electrons.

The above classification is for *electrode discharges*. By using alternating current power supplies of frequency much higher than normal mains frequency it is possible to dispense with electrodes and create *electrodeless discharges*. These will be discussed in Chapter 10.

6.2 Ignition—Producing a Self-sustaining Discharge

If the current through a discharge lamp is increased from a very low value up to about 10 A, the variation in voltage across the electrodes follows a complicated pattern as the lamp passes through the various discharge regions. This is shown in Fig. 6.1.

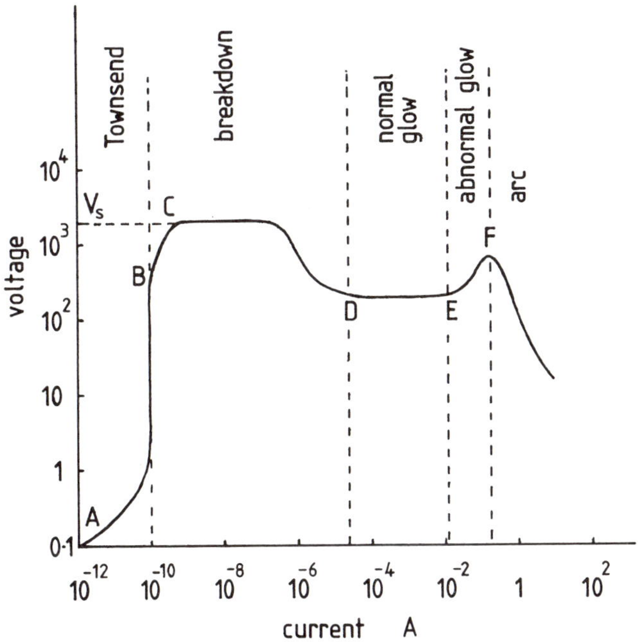

Fig. 6.1. Gas discharge characteristic (schematic).

Let us consider a tubular lamp of length about 20 cm with electrodes at each end, cathode and anode, and filled with gas at a pressure of a few torr. There will be a few electrons and ions in the tube caused by cosmic rays or natural radioactivity. More can be introduced if the cathode is heated or irradiated. If a voltage is established between the electrodes the electrons will flow towards the anode. If the voltage is high enough the rate at which the electrons reach the anode will equal the rate at which they are introduced into the tube. Point B in Fig. 6.1 represents the saturation current. The region between A and B represents the *Townsend discharge*, a steady state discharge which is not self-sustaining because the current carriers are introduced into the tube by an external agency. This type of discharge, which involves very small currents, is familiar in the vacuum photodiode (Section 4.3.2.3). When the voltage is increased to V_s the gas atoms in the tube begin to play a part. The potential gradient is now sufficient for the electrons to achieve enough kinetic energy to cause some ionization of the gas atoms when collisions take place. This causes a multiplication of the number of current carriers in what is known as the *Townsend avalanche*, with a rapid increase in the current flowing and a reduction in the impedance between the electrodes and the voltage across the lamp. The region from C to D represents the *ignition*, or *breakdown* region. At point D the number of positive ions striking the cathode is sufficient to cause enough electron ejection to eliminate the need for any external source of electrons, and we now have a *self-sustaining glow discharge*. At point D, only a small part of the cathode area is in use. The current can then be increased without any change in voltage simply by increasing the area of the cathode being used—the current density remains constant. The region from D to E represents the *normal glow*. At point E the whole of the cathode area is in use, and further increase in current can only be obtained by increasing the current density at the cathode. In order to do this the voltage must be increased so that the positive ions strike the cathode with greater momentum and therefore eject more electrons. At point F, however, the bombardment of the cathode is such that it becomes so hot that thermionic emission becomes the dominant process by which electrons are ejected into the tube, and the voltage requirement is then reduced. Point F represents the *glow to arc transition*. The region from E to F is the *abnormal glow discharge*, while the region beyond F represents the *arc discharge*. It is worth bearing in mind that an electric field normal to the surface of a cathode will reduce the work function by the Schottky effect.

Steady state discharges which are practical sources of light are almost all either abnormal glow discharges or arc discharges.

The voltage required to start a lamp is that which will produce a rapid build up of the concentration of ion–electron pairs. The starting voltage will be reduced if an artificial means is used for introducing electrons into the tube, e.g. preheating the cathode, or using a Tesla coil to produce local ionization. A third electrode close to the cathode is sometimes used to produce a high local field which may even allow electrons to be produced by field emission. There has even been some use of β emitters (e.g. ^{204}Tl, ^{85}Kr, ^{3}H) in the tube (Summer, 1962, p. 7).

In principle, if the voltage between the electrodes is equal to the ionization potential of the gas, an electron can gain enough energy by accelerating between the electrodes to create an ion–electron pair in a single collision. In practice, however, because of collisions, mainly elastic, the electrons follow a zig-zag path between the electrodes with only a general drift in the direction of the field, and the minimum starting voltage is much higher than the ionization potential of the gas.

6.3 Potential Gradients in the Discharge

In an electrode lamp the electric field is not uniform between the electrodes. In the low pressure d.c. mercury discharge lamp several distinct regions can be identified between the electrodes as shown in Fig. 6.2. Most

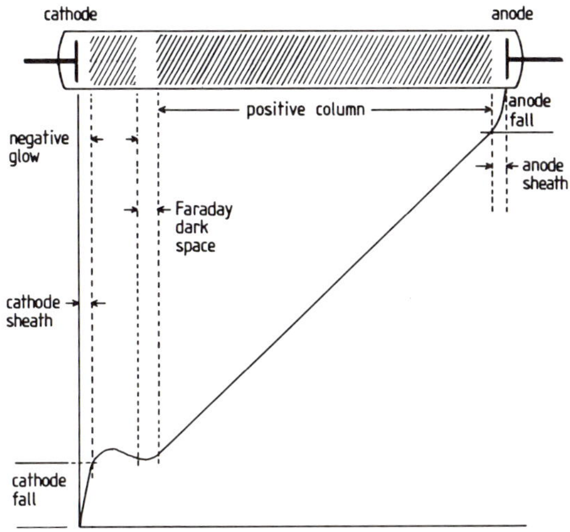

Fig. 6.2. Discharge regions in the low pressure arc (schematic).

of the tube is occupied by the "*positive column*" which has a uniform potential gradient, no net space charge, and is the prime source of radiation. In the vicinity of the electrodes, space charge, positive in the case of the cathode and negative in the case of the anode, causes an increase in the potential gradient, known as the *cathode fall* and the *anode fall*. The anode fall is usually quite small if the current is low, between 1 and 10 V, but the cathode fall may be much higher with a cold cathode lamp, e.g. 100 V. The cathode fall adjusts itself to provide enough positive ion momentum to give the required electron emission from the cathode. Hence, in the region of the abnormal glow discharge, increasing the current causes an increase in cathode fall. However, after the glow to arc transition takes place the cathode fall drops and may be lower than 15 V in the mercury discharge.

The cathode and anode falls occupy only a fraction of a millimetre. In the low pressure mercury lamp there is an area around the cathode called the *negative glow* which is about 3 cm in length (Waymouth, 1971, p. 73). This is separated from the end of the positive column by a non-radiating region of about 1 cm in length known as the *Faraday dark space.* When the lamp is operated from an a.c. supply both electrodes are anode and cathode in turn, and the negative glow and the Faraday dark space switch continually from one end of the lamp to the other. This can be demonstrated by placing a photosensitive detector near one end of a lamp and observing by use of an oscilloscope that alternate light pulses are weak.

If the pressure in the lamp is kept constant, and the length of the lamp is varied, it is found that the positive column contracts or expands accordingly while the other regions remain the same size. If the lamp is made very short the positive column may disappear entirely. The length of the cathode region, i.e. from the cathode to the cathode end of the positive column, is inversely proportional to the pressure, and if this is increased to near atmospheric, the whole cathode region becomes invisibly small.

6.4 Gas Discharges as Circuit Elements

In a tungsten filament the current is carried by the free electrons, the concentration of which is constant. If the voltage across the filament is increased, the electrons accelerate more rapidly in the electric field and an increased current flows. The filament is said to have a "positive voltage–current characteristic", with the complication that increasing the current causes the filament to heat up and its resistance to increase. A tungsten filament at normal operating temperature has a resistance about 16 times its value at room temperature.

The situation in the positive column of a gas discharge is rather more complicated, because the concentration of current carriers can vary over a wide range. The acceleration of electrons leads to the creation of new current carriers by collision. However, current carriers are continually being lost by recombination of electrons and ions. In a stable d.c. discharge the processes producing current carriers are exactly balanced by those by which they are lost. There are certain combinations of voltage and current which allow this balance. These are shown for an arc discharge in Fig. 6.3, which is derived from the arc region of Fig. 6.1.

It will be noticed that the graph has a negative slope. For this reason gas discharges are said to have a "negative characteristic". Unlike a tungsten filament, the discharge is not self-limiting, and the negative characteristic makes it inherently unstable. Suppose that in Fig. 6.3 there is a voltage V_1

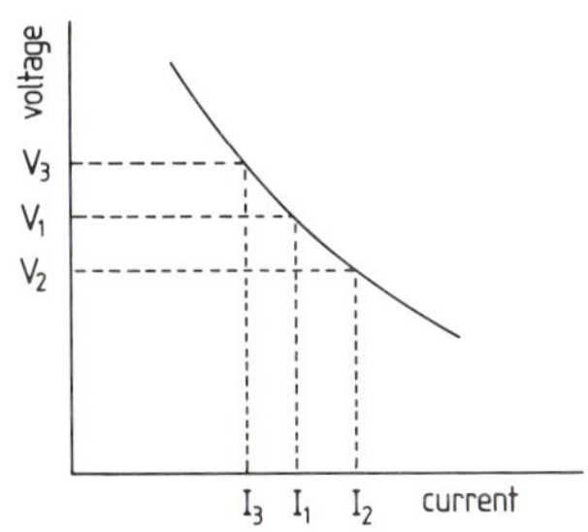

Fig. 6.3. Negative characteristic of an arc discharge.

across the discharge, with a current I_1 flowing. If the current increases momentarily to I_2 there will be an increase in current carriers which will not be balanced by a corresponding increase in recombination. The impedance of the tube will drop, and therefore the voltage required to maintain the arc will drop to V_2. This leaves an excess voltage of $V_1 - V_2$ which will cause further ionization and a runaway situation with current increasing until a fuse blows or the lamp is destroyed. On the other hand, if the current decreases momentarily to I_3, a chain of events will ensue leading to the extinction of the arc. It is also easy to see that if a voltage sufficient to start the lamp is placed across the electrodes the current will increase until catastrophe occurs.

Auxiliary circuit elements known as "ballast" are required to "correct" the negative characteristics of arcs. A suitable impedance is placed in series with the lamp, so that the impedance/lamp *combination* has a *positive* characteristic. If the lamp is operated using d.c. this must be a resistor. If the current increases, the voltage drop across the resistor increases, and therefore the voltage available to the lamp decreases. If the lamp is operated using a.c. the impedance can be a resistor, an inductor or a capacitor. Resistors are not normally used because they consume power.

The characteristic of a lamp is the sum of the characteristics of the individual discharge regions. In Fig. 6.1 it can be seen that the abnormal glow discharge has a positive characteristic whereas the arc discharge has a negative characteristic. However, if one takes the case of a very long cold cathode low pressure mercury lamp, the overall characteristic may be negative. This is because although the cathode fall has a positive characteristic, the positive column has a negative characteristic, and if the positive column is long enough its characteristic will predominate. With the arc discharge the cathode fall is usually quite small compared with the voltage drop across the positive column and therefore the overall characteristic of the arc is very much dominated by that of the positive column. The characteristic of the cathode fall of an arc is negative (Elenbaas, 1959, p. 77).

The reasons for the negative characteristic of the positive column are quite complex and have been explained for the low pressure mercury lamp by Elenbaas (1959, p. 76) and Waymouth (1971, p. 26).

In a stable d.c. discharge the rate of formation of new ion–electron pairs is exactly matched by the rate of recombination of ions and electrons. The positive column is a plasma, with no net space charge, i.e. the number of electrons n_e is equal to the number of ions n_i. $n_e = n_i = n$.

Suppose that the rate of recombination of ion–electron pairs is proportional to the xth power of n, while the formation of new ion–electron pairs is proportional to the yth power of n. Suppose that the current is increased at constant voltage and pressure. This implies an exactly proportional increase in the value of n. The discharge will continue to be stable only if the rate of recombination of ion–electron pairs remains equal to the rate of their formation. This will be the case only if $x = y$. If $y > x$ a stable value of n cannot be maintained at the higher current because the rate of formation of new current carriers will exceed the rate of recombination, and the current will continue to increase. The only way of stabilizing the discharge at the higher current is to reduce the voltage. Hence the negative characteristic.

In the case of the low pressure mercury lamp, the recombination of ions and electrons takes place almost entirely at the walls of the tube. The rate of recombination is determined by the diffusion of ions and electrons to the walls (ambipolar diffusion), and is proportional to the first power of the electron and ion density, i.e. $x = 1$. The value of y depends on the mechanism of ionization. The simplest mechanism involves an inelastic collision between an electron having a kinetic energy higher than the ionization energy of mercury:

$$Hg + e \rightarrow Hg^+ + 2e$$

At a particular potential gradient there will be a distribution of electron energies, with a particular fraction of the total number of electrons having sufficient energy to ionize a mercury atom in the *single stage* process shown above. It is not difficult to see that the rate of formation of ion–electron pairs by this process will be proportional to the electron density and therefore $y = 1$. This implies that the current can be varied at will while the potential gradient remains constant, i.e. a "neutral" characteristic.

However, the ionization of gas atoms may involve two or more stages. For instance, an inelastic collision between an electron and an atom may result in the atom being excited. This excited atom may then collide with another electron which can provide the extra energy required for ionization. Vriens *et al.* (1978) have shown that in the low pressure mercury discharge an important ionization process involves collisions between mercury atoms in 6^3P excited states. It is not difficult to show that when the ionization of an atom involves more than one electron collision, an increase in current at constant voltage will give a more than proportional increase in the rate of production of new current carriers, i.e. $y > 1$ and therefore $y > x$, and therefore the characteristic of the discharge is negative.

The voltage–current characteristics of a.c. operated lamps are more complex than those of d.c. operated lamps, because the discharge may be extinguished during each half cycle and the tube may have a positive characteristic during part of the cycle. The situation becomes very complicated at high frequencies, when the rates of ionization and recombination become slow compared with the rate of fluctuation of the electric field. Under these circumstances discharges can assume almost ohmic character, because the electron density cannot follow the changes in the field, and remains more or less constant.

6.5 Photon Production in Gas Discharges

(See also Section 2.10.)

When a free electron is accelerated in the electric field and then collides with a ground state gas atom, there are three possibilities:

(1) An elastic collision may take place, with little exchange of energy because of the great difference of masses of the participants (Howatson, 1965, p. 22).
(2) The atom may be excited, with a corresponding loss of kinetic energy of the electron.
(3) If the kinetic energy of the electron is sufficient, the atom may be ionized, and lose one of its electrons.

In general collisions can be classified as "elastic", in which the participants simply exchange kinetic energy, or "inelastic", in which the *internal* state of the atom is changed, either by excitation to a higher electronic energy level, or by ionization.

At *low pressures*, e.g. 10^{-3}–10^{-2} T, the mean free path of electrons will be high and the frequency of collisions with atoms will be low. There will be little exchange of energy between electrons and atoms by elastic collision. Hence the distribution of kinetic energies for the atoms will be affected only to a small extent by the presence of the electric field. On the other hand the electrons can achieve very high kinetic energies. The distribution of electron velocities follows approximately the Maxwell–Boltzmann distribution law (Easley, 1951). The distribution is characterized by an "electron temperature" which is typically in the region between 10 000 and 15 000 K. In the low pressure mercury discharge the distribution of energies is such that process (2) above favours excitation to the lowest excited states, and therefore the radiation emitted is predominantly derived from transitions between the lowest non-metastable excited states and the ground state.

At *higher* pressures, e.g. 1 atm, the frequency of collisions is much greater, with the result that the electrons, which gain energy from the electric field, share it to a much greater extent with the atoms. The result is that a state of "local thermal equilibrium" is reached when all the components in the discharge—electrons, ions and neutral atoms—are effectively at the same temperature. This is further discussed in Section 8.1. The electrons will always be at a slightly higher temperature than the atoms and ions, as they receive the energy from the field and then transfer it to the atoms and ions by collisions which due to the disparity of mass are inefficient. Also, of course, perfect equilibrium is not possible in a plasma that is continually losing energy by radiation.

The maximum temperature of the plasma will typically be in the region of 6000 K, which means that a small but significant fraction of the atoms will be in the various excited states. They can lose their energy by a variety of processes, the most important of which is radiative decay. This leads to a line spectrum with many of the possible transitions represented. The recombination of ions and electrons takes place in the body of the plasma, with the emission of a small amount of continuum radiation, rather than non-radiatively at the walls, as in a low pressure discharge. Also, collisions between atoms can give "molecular radiation" which again enhances the continuum.

At *very high pressures*, e.g. 100 atm, the lines broaden and the processes producing continuum radiation become very important. Because of the high pressure, a photon emitted from an atom is quite likely to be absorbed by another atom before it can escape from the tube. This is particularly true of resonance radiation, i.e. radiation for which the lower energy level of the transition is the ground state. Whereas very low pressure discharges give almost entirely resonance radiation, very high pressure discharges give a continuum with a gap at the resonance wavelength. At a pressure of several hundred atmospheres the discharge becomes sufficiently opaque that its output approximates to that of a black body.

6.6 Envelope Materials

Enclosed discharges require an envelope. If the discharge is to be a viable source of UV radiation the envelope material must transmit UV. The envelope must be impermeable to gases and chemically resistant to lamp filling materials at the temperature of operation. The envelope must have good degassing characteristics to prevent unwanted gases being evolved during lamp operation. The envelope material must be an electrical insulator. Medium pressure mercury and metal halide lamps operate with an envelope temperature in the region 600–800 °C, and the envelope must retain its integrity for long periods at this temperature, and show a high resistance to thermal shock. It must be able to resist a differential pressure of several atmospheres.

Low pressure mercury lamps used as sources of 254 nm radiation often have envelopes made from special grades of UV-transmitting glass. Medium and high pressure lamps use transparent vitreous silica.

Silica is silicon dioxide, SiO_2. It occurs profusely in nature. There are several crystalline forms, the main ones being quartz, tridymite and cristobalite. Quartz is the form which is stable at ambient temperature, while tridymite and cristobalite are metastable and much rarer, although the reversion to quartz is very slow at normal temperatures. Apart from the various crystalline mineral forms, found in various states of purity, there also occur vitreous forms, e.g. silica sand.

When silica of high purity is melted (m.p. $\sim$ 1700 °C), and then cooled, vitreous silica is formed. If the silica is rock crystal, a high purity form of crystalline quartz, the vitreous silica is transparent. On the other hand, if silica sand is used, the vitreous silica is translucent, due to the presence of tiny bubbles in the material.

The properties of transparent vitreous silica depend not only on the quality of the rock crystal used, but also on the method of melting. There are two principal methods. The first uses a graphite resistance furnace to melt the mineral and the transparent vitreous silica produced is referred to as *Type I*. The second uses an oxy-hydrogen flame to melt the powdered mineral and gives *Type II* material.

The impurities in the original rock crystal are also present in the vitreous silica. However, the flame method of melting does cause the partial evaporation of some impurities, particularly aluminium and the alkali metals, so that the metallic impurity of Type II material is somewhat lower than that of Type I material. On the other hand the flame introduces hydroxyl groups into the silica. The result is that Type II vitreous silica has better transmission in the short wavelength UV than Type I, but the hydroxyl groups in Type II give rise to very strong absorption in the infra-red at 2·73 μm, with lesser absorptions at 2·22 μm and 1·38 μm.

Some grades of transparent vitreous silica show an absorption at 242 nm with associated fluorescence at 280 and 396 nm. This is probably due to aluminium impurity being substituted for silicon in the lattice (Browell and Hetherington, 1965). The absorption can be eliminated by heat treatment in an oxidizing atmosphere. Figures 6.4 and 6.5 show spectral transmission curves for Type I and Type II silicas.

For some purposes the impurity level in vitreous silica derived from natural sources is too high. Very pure silica can be synthesized from $SiCl_4$ or other volatile silicon compounds which can be readily purified by fractional distillation. The compound is oxidized and the oxide fused in an oxy-hydrogen flame. This gives *Type III* transparent vitreous silica. Although it has very low levels of metallic impurity, and therefore excellent transmission in the short wave UV, hydroxyl groups are introduced into the lattice by the oxy-hydrogen flame with the result, as with Type II, being absorption in the infra-red. It is possible to make vitreous silica synthetically either by heating electrically or with a hydrogen-free flame, and this method gives *Type IV*. Spectral transmission curves for the two types of synthetic silica are shown in Fig. 6.6

Vitreous silica is thermodynamically unstable at temperatures below 1710 °C. At normal temperatures, however, the viscosity is so high that the rate of transformation to crystallinity is infinitesimal and cannot be detected. Even at temperatures up to 1200 °C large pieces of pure transparent vitreous silica kept free of surface contamination show little sign of crystallization (or as it is normally called, "devitrification"). It might be thought that, as lamp envelopes will seldom be required to operate at temperatures above 800 °C, devitrification will be of little concern. However, devitrification starts at surfaces and is strongly catalysed by certain contaminants. Perspiration is notorious, and opaque areas resembling fingerprints are commonplace on silica which has not been carefully handled. Alkali and alkaline earth ions are powerful catalysts. Devitrification is accelerated by the thermal cycling to which a lamp envelope is subjected in normal use (Adams, 1978). A high hydroxyl content promotes devitrification, and indeed, despite its very low level of metallic impurity,

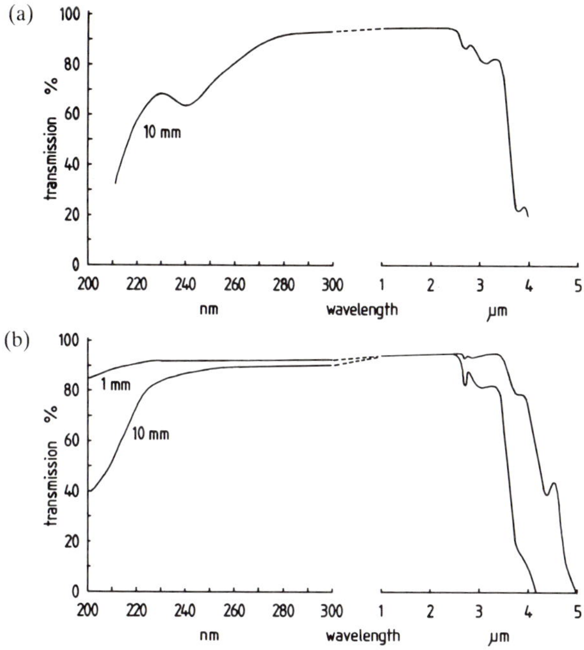

Fig. 6.4. Spectral transmission of Type I transparent vitreous silica: (a) I. R. Vitreosil® (Thermal Syndicate Ltd) (note *presence* of absorption at 242 nm); (b) Infrasil® (Heraeus Quarzschmelze GmbH) (note *absence* of absorption at 242 nm). Transmission = $(1 - R)^2 \times 10^{-kd}$. Transmittance = 10^{-kd}. R is the reflection loss at each surface; k is the coefficient of absorption; d is the path length. ® indicates a registered trademark. (Data taken from manufacturers' publications with permission.)

Type III silica devitrifies more rapidly than the other types at high temperatures (Hetherington *et al.*, 1969).

When devitrification takes place cristobalite is formed, even at low temperatures, despite the fact that cristobalite is the thermodynamically stable crystalline form only between 1470 and 1710 °C. Being opaque it cuts down the transmission of light from the arc. Absorption of the radiation causes the tube to heat up, causing further devitrification and so on. Cristobalite has a much higher coefficient of expansion than vitreous silica and therefore thermal cycling causes stresses in a lamp envelope which has partly devitrified.

Synthetic silicas are much more expensive than those derived from rock crystal, and are therefore seldom used for lamp manufacture. A very low hydroxyl content is necessary for the envelopes of medium pressure

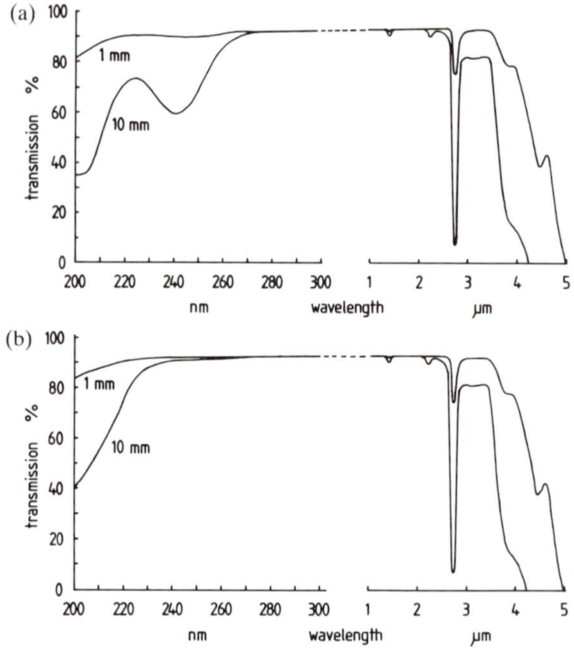

Fig. 6.5. Spectral transmission of Type II transparent vitreous silica: (a) Homosil®, Herasil® (Heraeus Quarzschmelze GmbH) (note *presence* of absorption at 242 nm); (b) Ultrasil® (Heraeus Quarzschmelze GmbH) (note *absence* of absorption at 242 nm). ® indicates a registered trademark. (Data taken from manufacturer's publications with permission.)

mercury and metal halide lamps, and therefore the vitreous silica most appropriate for the manufacture of these lamps is Type I.

Devitrification is not the only deterioration process which can take place. Ultraviolet radiation itself can cause some changes in the silica which result in a loss of transmission. Even the purest silica develops an absorption centred at 215 nm, which is thought to be caused by the loss of oxygen from the silica lattice into interstitial positions. The development of this absorption band in all types of vitreous silica implies that for lamps operating at near ambient temperature the benefit derived from choosing a grade of silica with particularly good short wavelength transmission is eventually lost. However, the absorption can be removed by heating at above 500 °C, and therefore may not develop in a lamp having an envelope temperature higher than 500 °C.

The presence of certain metal ions, e.g. Fe^{2+} and Mn^{2+}, can cause

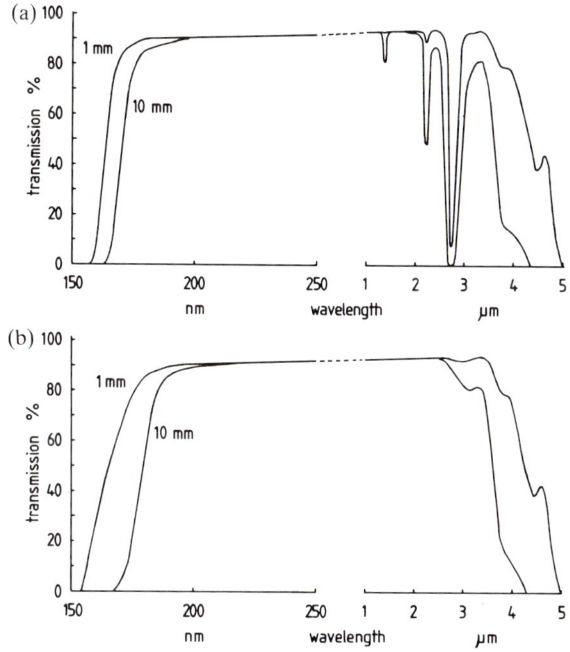

Fig. 6.6. Spectral transmission of synthetic transparent vitreous silica. (a) Type III, Suprasil® (Heraeus Quarzschmelze GmbH); (b) Type IV, Suprasil W® (Heraeus Quarzschmelze GmbH). Ⓡ indicates a registered trademark. Data taken from manufacturer's publications with permission.

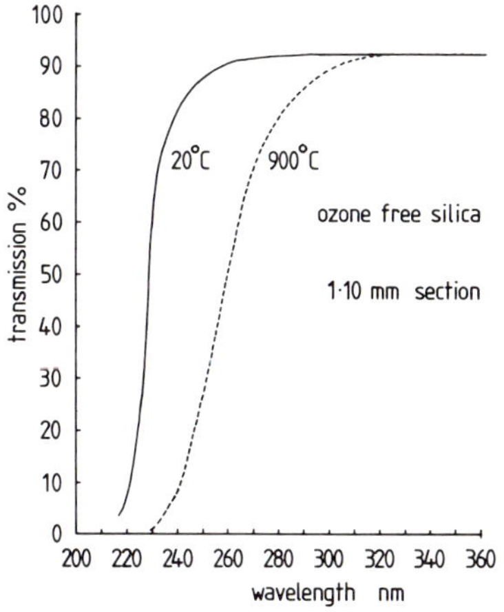

Fig. 6.7. Effect of temperature on the transmission of ozone-free silica. (Adapted from the data of Gion *et al.*, 1974.)

"solarization" and loss of transmission on irradiation (Koller, 1965, p. 169), but again, counteractive processes occur at high temperatures.

If a lamp is operated in air and emits radiation below about 224 nm, oxygen in the air will be converted to ozone, which is highly toxic and corrosive. The formation of ozone can be prevented by the use of special grades of silica which are doped, for instance with titanium dioxide, to reduce the transmission at short wavelengths. "Ozone-free" medium pressure mercury lamps are discussed in Section 8.8.

The spectral transmission data shown in Figs 6.4–6.6 are for ambient temperature. Rather limited data in the literature (Edwards, 1966; Vorob'ev and Daniel, 1970) indicate that short wavelength transmission is somewhat temperature dependent, and at the normal operating temperature of a medium pressure mercury or metal halide lamp, transmission will be somewhat reduced. This is particularly true of "ozone-free silica". Figure 6.7 shows transmission curves for 1·1 mm thick ozone-free silica at 20 °C and 900 °C, derived from data published by Gion *et al.* (1974).

References

Adams, O. (1978). *Light. Res. Technol.* **10,** 83–93.

Browell, T. P. and Hetherington, G. (1965). *Hilger J.* **9** (3), 41–48.

Easley, M. A. (1951). *J. Appl. Phys.* **22,** 590–593.

Edwards, O. J. (1966). *J. Opt. Soc. Amer.* **56,** 1314–1319.

Elenbaas, W. (1959). "Fluorescent Lamps and Lighting", Philips Technical Library. Cleaver-Hulme Press, London.

Elenbaas, W. (1972). "Light Sources", Philips Technical Library. Macmillan, London.

Gion, H., Okumura, H. and Nakamura, N. (1974). *Natl. Tech. Rep. Matsushita Electr. Ind.* **20,** 332–339.

Hetherington, G., Stephenson, G. W. and Winterburn, J. A. (1969). *Electron. Eng.* **41** (495), 52–55.

Howatson, A. M. (1965). "An Introduction to Gas Discharges". Pergamon, Oxford.

Koller, L. R. (1965). "Ultraviolet Radiation", 2nd edn. John Wiley & Sons, New York.

Summer, W. (1962). "Ultraviolet and Infra Red Engineering". Pitman, London.

Vorob'ev, A. N. and Daniel, E. V. (1970). *Zh. Prikl. Spektrosk.* **12,** 347–349.

Vriens, L., Keijser, R. A. J. and Ligthart, F. A. S. (1978). *J. Appl. Phys.* **49,** 3807–3813.

Waymouth, J. F. (1971). "Electric Discharge Lamps". MIT Press, Cambridge, Mass.

7 The Low Pressure Mercury Lamp

7.1 Why Mercury?

As was stated in Chapter 6, a low pressure discharge lamp produces predominantly resonance radiation. The intensity is low but the efficiency is often high. The ideal filling gas for a low pressure lamp for efficient UV production must have the following properties:

(1) It must have a fairly low ionization energy, so that the avalanche effect will take place readily. This implies a metallic element.
(2) It must have sufficient vapour pressure at a reasonable temperature (i.e. at ambient or slightly above ambient) to provide the optimum pressure for production of resonance radiation.
(3) Although the ionization energy must be reasonably low, the lowest excited state which is not metastable must be at such a level above the ground state that the resonance radiation appears at a useful wavelength.
(4) The element must be chemically inert to envelope and electrode materials.

Mercury fulfils these requirements admirably, and uniquely. Figure 7.1 shows vapour pressure curves for various metals, and it can be seen that mercury is much more volatile than any other metal.

The low pressure mercury lamp forms the basis of the ubiquitous fluorescent lighting tube. In this case the envelope is of glass, which is coated on the inside with a fluorescent powder which converts the resonance mercury radiation at 253·7 nm into white visible light. The other principal use of the low pressure mercury lamp is as a source of germicidal radiation.

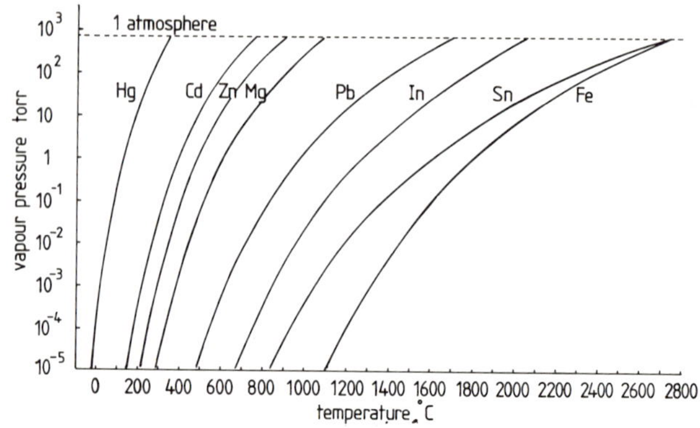

Fig. 7.1. Vapour pressure curves for various metallic elements.

7.2 Construction

The low pressure UV radiating mercury lamp consists of a transparent tube with an electrode at each end. The tube may be of pure transparent vitreous silica for maximum transmission of UV radiation, or may be of a UV-transmitting glass. The filling is not mercury alone, but a mixture of mercury and an inert gas, usually argon, at a pressure of a few torr. The mercury is introduced as a single drop, and almost all the mercury remains in the liquid form during the operation of the lamp. The pressure of mercury is just the vapour pressure of mercury at the operating temperature, which is in the region of 40 °C. The vapour pressure of mercury at this temperature is $\sim 7 \times 10^{-3}$ T, which, as will be explained later, is very conveniently the optimum for the most efficient production of resonance radiation. The inert gas is therefore at a much higher pressure than the mercury, but contributes almost nothing to the spectral output.

There are two basic types of low pressure mercury lamp, which may be distinguished by the form of the electrodes. The cathode of a "cold cathode" lamp is of fairly massive hollow cylindrical construction and may be made of iron or nickel. The cold cathode lamp is an example of an abnormal glow discharge, and depends on the bombardment of positive ions on the cathode to release electrons. It operates at a relatively high voltage and a relatively low current.

The cathode of the "hot cathode" lamp is of much more delicate construction, a prime requirement being a low heat capacity. It consists usually of a coiled coil of fine tungsten wire. Embedded between the turns is a mixture of alkaline earth oxides (calcium, barium and strontium), which reduces the work function and aids electron emission. The mechanism by which this happens is rather complex (Bouwknegt *et al.*, 1975).

The hot cathode lamp is an example of an arc discharge. Electrons are supplied to the discharge by thermionic emission from the hot cathode. Thermionic emission is discussed in more detail in Section 8.2.1. Hot cathode lamps in general operate at a lower voltage and a higher current than cold cathode lamps.

7.3 Circuits

The low pressure mercury lamp has a negative characteristic and therefore needs a ballast in series with it for stable operation. If the lamp is operated from a d.c. supply, this ballast must be a resistor. It is, however, more common to operate a lamp from an a.c. mains supply. In this case the two electrodes take it in turn to be the cathode. There are many types of circuit used with a.c. operated lamps. These have been described by Funke (1959) and Chapman (1972). Circuits are designed to achieve the following: (1) reliable and rapid starting; (2) stabilization of light output; (3) re-ignition of the discharge every half cycle; (4) a satisfactory lamp current waveform; (5) the presentation to the mains supply of a high power factor. (A low power factor means high line currents and the necessity for heavy conductors to minimize circuit losses. Also, for various reasons low power factor installations are frowned upon by those responsible for electricity supply.)

The ballast used with an a.c. operated lamp can in principle be a resistor, a capacitor or an inductor. Resistors are not used because a large power loss would result. Capacitors and inductors can provide a non-resistive impedance. The voltage across and the current through these components are 90° out of phase, so that the net power consumed is zero. In practice a purely capacitative ballast cannot be used at normal mains frequency because an unacceptable waveform would result. A purely inductive ballast in series with the lamp would give an unacceptably low power factor, so that in practice the ballast circuitry contains both inductance and capacitance. The most common arrangement is to have a choke in series with the lamp, and a power factor correction capacitor across the mains.

For reliable starting of a *cold cathode* lamp a fairly high voltage must be applied to the electrodes. In most cases the mains voltage will not be high enough, and therefore a step-up transformer is necessary. This may be of the leakage reactance type, so that the secondary provides all or part of the choke ballast.

Hot cathode lamps are of two types. The first type (instant start) has a single connection to each electrode, and as with the cold cathode lamp ignition is almost instantaneous. Lamp voltage is high at first, but the delicate electrodes rapidly heat up, causing a glow to arc transition, and a reduction in the cathode fall, as thermionic emission becomes the principal mechanism for ejection of electrons from the cathode. The disadvantages of this lamp are that a high starting voltage is required, and the high cathode fall during the glow period causes electrode deterioration (Waymouth, 1971, p. 59). The second type of hot cathode lamp (switch start) has a facility for preheating the electrodes for a short period before ignition. Each electrode has two connections for this purpose. A typical circuit is shown in Fig. 7.2. This incorporates a glow switch starter. This is a

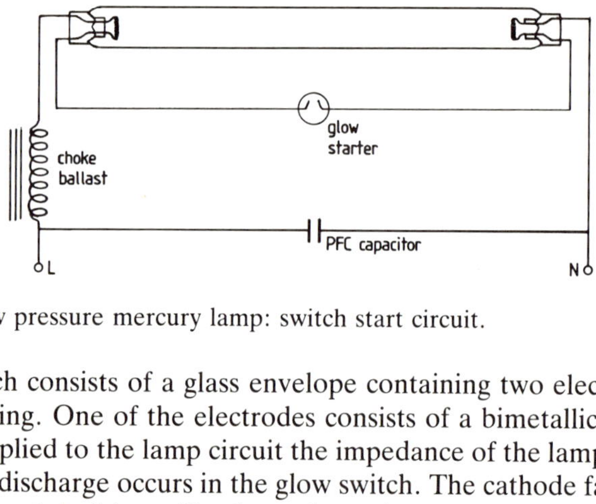

Fig. 7.2. Low pressure mercury lamp: switch start circuit.

device which consists of a glass envelope containing two electrodes and a rare gas filling. One of the electrodes consists of a bimetallic strip. When power is applied to the lamp circuit the impedance of the lamp is very high and a glow discharge occurs in the glow switch. The cathode fall causes the temperature of the bimetallic electrode to rise, and it bends to contact the other electrode, so causing a short circuit which allows the heating of the lamp electrodes. The bimetallic strip then cools and contact is broken, causing a voltage surge across the lamp electrodes sufficient to cause ignition. The discharge then maintains the temperature of the electrodes.

With preheated electrodes there is a delay of one or two seconds between switching the lamp on and ignition. The great attraction of this type of circuit is that the preheating of the electrodes greatly reduces the voltage required for ignition (Elenbaas, 1959, p. 83), and so there may be no need for a step-up transformer, so reducing the cost of the installation. Also, the glow stage is eliminated, so minimizing the sputtering of oxides from the electrodes. Some specialized lamps, e.g. for use in photocopying machines, have electrodes which are permanently heated by an external circuit. This allows the electrodes to be of more robust construction, and for almost instantaneous ignition.

Figure 7.3(a) shows lamp voltage and current waveforms, Fig. 7.3(b) mains voltage and light output waveforms, and Fig. 7.3(c) lamp input power and lamp resistance, for a Philips TUV 15 germicidal lamp. This was operated from a simple choke ballast with a glow starter switch. There was no power factor correction capacitor. The lamp current waveform is a distorted sine wave, while the lamp voltage waveform is a distorted square wave. The oscillations in the voltage waveform are due to fluctuations in the anode fall (Waymouth, 1971, p. 110). The lamp current lags behind the mains voltage. The true r.m.s. current was found to be 0·46 A, and the r.m.s. mains voltage 248 V. However, only 35 W were absorbed by the lamp–choke combination, indicating a very low power factor presented to the mains (about 0·3). This could be raised to about 0·85 by use of a power factor correction capacitor as in Fig. 7.2, with a large decrease in the current drawn from the mains. The time averaged power absorbed by the lamp alone was 21 W. The power factor of the lamp was found to be 0·87, lagging, so that the lamp has some inductive character.

7.4 The Role of the Rare Gas

Although the argon used in low pressure mercury lamps does not contribute directly to the production of photons, it plays a vital part in both the establishment and the maintenance of the discharge.

When a lamp is first switched on, electrons are accelerated in the electric field. The vapour pressure of mercury at ambient temperature is so low that there are few collisions between the few electrons initially present and mercury atoms. However, with a few torr of argon present, many more collisions take place between electrons and argon atoms. Argon, in common with the other rare gases, has quite a high ionization energy (15·76 eV). However, its lowest excited state (at 11·55 eV) is metastable, i.e. non-radiative. Collisions between energetic electrons and argon atoms lead to the excitation of significant numbers of argon atoms to this metastable state. The energy of the excited argon atoms can only be lost by collision. If this is with a mercury atom, ionization of the mercury atom can take place, because the energy of the metastable state of argon is higher than the ionization energy of mercury (10·43 eV). There is another possibility involving the metastable state of argon. If a collision takes place between an argon atom in its metastable state and an energetic electron,

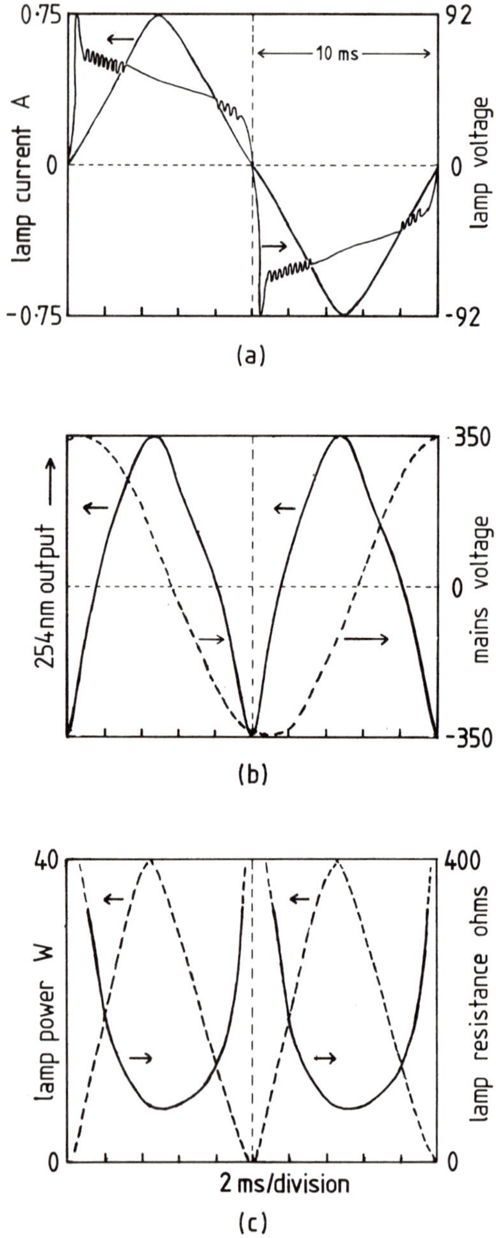

Fig. 7.3. TUV 15 lamp: (a) lamp voltage and current waveforms; (b) mains voltage and light output waveforms; (c) lamp input power and lamp resistance waveforms.

there is the possibility that the electron will contribute the extra 4·21 eV required to ionize the argon atom, so increasing the number of current carriers.

The energy of the metastable state of argon, which exceeds the ionization energy of mercury, enables argon to form what is known as a "Penning mixture" with mercury (Penning, 1927, 1928). This greatly reduces the starting voltage, and aids the formation of the discharge. At very low ambient temperatures, however, the vapour pressure of mercury is so low that there are insufficient collisions between argon atoms and mercury atoms for the "Penning effect" to be significant, and the starting voltage is then that for a pure argon discharge (Elenbaas, 1959, p. 84).

Helium (19·80 eV) and neon (16·62 eV) also form Penning mixtures, but krypton (9·91 eV) and xenon (8·32 eV) do not.

Although the prime purpose of the rare gas is to aid starting, it also performs several other functions. It reduces the mobility of electrons and ions and so reduces the recombination loss which takes place by "ambipolar" diffusion of ions and electrons to the wall. It also reduces the cathode fall and the loss of emissive materials from the electrodes. However, if the pressure of the rare gas is too high, the electrons lose too much energy by elastic collisions with the rare gas atoms, so reducing the electron temperature, and therefore the fraction of electrons with sufficient energy to excite mercury atoms to the 6^3P states.

Argon is used almost exclusively as the rare gas in the conventional low pressure mercury lamp. This is partly because argon is the most plentiful and cheapest of the rare gases. Also, of the three gases which form Penning mixtures with mercury, it is the one which is most effective at reducing ambipolar diffusion and gives the least elastic collision losses. The influence of the rare gas on the operating characteristics of the low pressure mercury lamp will be further discussed in later sections.

7.5 Processes in the Positive Column

The processes taking place in the positive column are of major interest because it is from the positive column that almost all the radiation is emitted, and, at least in the case of the arc discharge when the cathode fall is small, it is the positive column which largely determines the electrical characteristics of the lamp.

The important processes in the positive column are collisions between particles and the emission and absorption of photons. A distinction has already been drawn between elastic and inelastic collisions (see Section

6.5). Collisions are characterized by a parameter known as the "collision cross section", which is a measure of the probability not just of a collision taking place between two particles, but also of a particular result being obtained. The probability of a particular collision taking place when a particle of one type is introduced into a volume containing n particles per unit volume of another type, in a distance travelled dx, is

$$\mathrm{d}\bar{\omega} = \mathrm{n}\sigma(v)\, dx$$

(Papoular, 1965, p. 17). $\sigma(v)$ is the collision cross section and is a function of the kinetic energy of the particle. It has the dimensions of area.

Because the electrons can achieve high kinetic energies by acceleration in the electric field, most but not all the important collisions taking place involve an electron as one of the participants. Graphs showing the dependence of cross section on electron energy for many of the significant collisions have been given by Cayless (1959).

In the low pressure mercury lamp the electrons can be considered to behave like an ideal gas. Electron–electron collisions are common because although electrons are very small particles, because they are charged they exert an influence over a considerable distance, and therefore the cross section is quite high. Because the collision is between particles of the same mass, energy transfer is very efficient. The result is that the distribution of electron velocities approaches that expected for a state of thermodynamic equilibrium. Measurements by Easley (1951) showed that the distribution of electron velocities in the low pressure mercury discharge follows the Maxwell–Boltzmann law, and therefore the distribution can be described by an electron temperature T_e. T_e is typically about 10 000–15 000 K. If there are n electrons each of mass m, and dn have velocities in any direction between v and $v + \mathrm{d}v$, then

$$\frac{\mathrm{d}n}{n} = 4\pi \left(\frac{m}{2\pi kT}\right)^{3/2} \mathrm{e}^{-mv^2/2kT}\, v^2 \mathrm{d}v.$$

It is often more useful to have the distribution in terms of energy ε:

$$\frac{\mathrm{d}n}{n} = 2\pi \left(\frac{1}{\pi kT}\right)^{3/2} \mathrm{e}^{-\varepsilon/kT}\, \varepsilon^{1/2}\, \mathrm{d}\varepsilon.$$

In fact recent work (Ligthart and Keijser, 1980) has shown that there are fewer high energy electrons than would be expected from the Maxwell–Boltzmann distribution law.

In most collisions between electrons and atoms, the electron has more energy than the atom, and any exchange of energy is from the electron to the atom. This is called a "*collision of the first kind*". Sometimes, however, a slow moving electron will collide with an excited atom and the transfer of energy will be from the atom to the electron. This is known as a "*collision of the second kind*".

Most collisions between electrons and atoms are elastic, and involve a transfer of energy from the electrons to the atoms. This useless heating of atoms (mainly argon) dissipates a significant fraction of the input power, about 20% (Steele, 1955), and causes a radial distribution of gas temperature (Kenty *et al.*, 1951).

Ionization processes provide a supply of current carriers to replace those lost by *recombination*. During ignition, the Penning effect is important:

$$e + Ar \rightarrow Ar^* + e$$
$$Ar^* + Hg \rightarrow Hg^+ + e + Ar$$

When the arc is established, only a small fraction of the electrons have sufficient energy either to excite an argon atom or to ionize a mercury atom from the ground state. Therefore most ionizations take place in more than one stage. These may involve a first collision with an electron, giving an excited mercury atom, which is then ionized in a second collision with an electron (Waymouth and Bitter, 1956), i.e.

$$e + Hg \rightarrow Hg^* + e$$
$$Hg^* + e \rightarrow Hg^+ + 2e$$

The ionization process may involve the collision between two excited mercury atoms (Tan and von Engel, 1968; Franck and Ludemann, 1972; Vriens *et al.*, 1978), i.e.

$$Hg^* + Hg^* \rightarrow Hg_2^+ + e$$
$$Hg^* + Hg^* \rightarrow Hg^+ + Hg + e$$

This is known as "*associative ionization*". It is of interest that Hg_2^+ may be formed. This has a lower energy (9·6 eV) than Hg^+ (10·4 eV). Vriens *et al.* (1978) have shown that as the mercury pressure is increased, associative ionization becomes increasingly important, and becomes predominant at wall temperatures above 50 °C.

Recombination of Hg^+ ions and electrons takes place almost entirely at the wall in the low pressure lamp, with volume recombination being negligible (Waymouth and Bitter, 1956). Ions and electrons migrate to the walls by ambipolar diffusion. A radial electric field is developed which ensures that ions and electrons migrate at the same rate (Waymouth, 1971, p. 336).

Most of the radiation emitted from the low pressure mercury lamp is at 253·7 nm and produced by the transition $6^3P_1 \rightarrow 6^1S_0$. The creation of the 6^3P_1 excited state is often a multistage process with other excited states, and in particular the metastable states 6^3P_0 and 6^3P_2, playing an important role (Kenty, 1950).

Mathematical treatments of the processes taking place in the positive column, with the intention of calculating the voltage across the column and the concentrations of electrons, ions and excited and ground state atoms, have been made by several authors, the most significant being those of Kenty (1950), Waymouth and Bitter (1956) and Cayless (1963).

7.6 Spectral Output

Almost all the radiation emitted by the low pressure mercury lamp is resonance radiation at 253·7 and 185·0 nm, arising from the transitions $6^3P_1 \rightarrow 6^1S_0$ and $6^1P_1 \rightarrow 6^1S_0$. Few mercury atoms are excited to levels above 6^1P_1, and therefore the other mercury lines are very weak (Childs, 1962). Much of the radiation from the arc at 185·0 nm is absorbed by the envelope material or by atmospheric oxygen surrounding the tube, with the formation of ozone (Koller, 1945).

The main item of interest in the spectral distribution of radiation from the low pressure lamp is the 185·0 nm : 253·7 nm output ratio. The 253·7 nm output is always much greater than that at 185·0 nm, despite the fact that the $6^3P_1 \rightarrow 6^1S_0$ transition is "forbidden" according to the selection rules based on Russell–Saunders coupling, because the transition involves a change of multiplicity. Several authors have investigated the factors which influence the 185·0 nm : 253·7 nm ratio. Barnes (1960) examined the ratio as a function of current, wall temperature, inert gas (Kr, Ar or Ne) and inert gas pressure. He found values between 0·12 and 0·34, having allowed for the transmission of the lamp envelope. The ratio was found to increase with current and wall temperature, and decrease with the atomic weight and pressure of the inert gas. The influence of current and mercury pressure led Barnes to conclude that the population of the 6^1P_1 state involved multistage excitation, with the transition $6^3P_2 \rightarrow 6^1P_1$ becoming increasingly important at higher currents and pressures.

Barnes' measurements were made with a.c. lamps. Read (1964) examined d.c. lamps, and obtained results in general agreement with those of Barnes. He examined the influence of tube diameter, and found that decreasing the diameter from 36 to 24 mm produced a marked increase in the 185·0 nm : 253·7 nm ratio, ascribed to an increase in electron temperature. Johnson (1971a) examined the influence of current density, mercury pressure and discharge tube diameter, for high current discharges. He found, as did Barnes, that the 185·0 nm : 253·7 nm ratio increased with pressure, and concluded that this was because there was more line broadening and less *self-absorption* for the 185·0 nm line (see Section 7.7).

When a low pressure mercury lamp is operated under normal conditions, i.e. with a tube diameter of about 36 mm and a current of about 0·4 A, in the region of 95% of the radiation emitted is the resonance radiation of mercury *atoms*. This is because the ratio of mercury ions to mercury atoms in the discharge is very small, about 0·001. At very high current densities, however, the resonance radiation of mercury *ions* becomes significant, with a line at 194·2 nm (Johnson, 1971b).

7.7 Self-absorption

The process by which a photon is emitted from an excited atom is reversible, i.e.

$$\text{Hg}^* \text{ (more excited)} \leftrightarrows \text{Hg}^* \text{ (less excited)} + h\nu$$

This means that if a photon is emitted from an excited atom in the centre of the lamp it can, in principle, be reabsorbed by another atom before escaping from the tube. The probability of this occurring depends on the concentration of atoms in the lower excited state of the transition producing the photon. "Self-absorption" or "imprisonment" is most important for the 185·0 nm and 253·7 nm lines of the mercury lamp.

The phenomenon of self-absorption was first put forward by Compton (1922) as one of a number of mechanisms of "cumulative ionization". Self-absorption was demonstrated experimentally by Zemansky (1927) who irradiated mercury vapour with 253·7 nm radiation and measured the decay time of the re-emitted radiation after the primary source was switched off. He concluded that a photon could be absorbed and re-emitted up to a thousand times before escaping from the discharge. His experiments were repeated by Albert *et al.* (1947). Both sets of results indicated that at low mercury pressures the degree of self-absorption increases with pressure, but beyond a certain pressure (equivalent to a tube temperature of about 120 °C) the trend was reversed and self-absorption then decreased. This was because of pressure broadening (see Section 8.6.8). The influence of pressure broadening on self-absorption has been more recently examined by Hammond and Gallo (1971, 1972, 1974). A theoretical treatment of self-absorption was made by Holstein (1947, 1951), who pointed out that as mercury has five isotopes of approximately equal abundance, the absorption is only a fifth of what it would otherwise be, because each isotope emits and absorbs at a wavelength slightly different from the others. Franck and Schipp (1975) examined a lamp containing only the single isotope ^{202}Hg and found lower efficiency because of increased self-absorption. See also Walsh (1959) and Waymouth (1971, p. 24).

Self-absorption is important because although the most likely fate of a mercury atom in the 6^3P_1 state is that it will emit a photon, there is the possibility that a collision will occur which will prevent this happening. As a photon emitted near the centre of the tube must undergo many absorptions and re-emissions before escaping the tube, there is a significant possibility that the chain will be broken at some point.

Hammond and Gallo (1971) showed that increasing the argon pressure in the low pressure mercury lamp decreased the loss of 253·7 nm radiation by self-absorption, because of the broadening of the resonance line. However, to eliminate self-absorption required an argon pressure of at least 100 T, and any benefits in efficiency given by the elimination of self-absorption were outweighed by the increased elastic losses.

7.8 Efficiency

The construction, filling and operating conditions of the low pressure mercury lamp are in general chosen to maximize the efficiency with which electrical input energy is converted into resonance radiation. For a typical hot cathode fluorescent lamp, about 60% of the power input into the lamp is radiated by the arc as resonance radiation at 185·0 and 253·7 nm. About 3% appears at other wavelengths, mainly in the visible. About 15% is lost at the electrodes, and the remainder (22%) is lost from the positive column by elastic collisions with the tube wall and with argon atoms. It is is worth bearing in mind that in using the low pressure mercury lamp for photopolymerization purposes it may be necessary to accept a lower efficiency in return for a higher intensity.

The influence of the lamp characteristics and operating conditions on efficiency of the low pressure mercury lamp can be summarized under six headings:

7.8.1 Temperature

The temperature of the lamp (i.e. of the lamp tube) has little influence *per se* on the efficiency of the lamp. However, as the mercury is present in excess, a rise in the tube temperature causes an increase in the mercury pressure. There is an optimum pressure for maximum efficiency. If it is too low there is insufficient likelihood of collisions between electrons and mercury atoms. Starting becomes difficult. If the pressure is too high, self-absorption reduces the output of resonance radiation and elastic losses due to electron–mercury atom collisions become significant (Verbeek and Drop, 1974). The optimum pressure is about 7×10^{-3} T, which corresponds to a tube temperature of 40 °C (Jerome, 1956), easy to maintain under most conditions. The exact temperature for optimum output of 253·7 nm radiation depends on the tube diameter (Hammond and Gallo, 1974). The optimum temperature for 185·0 nm radiation is generally higher (Johnson, 1971a). The effect of temperature on the output of a lamp can be reduced by introducing the mercury into the lamp in the form of an amalgam (Jack and Vrencken, 1980).

7.8.2 *Tube diameter*

If the tube diameter is too small, there will be excessive losses by elastic collisions and ion–electron recombination at the tube wall. If the diameter is too large, self-absorption becomes excessive.

7.8.3 *Tube length*

The electrode losses are more or less independent of the tube length, and therefore an increase in the length of the positive column, which implies an increase in tube length, gives an increase in efficiency.

7.8.4 *Inert gas pressure*

If the argon pressure is too low, starting will be impaired, wall losses will be increased, and electrode damage will be enhanced. If the argon pressure is too high, although there will be a reduction in self-absorption losses (Hammond and Gallo, 1971, 1972), this will be outweighed by an unacceptable lowering of the electron temperature caused by elastic collisions between electrons and argon atoms.

7.8.5 *Current*

If the mercury pressure is kept constant, and the current through the lamp increased, the UV output will increase until a plateau is reached

(Waymouth *et al.*, 1957; Waymouth, 1971, p. 25). The efficiency with which UV is produced decreases with increasing current. This is a major shortcoming of the low pressure lamp, as it severely limits the intensity of radiation that can be achieved. The reasons are quite complicated. Because of the negative characteristic of the arc, an increase in current density decreases the voltage across the positive column and therefore the electron temperature. This means that a smaller fraction of the total number of electrons is capable of exciting mercury atoms to the 6^3P levels. However, the increased electron density implies an increased frequency of collision of electrons with mercury atoms in the 6^3P states. Collisions of the first kind result in excitation to higher levels and the ultimate emission of non-resonance radiation, while collisions of the second kind result in de-excitation without photon emission.

The plateau of light output of the common fluorescent tube is about 2·5 times its rated value, if the mercury pressure is held constant (Waymouth, 1971, p. 26).

7.8.6 *Supply frequency*

The influence of supply frequency on the behaviour and efficiency of the low pressure lamp has engaged the attention of numerous workers (e.g. Read and Kerry, 1965; Hasebe, 1969; Drop and Polman, 1972; Polman *et al.*, 1972, 1975).

When ionization takes place in the low pressure lamp, the electron–ion pair has a limited lifetime, determined by the time taken to diffuse to the walls of the tube where recombination takes place. The lifetime is of the order of 1 ms. This means that if the electric field is suddenly reduced to zero the electron concentration will reduce to a low value in about 1 ms. Therefore if a lamp is operated from a low frequency a.c. supply, e.g. 5 Hz, such that $f^{-1} \gg 1$ ms, when the voltage across the lamp is reduced to zero at the end of each half cycle the discharge is extinguished, and must be re-ignited at the beginning of the next half cycle. In principle the lamp passes through the entire voltage current characteristic shown in Fig. 6.1 at the beginning of each half cycle. Once the arc is established, the voltage–current characteristic is the same as the static characteristic of a d.c. discharge.

If the frequency is such that $f^{-1} \ll 1$ ms, the electron concentration can no longer respond effectively to the fluctuating field and remains almost constant. This implies an unchanging concentration of current carriers and therefore a constant impedance. The discharge no longer has a negative characteristic, and behaves like a resistor.

Although when the field is reduced to zero the lifetime of the electrons is about 1 ms, the reduction in the electron temperature is much more rapid, and the electrons lose most of their kinetic energy within a few microseconds. If the frequency is very high, e.g. 1 MHz, the electron temperature cannot respond quickly enough and remains effectively constant. This means that the light output no longer responds to the fluctuations in electric field. Hasebe (1969) has demonstrated how the modulation of both electron temperature and light output is reduced as the frequency is increased in the kilohertz range. The modulation in light output is actually dependent on the lifetime of the 6^3P excited states. This is rather longer than for the electron temperature (Drop and Polman, 1972).

It has generally been found that the efficiency of the low pressure mercury lamp is greatest for the d.c. discharge, and for a.c. discharges increases with frequency (Campbell, 1960). It is important to distinguish between the effects at the electrodes and those in the positive column (Read and Kerry, 1965). Read and Kerry showed that efficiency was not only dependent on the frequency of the supply, but also on the type of ballast employed, if the frequency was below 500 Hz. A series inductance gave both better efficiency in the positive column and lower electrode losses than a series capacitance. This is probably due to effects of the current waveform. It was found that a 1 kHz square wave supply gave positive column efficiency equal to that of a d.c. discharge, and even lower electrode losses.

Drop and Polman (1972) have given a mathematical treatment of the positive column at different supply frequencies, and have predicted a minimum efficiency at 500 Hz, in general agreement with experimental results.

Although both increased positive column efficiency and decreased electrode losses can be obtained by operating a lamp on either a d.c. supply or a high frequency a.c. supply, in practice this is seldom feasible, because any increase in lamp efficiency is offset by the losses in a resistive ballast (d.c. operation), or in frequency converters (high frequency a.c.).

7.9 Lamps with Enhanced Output

The striking aspect of the low pressure mercury lamp is that approximately 60% of the lamp input power can be radiated at a wavelength, 253·7 nm, which is highly effective in the excitation of many of the common photoinitiators employed for photopolymerization. Unfortunately, although the efficiency is high, the intensity available from the conventional low pressure lamp is very low, and for reasons mentioned in the previous section, increasing the current through the lamp causes a rapid saturation of the UV output. A great amount of ingenuity has been displayed in attempts to delay the onset of saturation to higher current densities, and it has been possible to maintain reasonable efficiency up to tube loadings of $10\,W\,cm^{-1}$, i.e. about 20 times that of the conventional lamp.

The conventional low pressure mercury lamp has a tube diameter of 36 mm, an inert gas fill of 3 T of argon, and operates at a current of 400 mA, and a tube loading of about $0{\cdot}5\,W\,cm^{-1}$, with a wall temperature of 40 °C. These are the conditions giving maximum overall efficiency, and most fluorescent lamps are based on this design. Waymouth (1971, p. 31) has suggested that the basic reason for the early saturation of the UV output of the arc of the conventional lamp is that 6^3P levels of mercury approach a state of "local thermal equilibrium" with the electron gas, and that the way to increase the saturation value of UV output is to increase the electron temperature. This can be done by increasing the rate of ion–electron recombination. The electron temperature must then increase so that the rate of formation of new current carriers balances the increased rate of loss. The rate of loss can be increased by (1) decreasing the distance the ions and electrons must travel to reach the wall, and (2) increasing the ambipolar diffusion coefficient, either by decreasing the rare gas pressure, or changing to a rare gas in which the mobility of ions and electrons is higher.

The General Electric "Power Groove®" lamp (Aicher and Lemmers, 1957) exploits (1) above by changing the geometry of the tube cross section to a "U" shape. A theoretical treatment of lamps of non-circular cross section has been given by Cayless (1960). The Power Groove fluorescent lamp operates at 1·5 A, and gives a significantly higher lumen output than circular cross section lamps operated at the same current (General Electric Lamp Catalogue).

A variation on the theme of (1) above is the introduction of a recombination structure into the lamp, made for instance of glass wool (Hasker, 1976).

Sadoski and Roche (1976) have designed a lamp specifically for photopolymerization which is capable of giving six times the output at 253·7 nm of a conventional germicidal lamp. They adopted a fairly narrow tube with a rather low pressure (2·5 T) of neon with 0·5% of argon. The ambipolar diffusion coefficient is higher in neon than in argon. The small amount of argon is included to create a Penning mixture and so provide easier starting. The lamp operates at 3·5 A. In order to maintain the optimum mercury pressure of 6 mT condensation chambers were created behind each electrode, shielded from the arc by metal heat shields.

A particularly interesting recent development has been the Brown Boveri high intensity UVC lamp, designed mainly for sterilizing and photochemical applications (British Patent 1,465,242; US Patent 3,971,968). This has been described by Braendli (1975) and its use by Bachmann (1975), Cerny (1977) and Maunder (1977). The development of the lamp was based on the high current discharge technique used in mercury ion lasers. The lamp has a narrow bore (1 cm diameter) and is operated from a stabilized d.c. supply providing a current density of about $10\,A\,cm^{-2}$. This is about 200 times that of a conventional low pressure lamp. The cathode is permanently heated to provide the thermionic emission necessary. The optimum mercury pressure at these very high current densities is obtained with a cold spot temperature of 70 °C, and condensation chambers and a cooling fan are used to maintain this. It is remarkable that this lamp produces 253·7 nm radiation with 30% efficiency, with smaller amounts of the other mercury lines, including the 194·2 nm Hg^+ resonance line, and that the UV emittance of the lamp (in watts per square centimetre) approaches that of a conventional medium pressure lamp. Braendli (1975) suggested that the high efficiency is due to a much higher discharge temperature and Doppler broadening of the emission lines, which reduces the self-absorption loss.

7.10 Fluorescent Lamps

Most low pressure mercury lamps are used as fluorescent lighting tubes, with a fluorescent powder coating on the inside wall which converts the 185·0 and 253·7 nm radiation from the arc into visible radiation suitable for illumination. Fluorescent tubes are also used for purposes other than illumination, with fluorescent coatings chosen to emit bands of radiation centering on wavelengths between about 300 and 450 nm. The most common emits a broad band of UVA, with peak emission at 360–370 nm. There are two versions of this lamp. One has a tube made from normal soda glass and the other a tube made from "Wood's" glass. Wood's glass contains oxides of nickel and cobalt, and is opaque to visible radiation but transmits UVA.

There is always an energy loss when a fluorescent wavelength converter is used because even if the quantum yield is unity, each energetic 253·7 nm photon is converted into a less energetic photon of a longer wavelength.

7.11 Uses

Low pressure mercury lamps have many attractions as radiation sources for the photoinitiation of polymerization. They are the most efficient converters of electrical energy into UVC radiation, and the 253·7 nm line corresponds well to the wavelength of greatest absorption of the majority of the photoinitiators used. They switch on and off almost instantaneously; they run cool so that they do not affect heat-sensitive materials or cause reflector tarnishing; they have long lifetimes and use very cheap and simple control gear. However, they do have two limitations. Their output is virtually monochromatic unless a fluorescent coating is used to convert the 253·7 nm radiation into longer wavelengths. Monochromatic 253·7 nm radiation is absorbed so strongly by most organic systems that penetration is very limited. Also, they provide a radiant emittance which is low compared with that of a medium pressure mercury lamp.

The kinetics of free radical polymerization, *in the absence of oxygen*, are favoured by a low intensity of irradiation (see Section 3.2.4). However, if a UV-curing coating being irradiated is not inerted, and the intensity of irradiation is low so that the creation of initiator radicals is slow, oxygen molecules have plenty of time to mop up initiator radicals as they are formed, and to terminate growing chains. If low pressure lamps are to be used to cure free radical systems it is necessary to protect the surface from attack by oxygen. Union Carbide have patented a system (US Patents 3,807,052 and 3,840,448) in which a coating is irradiated by a bank of germicidal lamps in an atmosphere of nitrogen to provide a "preferential surface cure" (PSC). The cure is completed by irradiation in air with medium pressure lamps. The system has been described by Krajewski *et al.* (1976), Troue (1976) and Pelgrims (1978).

The use of low pressure fluorescent UVA lamps for curing of coatings has been advocated by Levin and Homer (1972). These lamps are sometimes used in the curing of styrene–polyester wood finishes, to bring wax in the formulation to the surface. This reduces the evaporation of the volatile styrene and minimizes the effects of oxygen inhibition during subsequent irradiation with medium pressure lamps. Fluorescent tubes are widely used for exposing photopolymer printing plates. Oxygen inhibition is minimized by a protective plastic film which covers the surface of the photosensitive resin. Blue light-emitting ("super-actinic") fluorescent tubes are suitable for curing thick sections of glass-reinforced polyester resins. They are also used in diazo photocopying machines.

Low pressure mercury lamps with vitreous silica or UV transmitting glass envelopes are used for the sterilization of air, liquids and surfaces.

Fluorescent low pressure lamps are used for tanning, and for a variety of therapeutic uses in medicine, including the treatment of psoriasis and hyperbilirubinemia (Wanmaker *et al.*, 1975; Jewess, 1978; Parrish *et al.*, 1978, p. 19).

Low pressure mercury lamps are used for the practical applications of fluorescence described in Section 1.2.

7.12 Life

Low pressure mercury lamps have very long life times—many thousands of hours. The principal lamp deterioration process is electrode failure, and this is aggravated by frequent on–off switching. Hot cathode lamps have shorter lives than cold cathode lamps. When a hot cathode lamp passes through the glow phase prior to establishment of the arc, the high cathode fall causes sputtering of the emissive material from the electrode. Eventually the lamp may refuse to start, or may "hang up in the glow". It might be thought that if a low pressure arc is operated from an a.c. supply and the arc is extinguished and re-ignited every half cycle, there will be a glow–arc transition every half cycle, with a consequent shortening of the life of the lamp. However, Waymouth (1971, p. 104) has shown that there is sufficient heating of the electrodes in the anode phase to ensure that the temperature of the electrodes remains high enough for copious thermionic emission, so that the glow phase is effectively eliminated.

The tube temperature is too low for any possibility of devitrification, and if vitreous silica is used there should be little solarization. UV transmitting glass is more prone to solarization, and there is evidence that sodium ions which migrate to the inner wall of the tube can combine with mercury to form a light-absorbing coating (Kuemmel, 1975). Some tungsten will evaporate from the electrodes during operation and form dark deposits around the electrodes.

References

Aicher, J. O. and Lemmers, E. (1957). *Illum. Eng.* **52,** 579–584.
Alpert, D., McCoubrey, A. O. and Holstein, T. (1949). *Phys. Rev.* **76,** 1257–1259.
Bachmann, R. (1975). *Brown Boveri Rev.* **62,** 206–209.
Barnes, B. T. (1960). *J. Appl. Phys.*, **31,** 852–854.
Bouwknegt, A., Nienhuis, H., Schipper, D. J. and Tielemans, P. A. W. (1975). *Philips Tech. Rev.* **35,** 356–359.

Braendli, G. (1975). *Brown Boveri Rev.* **62,** 202–205.
Campbell, J. H. (1960). *Illum. Eng.* **55,** 247–256.
Cayless, M. A. (1959). *Brit. J. Appl. Phys.* **10,** 186–190.
Cayless, M. A. (1960). *Brit. J. Appl. Phys.* **11,** 492–498.
Cayless, M. A. (1963). *Brit. J. Appl. Phys.* **14,** 863–869.
Cerny, G. (1977). *Verpack Rundsch.* **28** (10), Tech. Sci. Suppl., 77–82.
Chapman, D. F. (1972). *In* "Lamps and Lighting" (Eds S. T. Henderson and A. M. Marsden), pp. 315–340. Edward Arnold, London.
Childs, C. B. (1962). *Appl. Opt.* **1,** 711–716.
Compton, K. T. (1922). *Phys. Rev.* **20,** 283–299.
Drop, P. C. and Polman, J. (1972). *J. Phys. D* **5,** 562–568.
Easley, M. A. (1951). *J. Appl. Phys.* **22,** 590–593.
Elenbaas, W. (1959). "Fluorescent Lamps and Lighting", Philips' Technical Library. Cleaver-Hulme Press, London.
Franck, G. and Ludemann, H. (1972). *Z. Naturforsch. A* **27a,** 1278–1286.
Franck, G. and Schipp, F. (1975). *Light. Res. Technol.* **7,** 49–51.
Funke, J. (1959). *In* "Fluorescent Lamps and Lighting" (Ed. W. Elenbaas), pp. 121–156. Philips' Technical Library. Cleaver-Hulme Press, London.
Hammond, T. J. and Gallo, C. F. (1971). *Appl. Opt.* **10,** 58–64.
Hammond, T. J. and Gallo, C. F. (1972). *Appl. Opt.* **11,** 729–734.
Hammond, T. J. and Gallo, C. F. (1974). *Appl. Opt.* **13,** 2164–2170.
Hasebe, K. (1969). *Brit. J. Appl. Phys.* **2,** 611–612.
Hasker, J. (1976). *Appl. Phys. Lett.* **28,** 586–588.
Holstein, T. (1947). *Phys. Rev.* **72,** 1212–1233.
Holstein, T. (1951). *Phys. Rev.* **83,** 1159–1168.
Jack, A. G. and Vrenken, L. E. (1980). *IEE Proc. A* **127,** 149–157.
Jerome, C. W. (1956). *Illum. Eng.* **51,** 205–213.
Jewess, B. W. (1978). *Light. Res. Technol.* **10,** 184–188.
Johnson, P. D. (1971a). *J. Opt. Soc. Amer.* **61,** 1451–1454.
Johnson, P. D. (1971b). *Appl. Phys. Lett.* **18,** 381–382.
Kenty, C. (1950). *J. Appl. Phys.* **21,** 1309–1318.
Kenty, C., Easley, M. A. and Barnes, B. T. (1951). *J. Appl. Phys.* **22,** 1006–1011.
Koller, L. R. (1945). *J. Appl. Phys.* **16,** 816–820.
Krajewski, J. J., Packer, E. S. and Yamazaki, T. (1976). *Metal Finish.* **74** (11), 58–63; *J. Coat. Technol.* **48** (623), 43–49.
Kuemmel, U. (1975). *Appl. Phys.* **7,** 93–97.
Levin, R. E. and Homer, H. H. (1972). *J. Illum. Eng. Soc.* **1,** 290–299.
Ligthart, F. A. S. and Keijser, R. A. J. (1980). *J. Appl. Phys.* **51,** 5295–5299.
Maunder, D. T. (1977). *Food Technol.* **31** (4), 36–37.
Papoular, R. (1965). "Electrical Phenomena in Gases". Iliffe, London.
Parrish, J. A., Anderson, R. R., Urbach, F. and Pitts, D. (1978). "UVA: Biological Effects of Ultraviolet Radiation with Emphasis on Human Responses to Longwave Ultraviolet". John Wiley & Sons, Chichester.
Pelgrims, J. (1978). *J. Oil Colour Chem. Assoc.* **61,** 114–118.
Penning, F. M. (1927). *Naturwiss.* **15,** 818.
Penning, F. M. (1928). *Z. Phys.* **46,** 335–348.
Polman, J. and Drop, P. C. (1972). *J. Appl. Phys.* **43,** 1577–1580.
Polman, J., van der Werf, J. E. and Drop, P. C. (1972). *J. Phys. D* **5,** 266–279.
Polman, J., van Tongeren, H. and Verbeek, T. G. (1975). *Philips Tech. Rev.* **35,** 321–330.

Read, T. B. (1964). *Brit. J. Appl. Phys.* **15,** 837–841.
Read, T. B. and Kerry, D. C. (1965). *Brit. J. Appl. Phys.* **16,** 453–455.
Sadoski, T. T. and Roche, W. J. (1976). *J. Illum. Eng. Soc.* **5,** 143–147.
Steele, H. L., Jr (1955). *J. Appl. Phys.* **26,** 1183–1184.
Tan, K. L. and von Engel, A. (1968). *J. Phys. D* **1,** 258–259.
Troue, H. H. (1976). SME Tech. Pap. (Ser.) FC76-519.
Verbeek, T. G. and Drop, P. C. (1974). *J. Phys. D.* **7,** 1677–1683.
Vriens, L., Keijser, R. A. J. and Ligthart, F. A. S. (1978). *J. Appl. Phys.* **49,** 3807–3813.
Walsh, P. J. (1959). *Phys. Rev.* **116,** 511–515.
Wanmaker, W. L., van Gils, J. F. and ter Vrugt, J. W. (1975). *Light. Res. Technol.* **7,** 19–22.
Waymouth, J. F. (1971). "Electric Discharge Lamps". MIT Press, Cambridge, Mass.
Waymouth, J. F. and Bitter, F. (1956). *J. Appl. Phys.* **27,** 122–131.
Waymouth, J. F., Bitter, F. and Lowry, E. F. (1957). *Illum. Eng.* **52,** 257–261.
Zemansky, M. W. (1927). *Phys. Rev.* **29,** 513–523.

8 The Medium Pressure Mercury Lamp

8.1 Local Thermal Equilibrium

In order for a mercury discharge lamp to give a high conversion of electrical energy into radiation it is necessary to have either a high electron temperature or alternatively a high mercury atom temperature. Figure 8.1

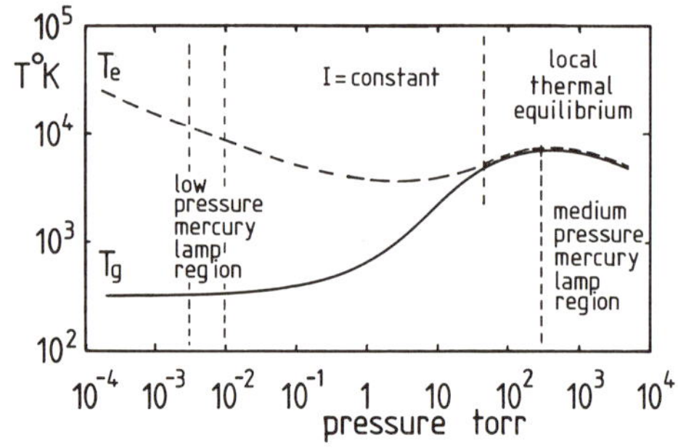

Fig. 8.1. Mercury discharges. Influence of pressure on T_e (electron temperature) and T_g (atom and ion temperature) showing the approach to local thermal equilibrium. Schematic, adapted from Elenbaas (1951, Fig. 7).

shows that the low pressure lamp (10^{-3}–10^{-2} T) provides a very high electron temperature. When the pressure is increased to about 10^2 T, the temperatures of the electrons, atoms and ions become almost identical, and a state of "local thermal equilibrium" or "local thermodynamic equilibrium" is said to exist. The concept of a "plasma temperature" and the laws of thermodynamics become applicable. This is the region of operation of the medium pressure mercury lamp, in which the plasma temperature is in the region of 5000–7000 K. There is only an approximation to equilibrium. Because the electrons transfer energy to the atoms they will necessarily be at a slightly higher temperature. Also, perfect equilibrium is not theoretically possible in a system which is continually losing energy by radiation. However, a model of the medium pressure mercury discharge based on the assumption of local thermodynamic equilibrium, and the use of the Boltzmann distribution law (to determine the populations of the excited states of the mercury atom) and the Saha equation (to determine the concentrations of ions and electrons) has led to a detailed understanding of the radiative and electrical characteristics of the medium pressure mercury discharge (Elenbaas, 1951, 1965a).

The medium pressure mercury lamp has become the most important radiation source for chemical applications, and in particular for the curing of surface coatings. It has a high radiative efficiency (percentage of input power converted to radiation) and quite good spectral efficiency (percentage of radiation appearing at wavelengths in the ultraviolet). The latter is a result of a very favourable system of electronic energy levels for the mercury atom, with many of the possible radiative transitions producing photons in the ultraviolet.

8.2 Construction

The construction of a typical linear medium pressure mercury lamp, as used for UV curing and other photochemical applications, is shown schematically in Fig. 8.2. The tube is of transparent vitreous silica,

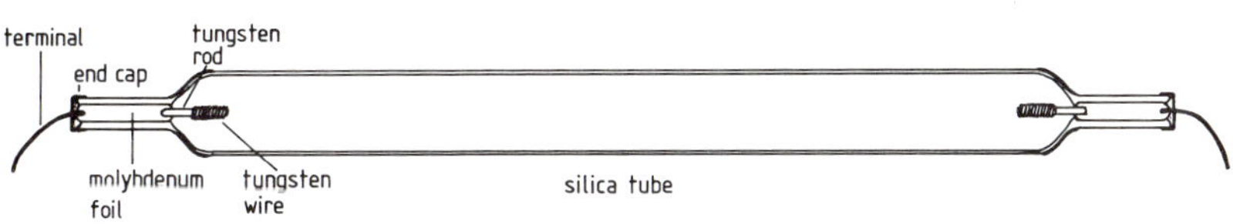

Fig. 8.2. The medium pressure mercury lamp: construction.

normally of internal diameter 10–30 mm, and wall thickness 1–2 mm, with electrodes at each end (Lienhard, 1975; Schooley, 1975). Vitreous silica has a very low temperature coefficient of expansion (5×10^{-7} K^{-1}), much lower than that of any electrical conductor, and therefore making an external connection to the electrodes presents a design problem. This is ingeniously overcome by using a piece of thin molybdenum foil, hermetically sealed in the silica. Although molybdenum has a coefficient of expansion (5×10^{-6} K^{-1}) which is much higher than that of silica, the thinness and ductility of the molybdenum allow it to accommodate the heating and cooling of the seal without cracking the silica. The thinness of the seal (less than 70 μm) limits the current which can be carried, and it is important that during the operation of the lamp the temperature of the molybdenum foil does not exceed 350 °C otherwise it will be oxidized and the seal will be destroyed (Ayotte and Hale, 1967). The ends of the tube are often coated externally with a reflective, heat-resistant material such as platinum, gold or zirconium oxide. This keeps the ends warm and prevents condensation of the mercury while the lamp is operating.

8.2.1 Electrodes

Almost all medium pressure mercury lamps operate from an alternating current supply. The electrodes act as anode and cathode in turn and are identical in construction (Bouwknegt *et al.*, 1975). The most common form of electrode consists of a tungsten rod around which is wrapped one or more coils of tungsten wire. Tungsten is chosen because, of all metals, it has the highest melting point and the lowest vapour pressure (Plante and Sessoms, 1973). The medium pressure mercury lamp is an arc discharge and when the lamp is running normally the electron current is supplied by thermionic emission from the hot electrodes. The thermionic emission from an electrode is given by the Richardson–Dushman equation:

$$J = AT^2 \exp(-e\Phi/kT),$$

where J is the current density (in amps per square centimetre), T is the absolute temperature, e is the electronic charge ($1{\cdot}60 \times 10^{-19}$ e.s.u.), Φ is the work function, an empirical constant for the emitting surface, with units of electron-volts, k is Boltzmann's constant ($1{\cdot}372 \times 10^{23}$ J K^{-1}) and A is a constant (the emission coefficient) for a particular surface, having a theoretical value for pure metals of 120 A cm^{-2} K^{-2}.

A high thermionic emission is favoured by a low work function and a high temperature. Unfortunately tungsten has quite a high work function (4·5 eV) and therefore a pure tungsten electrode needs to operate at a very high temperature to sustain the thermionic emission required—a temperature at which significant evaporation of tungsten takes place. Also a very high electrode temperature implies high electrode energy losses. Therefore, as with the low pressure mercury lamp, emissive materials are used to reduce the electrode temperature required for adequate thermionic emission. These materials may be alkaline earth or thorium oxides embedded between the turns of the inner winding (Martt *et al.*, 1960; Till and Unglert, 1960). It is also common practice to fabricate the electrode rod not from pure tungsten but from tungsten containing 1 or 2% of thorium oxide. Typical electrode construction is shown in Fig. 8.3. Electrodes must be

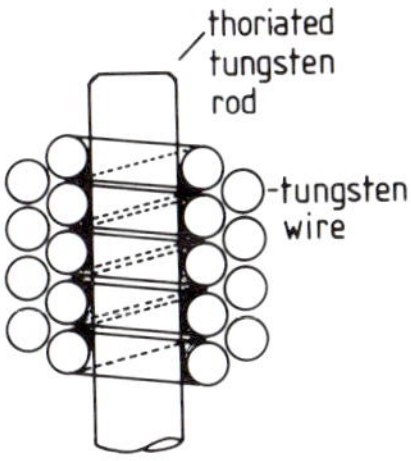

Fig. 8.3. Typical electrode construction for the medium pressure mercury lamp.

designed carefully to ensure efficient and reliable operation and a long lamp life. They must warm up quickly to allow a rapid glow to arc transition, and to reduce the contribution of secondary emission to the generation of the electron current. Secondary emission, which results from the bombardment of the electrodes by energetic electrons and ions, tends to cause sputtering of the tungsten and darkening of the inner wall of the tube. It is important in the first few seconds after a lamp is switched on (Keeffe, 1980), before the electrodes reach a stable operating temperature. The operating temperature must be high enough to allow copious thermionic emission but low enough to prevent rapid evaporation of tungsten and other electrode materials. Electrodes are heated by electron bombardment in the anode phase, ion bombardment in the cathode phase, and Joule heating. Heat is lost by electron emission, radiation and conduction. The electrode must be designed to allow proper functioning both during the low pressure warm-up period and the medium pressure stable running (Nelson, 1961). If the dimensions of the electrode have been correctly chosen for the lamp current, then when the lamp is first switched on and the mercury pressure is low, the winding starts to glow and a diffuse discharge emanates from it. As the mercury pressure rises the glowing area reduces in size, intensifies, and moves towards the end of the electrode and finally becomes an intense hot spot on the end of the rod. Regulation of the temperature profile of the electrode is one of the keys to successful operation. The tungsten wire coil acts as a radiator, and is several hundred degrees Celsius cooler than the arc spot on the tip of the rod (van de Weijer, 1965). Typical electrode tip temperatures lie between 1500 and 2000 °C, while the wire overwinding will be between 1000 and 1500 °C, for alkaline earth oxide activation.

The oxides which activate the electrode are stored between the turns of the inner winding. During operation the oxide is slowly reduced by tungsten (Waymouth, 1971, p. 178), releasing the metal, which migrates to the end of the electrode rod where it forms a monolayer, so reducing the effective work function. The surface of pure bulk barium has a work function of 2·1 eV. The value for thorium is 3·4 eV. However, barium as a monolayer on tungsten has a work function of only 1·56 eV, while thorium on tungsten has a work function of 2·63 eV (Smithells, 1976, p. 1027). Emission coefficients for real surfaces differ greatly from the theoretical value of 120 A cm^{-2} K^{-2}, and are usually lower. Tungsten, barium and thorium have values of 60, 60 and 70 A cm^{-2} K^{-2} respectively for the bulk metals. The values for monolayers are lower still, 1·5 A cm^{-2} K^{-2} for Ba/W and 3·0 A cm^{-2} K^{-2} for Th/W (Smithells, 1976, p. 1027). It is important that when the lamp has reached stable operation the arc spot does not remain on the winding. This could cause rapid loss of the emissive material.

In an operational lamp there is an axial electric field which by the Schottky effect reduces the work function of the cathode, and allows a field emission contribution, or an enhancement of thermionic emission.

8.2.2 Gas fill

Medium pressure mercury lamps operate at a pressure in the region 10^2–10^4 T. In the low pressure mercury lamp there is an excess of liquid mercury in the tube, and the mercury pressure is controlled by the ambient temperature. In the medium pressure lamp, however, all the mercury evaporates, and the pressure is therefore regulated by the careful measurement of the mercury dose. Figure 7.1 shows the vapour pressure curves for some common metallic elements, and it can be seen that mercury is by far the most volatile. However, to maintain the pressure required, the coolest part of the tube must be at a temperature in excess of 400 °C, and in practice the main body of the tube is maintained at 600–800 °C during stable operation.

As with a low pressure mercury lamp, a starting gas (usually argon) is incorporated at a few torr. The pressure of the argon has a considerable influence on the ignition voltage, warm-up characteristics and electrode deterioration, but the argon has little effect on the stable running of the lamp unless the argon pressure is abnormally high, in which case it increases the non-radiative losses and alters the spectral distribution of the light output.

Some of the characteristics of medium pressure mercury lamps will be illustrated in this chapter by experimental data. The more important of the lamp types used will be coded in the text "Lamp I", "Lamp II", "Lamp III", etc. All the lamps used had thoria-activated tungsten electrodes, and envelopes of 18 mm internal diameter. Variables were interelectrode distance, mercury dose (measured in milligrams per centimetre of lamp length), argon pressure (measured at ambient temperature), and silica type (ozone free, or ozone producing). The details are given in Table 8.1.

Table 8.1 Details of lamps used to supply experimental data. All lamps supplied by Primarc Jigs and Lamps Ltd, Reading, UK

Lamp no.	Interelectrode distance (cm)	Mercury dose (mg cm^{-1})	Argon pressure (torr)	Silica type
I	14·6	1·4	30	Ozone producing
II	14·6	1·4	40	Ozone producing
III	16·6	1·4	50	Ozone producing
IV	16·6	1·4	200	Ozone producing
V	22·5	4·8	30	Ozone producing
VI	22·5	8·0	30	Ozone producing
VII	22·5	10·1	30	Ozone producing
VIII	22·5	9·0	30	Ozone producing
IX	22·5	9·0	30	Ozone free
X	14·6	14·5	30	Ozone producing

8.3 Circuitry: The Medium Pressure Mercury Lamp as a Circuit Element

As an arc discharge lamp, the medium pressure mercury lamp has a negative resistance characteristic, and therefore stable operation requires the use of a suitable ballast. Lamps are almost always operated from an a.c. supply, for reasons which will become apparent, and the ballast may be either a capacitor or an inductor, or more commonly a combination of both.

The circuitry associated with the medium pressure mercury lamp has three principal functions:

(1) It must provide a sufficient electric field in the cold lamp to enable an arc to be established.
(2) It must control the current passing through the lamp.
(3) In the case of a.c. operated lamps, the circuitry must allow the arc to be re-established every half cycle when the current passes through zero.

8.3.1 Lamp voltage and lamp current

The high pressure of mercury during operation causes the tube to have a much higher impedance per unit length of arc than is the case with the low pressure lamp, and therefore a higher potential gradient is required to maintain the arc. This means that for all but the shortest lamps, mains voltage is insufficient to operate the lamp and a step-up transformer is required. Medium pressure mercury lamps operated from an a.c. supply have a potential gradient (r.m.s.) of about 5–30 V cm^{-1} of arc length, and it is normal for the open circuit voltage to be 50–100% higher than this.

Sometimes a lamp will *run* happily on mains voltage, but will not *start* on mains voltage. In this case it is possible to avoid the use of an expensive transformer by incorporating one or two extra electrodes in the lamp to create intense local fields, or alternatively to use a high voltage pulse unit to create a short-lived intense field between the electrodes.

The voltage at which a lamp operates (i.e. the potential difference between the electrodes) depends on the length of the lamp, the amount of mercury per unit lamp length, m (in milligrams per centimetre), the bore of the lamp, d (in centimetres), and (slightly) on the power consumed by the lamp. Elenbaas (1951, p. 30) has derived theoretically the following expression for a d.c. lamp:

$$E \propto \frac{P^{1/2}}{(P-A)^{1/3}} m^{7/12} d^{-3/2}, \qquad (8.1)$$

where E is the potential gradient in the positive column (in volts per centimetre), P is the power consumed by the lamp per unit length of the positive column (in watts per centimetre), and A is the conductive loss from the arc to the tube (in watts per centimetre). For the normal range of operation of the medium pressure mercury lamp, $P \gg A$ (see Section 8.7), and therefore expression (8.1) can be simplified to

$$E \propto P^{1/6} m^{7/12} d^{-3/2}. \qquad (8.2)$$

This expression indicates that the operating *voltage* of a d.c. lamp is almost entirely determined by the lamp itself, rather than the associated circuitry. The *current* through the lamp, however, is largely determined by the ballast. When a lamp is operated from an a.c. supply, the waveform of the current through the lamp responds to that of the supply voltage, although some distortion always takes place. If the supply frequency is low (50 or 60 Hz) the lamp voltage waveform can be viewed as a distorted square wave, with the voltage being more responsive to the direction of the current than to its magnitude.

There are a number of different types of ballast in common use. The advantages and disadvantages of each have been described in a recent paper by Majewski (1978). Ballasts fall into two main categories, those which are predominantly capacitative, and those which are predominantly inductive. I have examined an example of each of these types.

8.3.2 The constant wattage ballast

The constant wattage ballast is *predominantly capacitative*. The version I have used is shown in Fig. 8.4, described by Majewski as the "iso-stabilized" premium constant wattage ballast circuit. The capacitance in series with the lamp could be varied in steps between 12·2 and 20·3 μF. The ballast circuit was suitable for lamp voltages up to about 580 V, and lamp currents between 3·6 and 6·6 A.

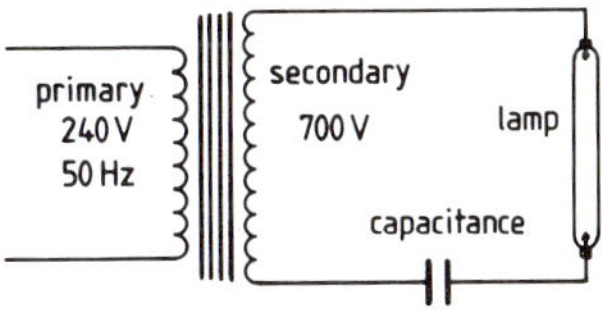

Fig. 8.4. The iso-stabilized constant wattage ballast circuit.

Lamps V, VI and VII (Table 8.1) were used to examine the operational characteristics of the constant wattage ballast. These lamps were of the type normally used for UV curing, and the range of mercury pressures and operating conditions were chosen to cover the normal range of operation of these lamps when a constant wattage ballast is used.

The primary (i.e. mains) voltage and current, and the lamp voltage and current (i.e. the voltage *across* the lamp, and the current *through* the lamp) were measured with an Avometer (a rectifying moving coil multimeter), for various lamp–capacitance combinations. For some combinations, the voltage and current waveforms (half cycle only) were displayed using a current probe and oscilloscope. The lamp waveforms for the three lamps, with 14·5 μF series capacitance, are shown in Fig. 8.5.

The product of instantaneous values of voltage and current gave the instantaneous power consumption of the lamp, while the voltage divided by the current gave the resistance of the lamp. The waveforms for power and resistance are shown in Fig. 8.6.

Analysis of the power waveform allowed the time averaged power consumption of the lamp to be determined. Similarly, analysis of the

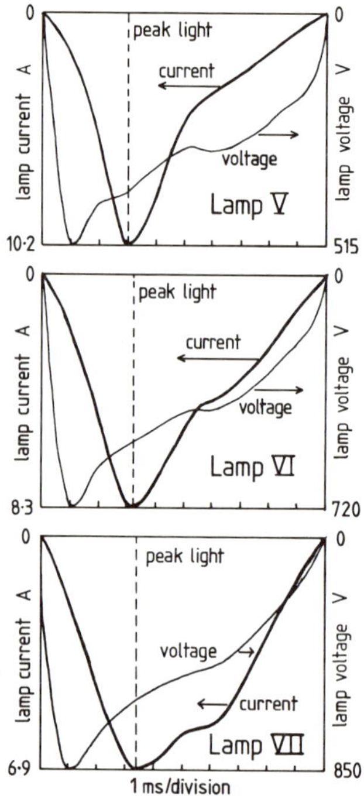

Fig. 8.5. Lamps V, VI and VII. Constant wattage ballast. Lamp voltage and lamp current waveforms.

primary waveforms (Fig. 8.7) allowed the time averaged power consumption from the mains to be calculated, and therefore the circuit losses. The results of the waveform analysis are shown in Table 8.2. Results for Lamp V operated with 19·3 μF series capacitance are also given. Changing the series capacitance was found to have little effect on the shapes of the waveforms.

The Avometer responds to the *mean* value of the waveform half cycle, and although it is *calibrated* for r.m.s. values, the calibration is only valid for a *sinusoidal* waveform. However, dividing the Avometer reading by the form factor for a sinusoidal waveform (1·11) gives the true mean, and this can be used to calibrate the oscilloscope trace, and hence calculate the true r.m.s. value.

A number of interesting conclusions can be drawn from the waveform analysis. The lower the mercury pressure, the lower the minimum resistance of the lamp, the lower the power factor of the lamp, and the peakier the lamp current waveform. Increasing the power for Lamp V decreases

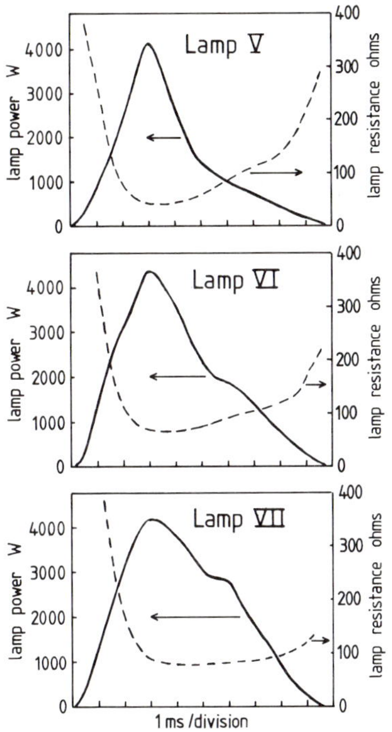

Fig. 8.6. Lamps V, VI and VII. Constant wattage ballast. Lamp power and lamp resistance waveforms.

the minimum resistance of the lamp. The constant wattage ballast controls the *mean* current, but not the *shape* of the current waveform. Although increasing the pressure increases the lamp voltage and the time averaged power consumed by the lamp, there is hardly any change in the *peak* power. For a particular input power, a higher pressure lamp, which operates at a higher voltage and a lower current than a lower pressure lamp, has associated with it a lower circuit loss.

Although the constant wattage ballast has a transformer which is designed to have a reactive secondary winding, it is a predominantly capacitative ballast. The observation that this type of ballast controls the *mean* but not the *instantaneous* current can be explained if one considers the way in which a capacitative ballast works. Ballasts in general operate by varying the voltage available to the lamp in such a way as to compensate for the negative characteristic and maintain a stable arc. The capacitor generates a voltage across itself proportional to the accumulated charge, and therefore the capacitative ballast controls only the amount of charge

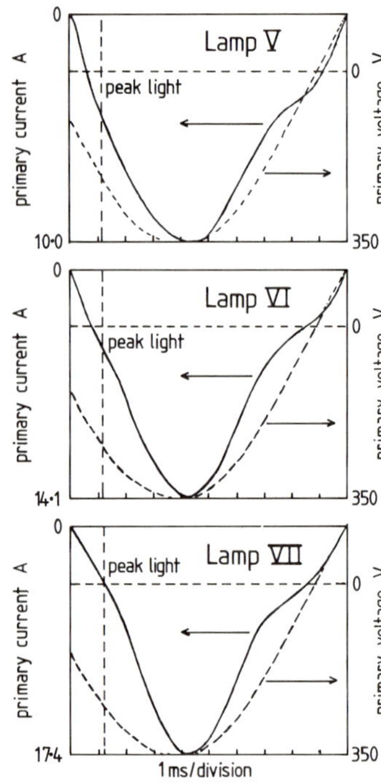

Fig. 8.7. Lamps V, VI and VII. Constant wattage ballast. Primary voltage and current waveforms.

passing through the lamp during each half cycle, i.e. the mean current. The rates of growth of ionization in the lamp are rapid compared with the duration of the normal mains cycle (50 or 60 Hz) and therefore rapid rates of current rise and excessive instantaneous values are possible, particularly when the mercury pressure and lamp impedance are low, as they are during the warm-up period. Peaky waveforms make large demands on the electrodes, and reduce lamp life.

Figure 8.8 shows the phase relationship between the open circuit voltage (i.e. that across the transformer secondary winding) and the lamp waveforms for Lamp V. This shows that the lamp current *leads* the open circuit voltage. During most of each half cycle the lamp voltage is relatively insensitive to lamp current. However, at the end of each half cycle, when the current changes direction, there is a very large and rapid change in lamp voltage. During this period, the open circuit voltage changes relatively little, and therefore there must be a large compensating change in the ballast voltage. As $V = Q/C$ for a capacitor (Q being the stored

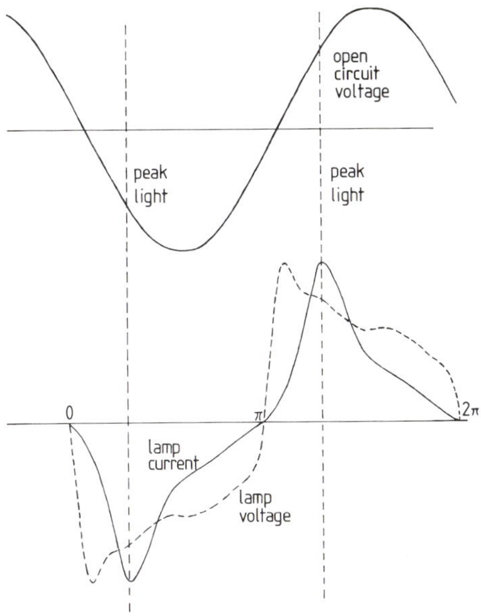

Fig. 8.8. Lamp V. Constant wattage ballast. Phase relationships.

charge), there must be a large current flow through the capacitor, and therefore the lamp. This explains the generally peaky waveforms found with the capacitative ballast. Note also that when the lamp current passes through zero, the open circuit voltage is in the wrong direction for re-ignition, and therefore the capacitative ballast is not particularly good for fulfilling requirement (3) listed above.

A power factor of less than unity implies that the lamp has reactance as well as resistance. This reactance is seen to be inductive, as the voltage leads the current during each half cycle. The voltage developed across the lamp is therefore to some extent dependent on the rate of change of current. This makes the mathematical analysis of the situation rather difficult.

Table 8.3 shows how the voltage and the current measured by the Avometer depend on the series capacitance and the mercury dose in the three lamps. It can be seen that the mean current is almost entirely determined by the series capacitance, and is almost independent of the mercury dose. The data in Table 8.3 can be used to check the validity of expression (8.1). The thermal conductive loss *A* is discussed in some detail in Section 8.7. It is surprisingly independent of mercury dose, tube diameter and power input, and is often taken to be a constant with a value of 10 W cm^{-1}. A more refined analysis (Elenbaas, 1965b) shows that *A* does

Table 8.2 Electrical characteristics of Lamps V, VI and VII. Constant wattage (predominantly capacitative) ballast

	Lamp V		Lamp VI	Lamp VII
Series capacitance (μF)	14·5	19·3	14·5	14·5
Time to reach $\frac{1}{2}$ max V (s)	80	65	108	118
Primary circuit				
Avometer current (A)	6·58	9·25	8·37	9·62
Mean current (A)	5·93	8·33	7·54	8·67
R.M.S. current (A)	6·64	9·52	8·69	10·3
Form factor (current)	1·12	1·14	1·15	1·19
Avometer voltage	247	247	247	247
Mean voltage	223	223	223	223
R.M.S. voltage	247	247	247	247
Form factor (voltage)	1·11	1·11	1·11	1·11
Time averaged power (W)	1600	2290	2070	2380
R.M.S. volts × r.m.s. amps	1640	2350	2150	2540
Avometer volts × Avometer amps	1620	2280	2070	2380
Power factor	0·98	0·97	0·96	0·94
Lamp				
Avometer current (A)	4·40	6·15	4·50	4·55
Mean current (A)	3·96	5·54	4·05	4·10
R.M.S. current (A)	5·00	6·98	4·73	4·55
Form factor (current)	1·26	1·26	1·17	1·11
Avometer voltage	340	352	461	548
Mean voltage	306	317	415	494
R.M.S. voltage	324	339	446	533
Form factor (voltage)	1·06	1·07	1·07	1·08
Time averaged power (W)	1390	2040	1890	2200
R.M.S. volts × r.m.s. amps	1620	2360	2110	2420
Avometer volts × Avometer amps	1500	2160	2070	2490
Power factor	0·86	0·86	0·90	0·91
Minimum resistance (Ω)	38	29	62	83
Ballast/transformer loss (W)	210	250	180	180

Form factor is the ratio r.m.s.: mean, and has a value of 1·11 for a sinusoidal waveform, and 1·00 for a square waveform.

Table 8.3. Potential gradient analysis

	Lamp V						Lamp VI					Lamp VII			
Series capacitance (μF)	12·2	14·5	16·6	18·2	19·3	20·3	12·2	14·5	16·6	18·2	19·3	12·2	14·5	16·6	18·2
Avometer voltage (V)	335	340	345	348	352	355	440	461	468	479	485	520	548	568	575
Avometer current (A)	3·65	4·40	5·12	5·70	6·15	6·58	3·65	4·50	5·18	5·70	6·20	3·73	4·55	5·20	5·75
Time average power (W)	1140	1390	1640	1840	2040	2170	1460	1890	2210	2480	2740	1660	2190	2600	2910

increase slightly with P, and that $A = 4{\cdot}5P^{1/4}$. In order to calculate E, the potential gradient in the arc, it is necessary to know the value of the sum of the cathode and anode falls. This value is about 15 V for the medium pressure mercury lamp with activated electrodes (Elenbaas, 1951, p. 145; Lowke *et al.*, 1975; Lowke and Zollweg, 1975). In order to calculate a value for E, 15 V was subtracted from the mean voltage measured between the electrodes, to give the voltage across the positive column, and this value was divided by the arc length to give E. A graph of E versus

$$[\mathrm{P}^{1/2}/(\mathrm{P} - 4{\cdot}5P^{1/4})^{1/3}]\ m^{7/12}\ d^{-3/2}$$

is shown in Fig. 8.9. It can be seen that the experimental data fit expression (8.1) quite well. However, the derivation of expression (8.1) from the "two-temperature" model of the discharge (see Section 8.5) involves a number of simplifying assumptions, for example that the width of the arc remains constant, which, as will be seen in later sections, is not strictly justified.

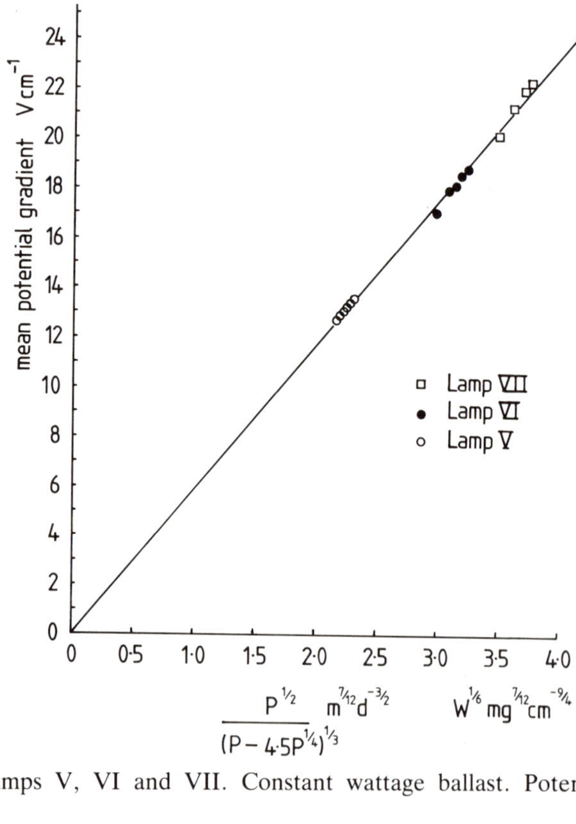

Fig. 8.9. Lamps V, VI and VII. Constant wattage ballast. Potential gradient analysis.

It was found that potential gradient measurements made with Lamps I, II, III and IV at high power inputs gave values which were slightly higher than would be predicted from Fig. 8.9. This is not surprising, as at low mercury pressures and high power inputs the positive ions have a significant effect on the impedance, and this is ignored in the derivation of expression (8.1) (Elenbaas, 1972, p. 146).

The constant wattage ballast is that most favoured for the operation of the long high intensity medium pressure mercury lamps used for UV curing. When a lamp is operated from a constant wattage ballast, the lamp current is rather insensitive to the primary voltage, and therefore the lamp output does not vary very much with the mains voltage. Hence the term "constant wattage". The power factor presented to the mains supply is quite high, as can be seen from Table 8.2. The circuit has good short circuit protection, as if the lamp terminals are shorted the capacitance limits the current. The primary current starts at a low value when the lamp is first switched on, and increases as the power consumed by the lamp increases. Therefore the rating of the input wiring, fusing and circuit breakers can be based on normal operating conditions.

NB: The proportionality factor derived from the gradient of Fig. 8.9 is 5.8. This agrees well with Elenbaas's experimental result of 6 (Elenbaas, 1951, p. 30). For arc power inputs between 60 and 120 W cm^{-1}, $P^{1/2}/(P - 4{\cdot}5P^{1/4})^{1/3}$ does not vary greatly and is $2{\cdot}23 \pm 0{\cdot}10$ W$^{1/6}$ cm$^{-1/6}$. Therefore $E \sim 13m^{7/12}d^{-3/2}$.

8.3.3 *The reactor ballast*

An example of an inductive ballast is shown in Fig. 8.10. This simple *reactor ballast* is suitable for operating short lamps which will run at less

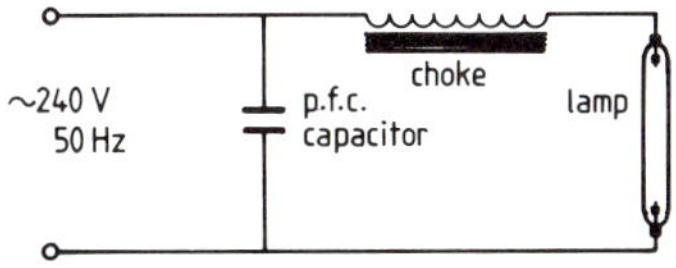

Fig. 8.10. The reactor ballast.

than mains voltage, and is a type commonly used with lamps designed for illumination purposes. Lamps I, II, III and IV were operated with a ballast of this type. The inductance consisted of four Thorn G53230T chokes connected in parallel. It was possible to reduce the lamp current by 25% or

50% by removing either one or two of these chokes. Six 25 μF capacitors connected in parallel were used for power factor correction. Although the lamps would run on mains voltage, they would not ignite on mains voltage, and therefore a high voltage (3–4 kV) pulse ignitor (Thorn G53279) was incorporated in the circuit.

Figure 8.11 shows lamp voltage and lamp current waveforms for Lamp I operated with the inductive ballast. The lamp current is seen to lag behind the mains voltage, and therefore the inductive ballast is sometimes referred to as a *lagging* ballast. The lamp voltage waveform is again a distorted

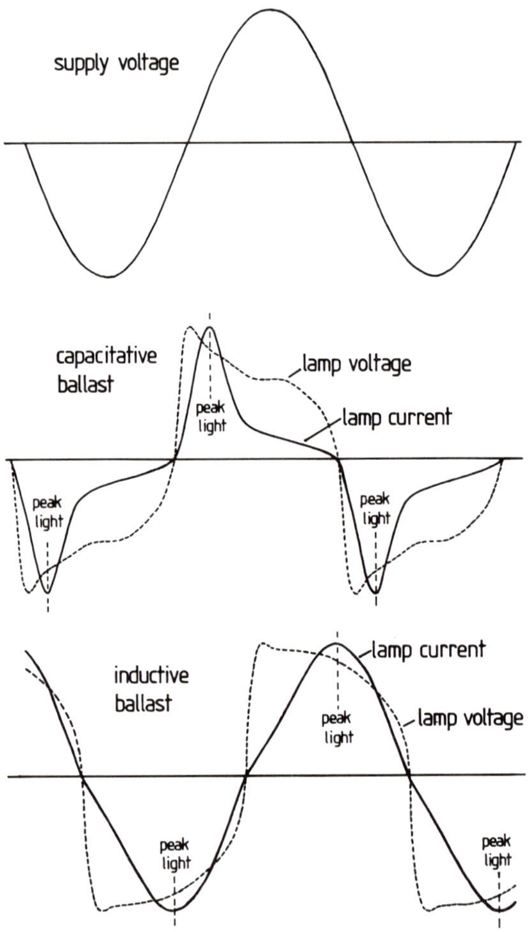

Fig. 8.11. Lamp I. Lamp voltage and lamp current waveforms for constant wattage (capacitative) and reactor (inductive) ballasts. Lamp current form factor: 1·33 (constant wattage), 1·13 (reactor).

square wave, and the lamp current waveform a distorted sinusoidal wave. The reason for the minimal current waveform distortion is that the choke generates a voltage across itself proportional to *the rate of change* of current through the lamp. Waymouth (1971, p. 312) has stated that "the effect of the inductance is to limit the growth of current during that part of the cycle when open circuit voltage is in the proper phase to cause the lamp current to increase. Once current has reached its maximum value and started to decrease, the inductive voltage reverses and prevents current from decreasing as fast as it otherwise would".

In order to make a direct comparison between the reactor (inductive) ballast, and the constant wattage (predominantly capacitative) ballast, Lamp I was operated on the same constant wattage ballast as was used for Lamps V, VI and VII. Lamp current and voltage waveforms were measured and are also shown in Fig. 8.11. Note that Lamp I, with an operating pressure at the lower limit of the medium pressure range, gives a very peaky current waveform with the constant wattage ballast. A comparison of the lamp power input waveforms for the two ballasts, for the *same time averaged power consumption*, is shown in Fig. 8.12. It can be seen that the peak power consumption with the constant wattage ballast is very much higher than it is with the reactor ballast.

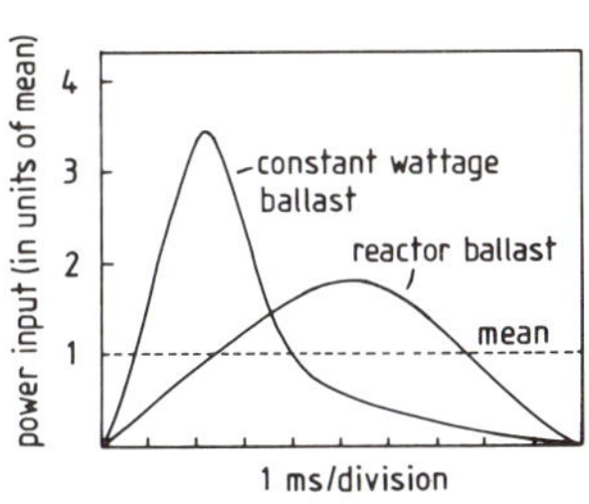

Fig. 8.12. Lamp I. Effect of ballast type on power input waveform.

It can be seen from Fig. 8.11 that when an inductive ballast is used, and the lamp current passes through zero, the open circuit voltage is in the right direction for re-ignition, and therefore the inductive ballast is relatively good in this respect.

The reactor ballast has several shortcomings. Even before the lamp is ignited, the power factor correction capacitor(s) draws a high current (albeit with a very low power consumption) and therefore wiring, fuses and circuit breakers must be specified with this in mind. The circuit provides no intrinsic protection against a lamp short. Unlike the constant wattage ballast, the reactor ballast provides poor regulation of lamp current and light output, both of which are quite sensitive to fluctuations in the supply voltage. The effect is demonstrated in Figs 8.13 and 8.14, which show how the lamp current, lamp voltage, lamp input power, and UV (250–400 nm) output vary with the input voltage for Lamp I.

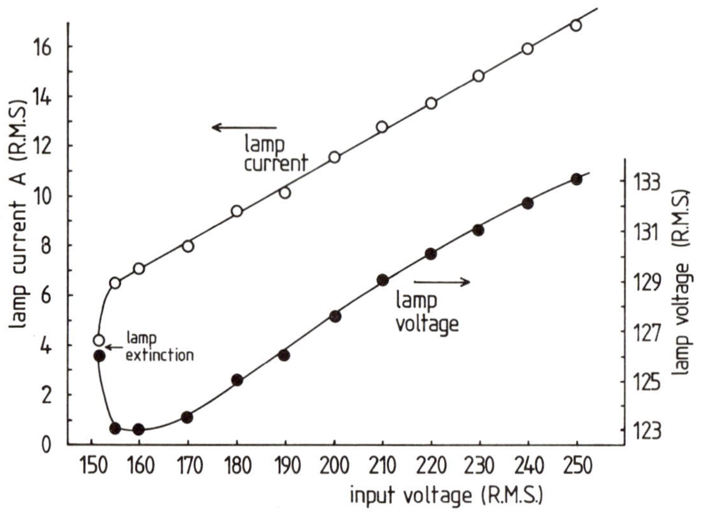

Fig. 8.13. Lamp I. Reactor ballast. Effect of input voltage on lamp current and lamp voltage.

When a lamp is operated from a constant wattage ballast, the mean lamp current is almost entirely dependent on the ballast, and is little influenced by the lamp itself. Expression (8.1) shows that an increase in mercury dose will increase the lamp voltage, and therefore the lamp power. The situation is rather more complex with a reactor ballast, because an increase in lamp voltage brings about a decrease in lamp current. Hehenkamp (1965) has discussed the quantitative relationship between lamp current and lamp voltage. If the lamp voltage waveform is taken to be a square wave, then it can be shown that the lamp current I is given by

$$I = \frac{\sqrt{V_{\mathrm{m}}^2 - 1{\cdot}09^2 V_{\mathrm{l}}^2}}{2\pi f L},$$

where V_{m} is the supply voltage (r.m.s.), V_{l} the lamp voltage, f the supply frequency, and L the inductance of the ballast. Figure 8.15 shows lamp current as a function of lamp voltage for five different lamps, varying in arc length and mercury fill. A reactor ballast with five different power settings

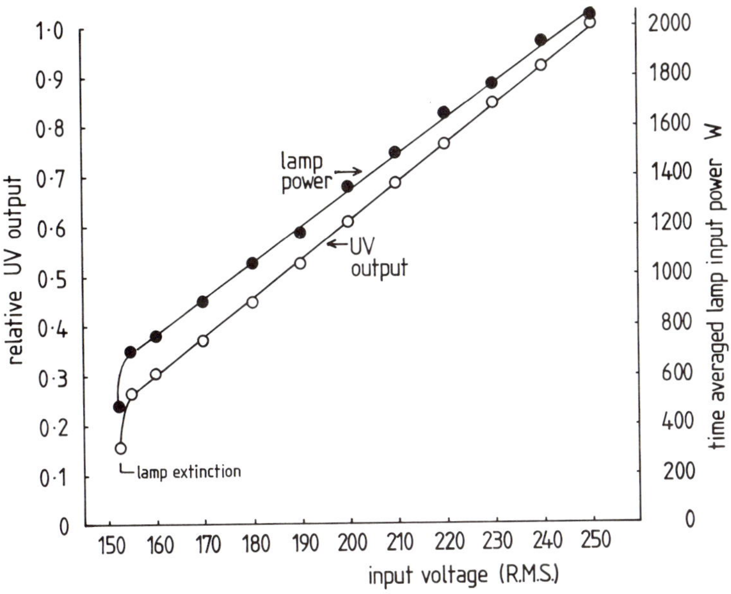

Fig. 8.14. Lamp I. Reactor ballast. Effect of input voltage on lamp power and UV output.

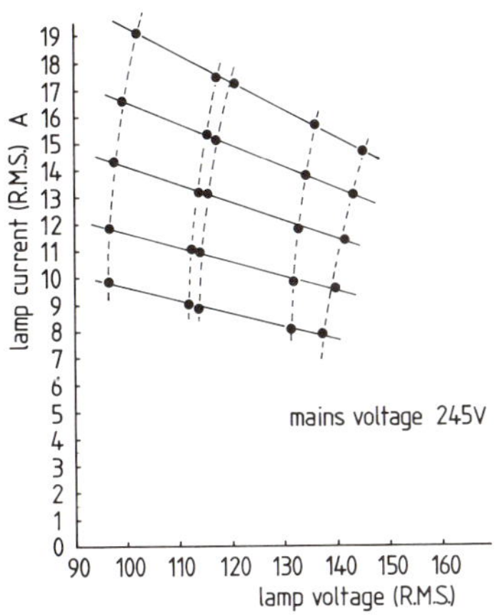

Fig. 8.15. The reactor ballast. Relationship between lamp voltage and lamp current (experimental results). Broken lines join data points for each lamp.

was used. Inductances were 50, 62·5, 75, 87·5 and 100% of a maximum value.

For the lamp power W_l, Hehenkamp gives

$$W_l = \frac{0{\cdot}9V_l}{2\pi fL}\sqrt{V_m^2 - 1{\cdot}11^2V_l^2}.$$

Graphs of $0{\cdot}9V_l\sqrt{V_m^2 - 1{\cdot}11^2V_l^2}$ against V_l for supply voltages of 220 and 240 V are shown in Fig. 8.16. It can be seen that if the lamp voltage is too high a fraction of the supply voltage the lamp power actually falls.

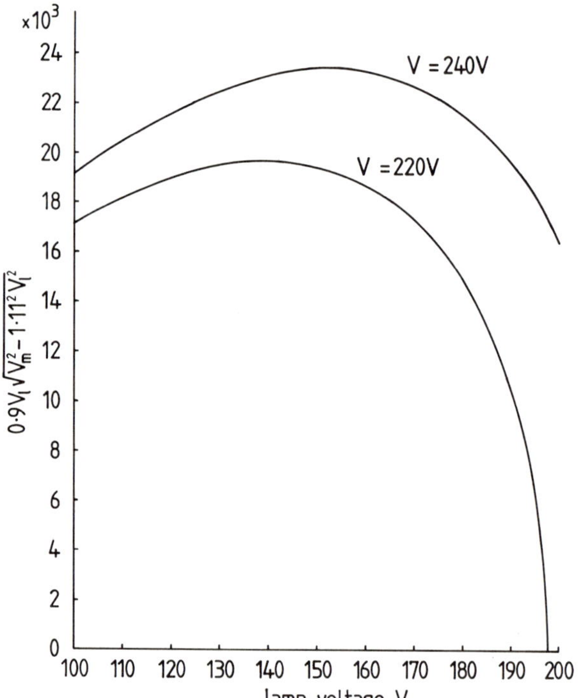

Fig. 8.16. The reactor ballast. Relationship between lamp voltage and lamp power. (Theoretical; after Hehenkamp, 1965.)

Variations in supply voltage have least effect on the lamp power when V_m is much larger than V_l. It can also be seen that over quite a large range of lamp voltage, the lamp power is almost constant. Although a reactor ballast gives poor regulation against variations in supply voltage compared with a constant wattage ballast, it does compensate for variations between individual lamps.

A point of general interest is that most ballast circuits contain series combinations of inductance, capacitance and resistance. The resistance is largely that of the lamp, and varies considerably during the warm-up period and during the mains cycle. There is therefore the possibility of conditions for resonance occurring, and the effects can sometimes be observed in the waveforms.

8.4 Warm-up and Restrike

The cold lamp behaves at first very much like a cold cathode low pressure lamp. Before ignition the lamp has a very high impedance because there are few current carriers present. A high enough open circuit voltage, or alternatively a high voltage pulse from a pulse ignitor, is sufficient to initiate the avalanche effect, and there is a rapid establishment of a low pressure glow discharge which fills the tube between the electrodes and radiates almost entirely at 185·0 nm and 253·7 nm. When the glow is established, the impedance of the lamp is quite low, because the pressure is low, and electrons and ions can move easily in the direction of the field. The glow to arc transition takes place within a few cycles, and visual examination of the lamp shows that the electrodes become incandescent very quickly.

Although the lamp radiates as a low pressure lamp in the early stages, the current, and therefore the power input, is much higher than would be normal for a low pressure lamp, and the result is that the tube heats up and the vapour pressure of the mercury increases. Note that during the early stages there is excess (i.e. liquid) mercury in the tube and the behaviour of the lamp does not reflect the final operating conditions. As the vapour pressure of the mercury increases, the impedance of the lamp increases, and therefore the voltage across the lamp and the power consumed. Conditions which favour rapid warm up are (a) a narrow bore tube, (b) a high lamp current, (c) a high argon pressure, (d) delaying the application of cooling air until the lamp voltage has risen to say one-half or two-thirds of the final value, (e) careful profiling of the tube behind the electrodes to prevent mercury evaporating from the centre of the tube and then condensing behind the electrodes. The current, and therefore the ballast, largely determines how rapidly a particular lamp warms up. As the power consumed by the lamp increases with its temperature, at a certain time after switch on there is a rapid increase in voltage to a maximum value (when all the mercury has evaporated), and a large increase in light output. The lamp volts versus time curves tend to be ∫-shaped. Maximum voltage is approached asymptotically, and the time to reach maximum voltage is somewhat indistinct. One can be more precise about the time taken to reach a certain fraction of the maximum voltage, or of the maximum light output. Figure 8.17 shows the time taken by Lamps I, V, VI and VII to reach half maximum voltage from cold, as a function of the stable operating current, as measured by an Avometer. The figure shows the very large influence of current on the warm-up characteristics.

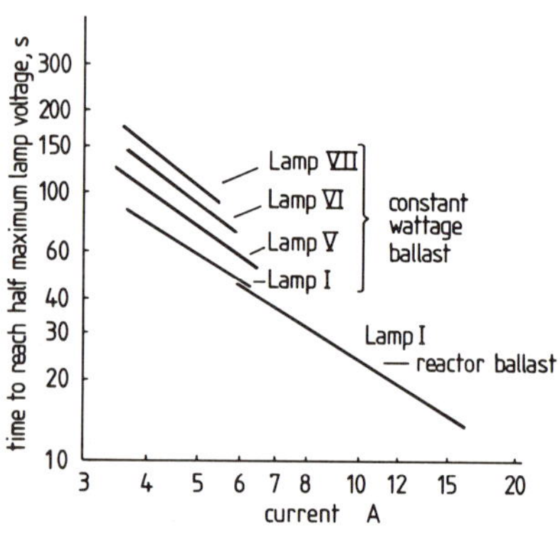

Fig. 8.17. Effect of lamp current on warm-up time.

It is of interest to examine how the lamp current waveform develops during the warm-up. This is shown for Lamp VII, operated from the constant wattage ballast with a series capacitance of 14·5 μF, in Fig. 8.18. Note that at the low mercury pressures during the initial stages of warm-up, the *peak* current is much higher than it is during stable running, but the *mean* current hardly changes. The phase of the current peak shifts during the warm-up, and the current waveform passes through shapes

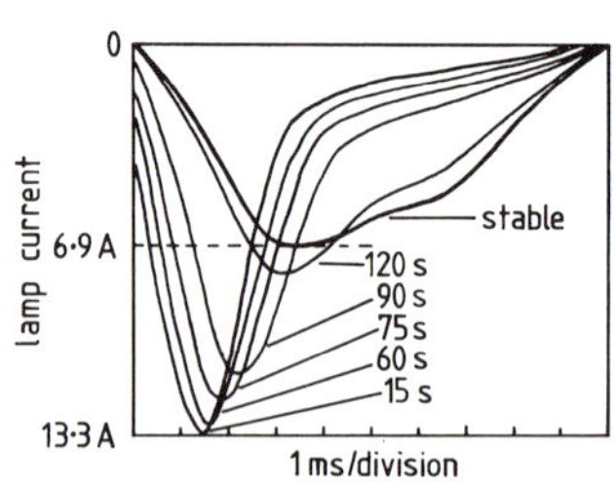

Fig. 8.18. Lamp VII. Constant wattage ballast. Development of lamp current waveform during warm-up.

which correspond to the *stable* operation of lamps with lower doses of mercury.

The waveform for the light output is influenced by the power input waveform, and in turn by the current waveform. Figure 8.19 shows the

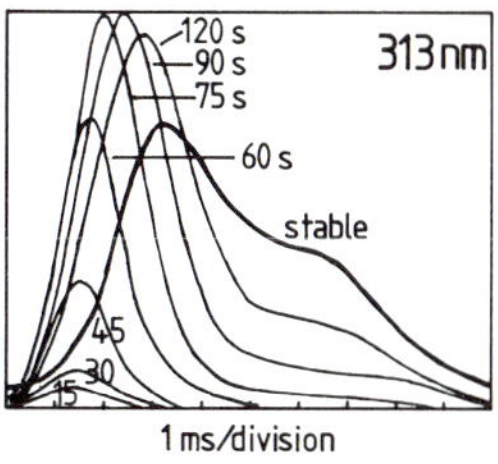

Fig. 8.19. Lamp VII. Constant wattage ballast. Development of light waveform (313 nm) during warm-up.

development of the light output waveform for the 313 nm line during the warm-up period of Lamp VII. The peak light output moves in phase with the current peak, and actually passes through a maximum before stable operation is reached. The light waveform begins as a small symmetric peak, but eventually takes on the asymmetric form of the current and power waveforms.

The warm-up characteristics of a lamp operated with a reactor ballast are rather different from those of the same lamp operated from a constant wattage ballast. The change in current waveform from immediately after ignition to stable operation, for Lamp I, is shown in Fig. 8.20. It can be seen that the mean current falls during warm-up. However, the shape of the waveform does not change greatly. For the same operating current, the

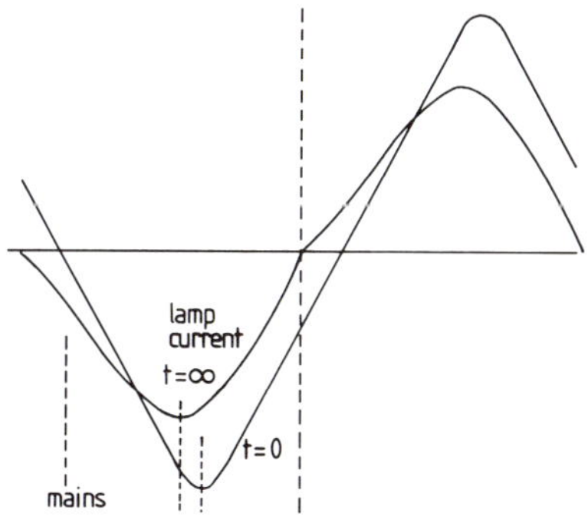

Fig. 8.20. Reactor ballast. Development of lamp current waveform during warm-up.

reactor ballast gives faster warm-up than the constant wattage ballast, and the less peaky waveform is less detrimental to lamp life.

The spectral distribution of the light output varies greatly during the warm-up period and eventually reaches the characteristic multiline spectrum superimposed on a small continuum. The changes which take place in the spectral distribution during warm-up will be further discussed in Section 8.6.5.

When the lamp is switched off, the current carriers (ions and electrons) recombine very quickly, so that within a fraction of a second the impedance of the tube increases to a high value. Under these circumstances it is not usually possible to restart the lamp until it has cooled sufficiently for most of the mercury to condense. The time needed is known as the "restrike time". The restrike time may be very short, particularly when a short lamp is ignited with a high voltage pulse, or may be several minutes. The ignition voltage for a lamp at any stage of cool down is determined by the inter-electrode distance and the mercury pressure (as indicated by the Paschen curve), and therefore the restrike time is determined by the voltage available and the method of cooling the lamp.

The time delay involved in warm-up and restrike is one of the major disadvantages of the medium pressure mercury lamp when electrodes are used for current coupling. Although the energy required to vaporize the mercury fill (only a few milligrams per centimetre of arc length) is very small, the lamp envelope and the electrodes, which have a much higher heat capacity, also have to be warmed up to a temperature above the boiling point of mercury. Even so, if the full rating of the lamp (e.g. $80\,W\,cm^{-1}$) could be absorbed throughout the warm-up period, the lamp could be raised to its operating temperature and full output within a few seconds. The problem lies in the current restriction imposed by the ballast.

High currents and narrow bore lamps allow fast warm-up. A high pressure of argon increases the impedance of the lamp, and therefore the voltage and power consumption immediately after the arc has been struck. External heating and cooling of the lamp has been advocated (DuRon, 1976). A compressed air blast on a small area of the lamp envelope provides a cold spot where mercury can condense, and so reduces the restrike time.

8.5 Arc Contraction

In a low pressure mercury lamp the discharge occupies the entire diameter of the tube. However, visual examination (with appropriate eye protection) of the medium pressure mercury lamp during the warm-up period reveals that, as the mercury pressure in the tube increases, the arc contracts away from the walls and forms a narrow bright cylinder between the electrodes. The contracted arc appears surprisingly well defined in photographs. Figure 8.21 shows the arc of Lamp I.

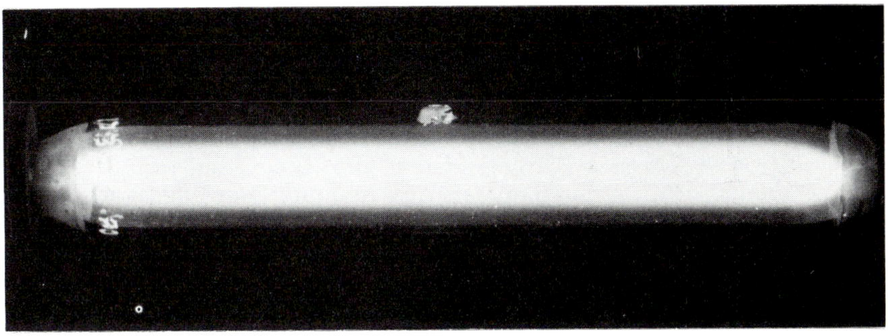

Fig. 8.21. Photograph of arc of Lamp I. The arc fills most of the tube diameter because the current is high (16·8 A r.m.s.) The lamp is horizontal, with the exhaust pip at the top.

The apparent width of the arc increases as the lamp current is increased (Burns and Nelson, 1964; Zollweg, 1979). When a lamp is operated horizontally, the arc may be displaced upwards from the axis of the tube by convection currents (see Section 8.7).

Because the centre of the arc is at about 6000 K, and the tube is at about 1000 K, there must be a radial temperature distribution. The form of the Boltzmann distribution law implies that as the plasma temperature drops from its central high value as one moves from the axis to the wall, the radiative output drops dramatically. Therefore almost all the radiation is emitted from the central portion of the arc, and a dark annulus surrounds the radiative column.

The concentration of electrons and ions in a plasma in thermodynamic equilibrium can be calculated from the Saha equation:

$$\frac{N_i N_e}{N_0} = 2\frac{g_i}{g_0}\left(\frac{2\pi m_e kT}{h^2}\right)^{3/2} \exp\left(-\frac{E_i}{kT}\right),$$

where N_i is the density of ions, N_e the density of electrons, N_0 the density of neutral atoms, g_i the statistical weight of the first ionization level, g_0 the statistical weight of neutral atoms, E_i the ionization energy for the first ionization level. For mercury, $g_i/g_0 = 2$. m_e is the mass of an electron.

The Saha equation shows that the degree of ionization, and therefore the concentration of current carriers, increases rapidly with increase in temperature. Hence the current density is highest at the axis of the arc, and this is where most power is absorbed from the field. Arc contraction is therefore a natural consequence of the radial temperature distribution, which itself follows from the distribution of current carriers.

The variation in temperature within the arc causes the radiation from different points in the arc to vary in spectral distribution.

The relationship between the temperature T and the radial distance from the centre of the tube, r, defines the radial temperature profile. If the axis of the arc coincides with the axis of the tube, then the gradient $\mathrm{d}T/\mathrm{d}r$ is always negative. In the medium pressure mercury lamp, the power input into any volume element is always greater than the radiative loss from that element, so that there is always a net production of heat (Waymouth, 1971, p. 174). This leads to an approximately parabolic radial temperature profile, with $\mathrm{d}^2T/\mathrm{d}r^2$ always being negative. This provides the condition for a stable arc (Zollweg *et al.*, 1975). In some metal halide arcs, $\mathrm{d}^2T/\mathrm{d}r^2$ may be positive, and this leads to "arc constriction", and an arc which does not remain stationary, but "snakes". Arc constriction will be further discussed in Chapter 9.

Elenbaas (1951, p. 24) suggested a simple model for the mercury arc which has proved to be very useful. The arc is considered to be a hot cylindrical core at a uniform, *effective temperature*, and having a well defined *effective radius*. The hot core is surrounded by gas at the temperature of the tube wall. All the current passes through the hot core, and none passes through the dark annulus. Similarly, all the radiation is emitted from the hot core. This model allowed Elenbaas, by rather complex reasoning, to derive an expression for the axial electric field. This is expression (8.1), given in Section 8.3.1.

8.6 The Spectral Distribution

The medium pressure mercury lamp operates in a region where local thermodynamic equilibrium can be assumed, and where the mercury atom temperature in the centre of the arc is near to a maximum value (Fig. 8.1). The optimum pressure for maximum plasma temperature depends on the diameter of the tube, d (in centimetres), and the current, I (in amps), and has been given by Elenbaas (1951, p. 7) as

$$\boxed{P_{\mathrm{m}} = 3d^{4/7}\, I^{-12/7} \quad \text{atm}} \tag{8.3}$$

If we take the example of a tube of 1·8 cm diameter, a current of 5 A gives a value for P_m of 0·27 atm, while a current of 10 A gives a value of 0·081 atm. These figures would appear to indicate that in general, with the medium pressure mercury lamp, we are working at pressures above that which would give the maximum plasma temperature.

The plasma temperature is in the region of 6000 K, and therefore kT is in the region of $8{\cdot}3 \times 10^{-20}$ J. One electron-volt is equivalent to $1{\cdot}6 \times 10^{-19}$ J, and therefore at 6000 K, kT is a substantial fraction of the differences in energies between excited and ground state levels in the mercury atom. Therefore, a small but significant fraction of the mercury atoms in the arc are in energy levels other than the ground state. If thermodynamic equilibrium is assumed, the Boltzmann distribution law may be applied to determine the ratio of the numbers of atoms occupying the two energy states ε_2 and ε_1:

$$N_2/N_1 = (g_2/g_1)\ e^{-(\varepsilon_2 - \varepsilon_1)/kT}.$$

g is the statistical weight of the state, and is equal to the number of distinct sub-states having the same energy, i.e. the "degeneracy". For a many electron atom, in the absence of an external magnetic field, g for a particular state is given by $2J + 1$ (Thorne, 1974, p. 91).

If ε_1 is taken as the ground state, the concentration of atoms in the various excited states shown in the Grotrian diagram for the mercury atom (Fig. 8.22) can be calculated if the plasma temperature is known. With known transition probabilities the rate of spontaneous photon emission at each wavelength can be calculated. However, the spectral distribution actually measured may not agree with this calculation, because some of the photons emitted will be absorbed by other mercury atoms and not escape the tube. While *emission* is proportional to the concentration of atoms in the upper state of the transition, *absorption* is proportional to the concentration of atoms in the lower state of the transition. Self-absorption is therefore most pronounced for the resonance lines (i.e. for transitions where the lower state is the ground state).

If we adopt the "two-temperature" model of the discharge (Section 8.5) it can be shown (Gallo, 1966) that the total power radiated in a spectral line is given by

$$P_\lambda = G_\lambda(g_2/g_0)N_0 \exp\,(-\ \varepsilon_2/kT_{eff})(hc/\lambda)\ AL\pi R_{eff}^2. \qquad (8.4)$$

G_λ is the escape factor for self-absorbed radiation, and is unity when no self-absorption takes place. N_0 is the number of ground state atoms confined within the effective radiating volume, which has a length L and a

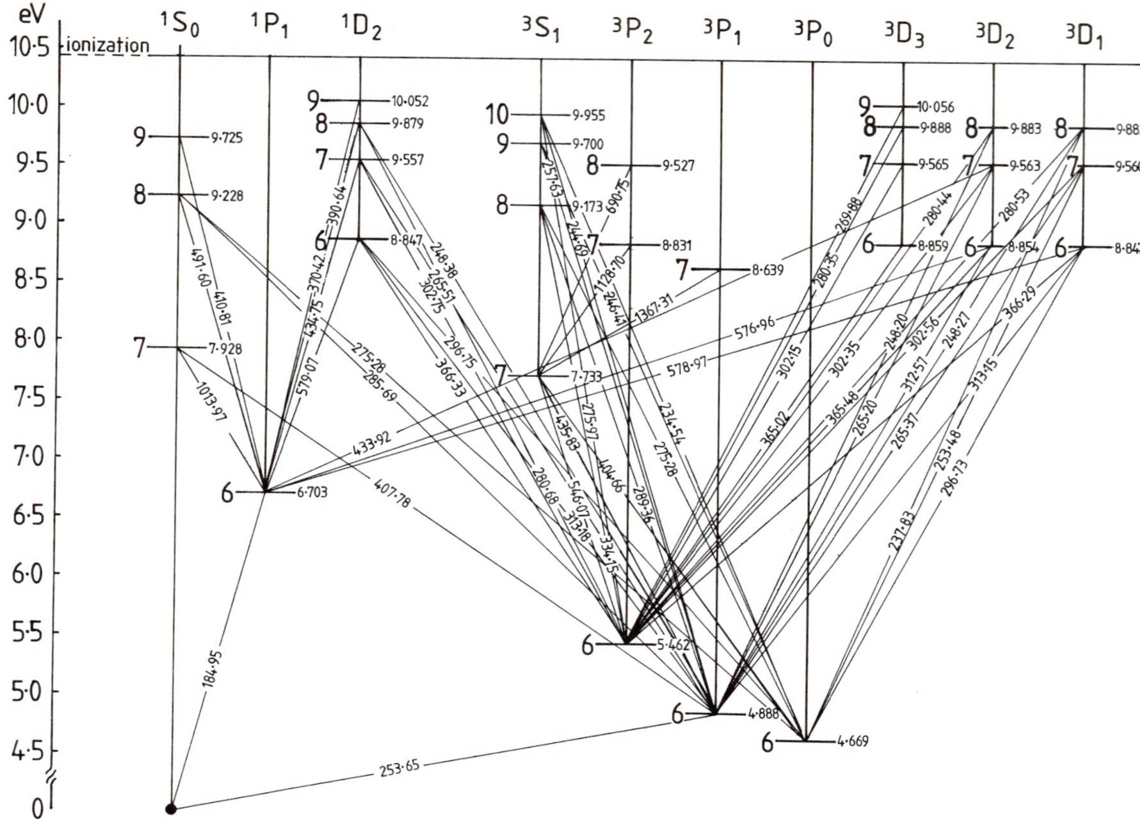

Fig. 8.22. The Grotrian diagram for atomic mercury. Transitions taken from The American Institute of Physics Handbook.

radius R_{eff}. g_2 is the statistical weight of the upper energy level. g_0 is the statistical weight of the ground state. ε_2 is the upper state energy. T_{eff} is the effective plasma temperature. A is the transition probability, or Einstein coefficient, for spontaneous emission.

Examination of the Grotrian diagram in Fig. 8.22 shows that the significant lines in the UV and visible regions result from transitions which terminate (1) in the ground state, or (2) in the 6^3P levels, or (3) in the 6^1P_1 level. The 185·0 and 253·7 nm lines are strongly self-absorbed. The many lines terminating in the 6^3P levels show moderate self-absorption, while those terminating in the 6^1P_1 level show little sign of self-absorption (Elenbaas, 1951, p. 16; Rautenberg and Johnson, 1964). It is difficult to predict spectral distributions because of uncertainties about values of G_λ. However, it is commonplace to assume that the two yellow lines of mercury (577·0 and 579·0 nm) have escape factors of unity, and therefore if their absolute intensity is known it is possible to calculate T_{eff}. Values of g_2A have been given by Corliss and Bozman (1962) and Mosburg and Wilke (1978).

It is important to realize that self-absorption of non-resonance radiation has little effect on the *total amount* of power radiated, but does have a strong influence on the *distribution* of line radiation.

The spectral output of a medium pressure mercury lamp can be varied by alteration of (a) the number of atoms in the arc, (b) the plasma temperature and (c) the spectral transmission of the tube. It is important to bear in mind that because of arc contraction the number of atoms which actually form the arc is only a small fraction of the total number of atoms in the tube, about $7\frac{1}{2}\%$, according to Elenbaas (1951, p. 60). Also, there is of course a radial temperature profile with a peak at or near the axis of the tube (Section 8.5). The spectral distribution of the light emitted will vary according to the part of the arc from which the light originates. The spectral distribution of the entire light output from the arc will reflect an "average" plasma temperature (i.e. T_{eff}) somewhat less than the peak value.

8.6.1 The region between 200 nm and 300 nm

The spectral distribution between 200 and 300 nm is of particular interest. Spectral distributions for this region for Lamps V, VI and VII, with conditions chosen so that the power consumed per unit arc length was about the same for all three lamps, are shown in Fig. 8.23. The experimental details for these spectra and for the others which will be presented in this chapter, may be found in Chapter 4. Points to note from

Fig. 8.23 are (1) the *broad hump which extends from below 200 nm to about 245 nm*, the shape of this depends on the pressure—as the pressure increases the hump narrows and the peak increases; (2) the *large broad asymmetric peak* extending between about 250 and 270 nm, with a small dip at 254 nm, the shape of this peak changes with pressure; (3) the *multitude of lines* superimposed on a small continuum, the continuum:line ratio increasing with pressure; (4) the area under the large broad asymmetric peak tends to increase with pressure, whereas the areas under the other peaks tend to decrease; (5) the *total* radiation between 200 and 300 nm varies little with pressure, but there are significant variations in spectral distribution.

8.6.2 *The short wavelength continuum*

(See also Section 8.6.8.)

The broad hump below 245 nm is something of a mystery. A tentative suggestion is that it is a result of ion–electron recombination:

$$Hg^+ + e \longrightarrow Hg^* + h\nu$$

Radiation produced in this way would be in the form of a continuum, because the kinetic energy of the participants in the collision, and the neutral atom produced, could vary continuously. If the excited state of mercury produced were one of the 6^3P levels, $h\nu$ would appear in the 200–245 nm region. The variation of the output in this region with the lamp current is consistent with this hypothesis.

Whereas in the low pressure discharge the recombination takes place mainly and non-radiatively at the wall, in the medium pressure discharge it takes place in the interior of the arc (Bleekrode *et al.*, 1975).

8.6.3 *The resonance transition* $6^3P_1 \longrightarrow 6^1S_0$

The "large broad asymmetric peak" is largely due to the resonance transition $6^3P_1 \longrightarrow 6^1S_0$, which is also responsible for almost the entire output of the low pressure mercury lamp. When the lamp is first switched on, the initial low pressure arc gives a sharp line at 253·7 nm, which actually decreases in intensity from an initially high value through a minimum in the intermediate pressure range. During the period of rapid voltage rise and pressure increase, the 253·7 nm line broadens and becomes asymmetric, the ultimate shape depending on the ultimate pressure in the

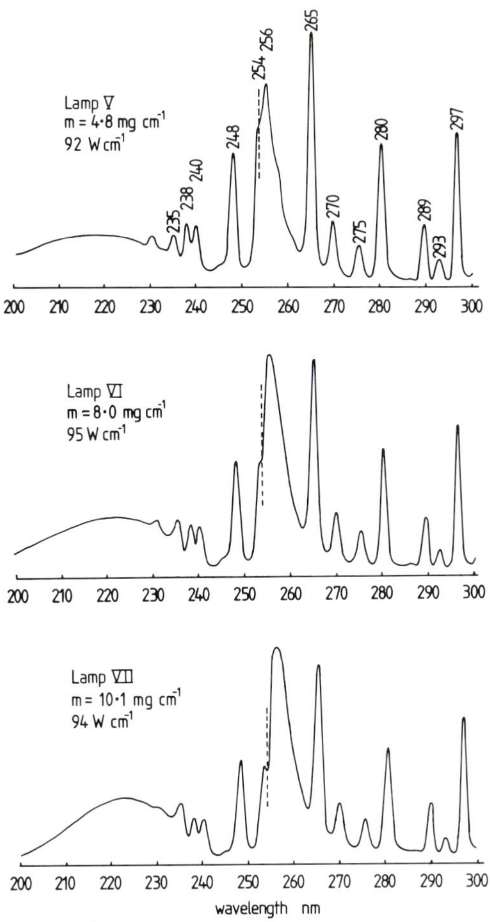

Fig. 8.23. Lamps V, VI and VII. Constant wattage ballast. Spectral output between 200 and 300 nm, showing influence of mercury pressure.

lamp. The wavelength of maximum output shifts to a longer wavelength, and a dip in output is observed at 253·7 nm. This is known as "reversed radiation" and is caused by self-absorption by the cool mercury around the arc. The broadening and asymmetry of the resonance peak is caused by "pressure broadening" (discussed later). It can be seen from Fig. 8.23 that the shifting and broadening of the resonance peak, and also the degree of reversal, increases with pressure. In fact the shape of the spectrum in this region is quite a good diagnostic tool for assessing the pressure in the lamp. Figure 8.24 shows the 246–272 nm region for Lamp II. It can be seen that while there is some asymmetry in the resonance peak, the peak wavelength

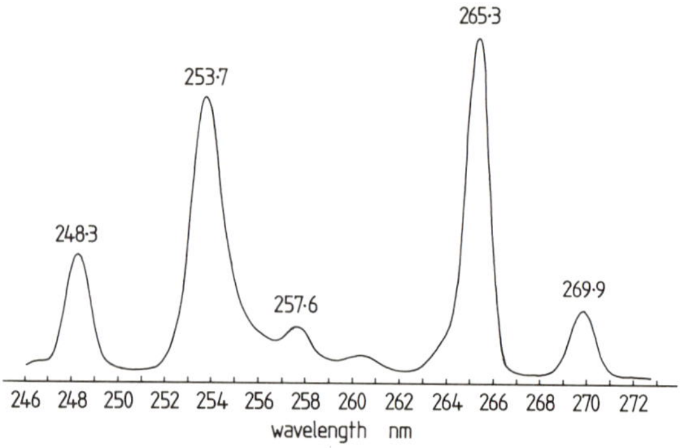

Fig. 8.24. 246–272 nm region for Lamp II.

is still at 253·7 nm, and while there is no doubt self-absorption, reversal is not observed (at this resolution). The 257·6 nm line, submerged under the broadened resonance peaks in Fig. 8.23, is now visible.

8.6.4 The line spectrum

Spectral distributions were measured for Lamps II, III, IV, V, VI and VII, each at different power input levels, for the region between 200 and 600 nm. Lamps III and IV were operated with a reactor ballast. Power input was varied both by varying the inductance in series with the lamp and by varying the supply voltage between 200 and 250 V. Lamps V, VI and VII were operated with a constant wattage ballast, and the power input was varied by changing the series capacitance. Lamp II was operated from both types of ballast to investigate the effects of the power waveform. The experimental arrangement shown in Fig. 4.23 was used for the measurement of spectral distribution.

Figure 8.25 shows spectral distributions for Lamps V and VII both operated with the same series capacitance. This shows how an increase in mercury pressure increases the power consumption and output, and alters the spectral distribution. Figure 8.26 shows spectral distributions (in the form of computer-generated 1 nm histograms) for Lamp V with different series capacitances. The same data are expressed in a cumulative form in Fig. 8.27. The output, in watts per centimetre of arc, for each of the significant peaks between 248 and 579 nm, for various combinations of lamp (V, VI or VII) and power input (again in watts per centimetre of arc), is shown in Table 8.4.

A peak sometimes represents a group of closely spaced lines which could

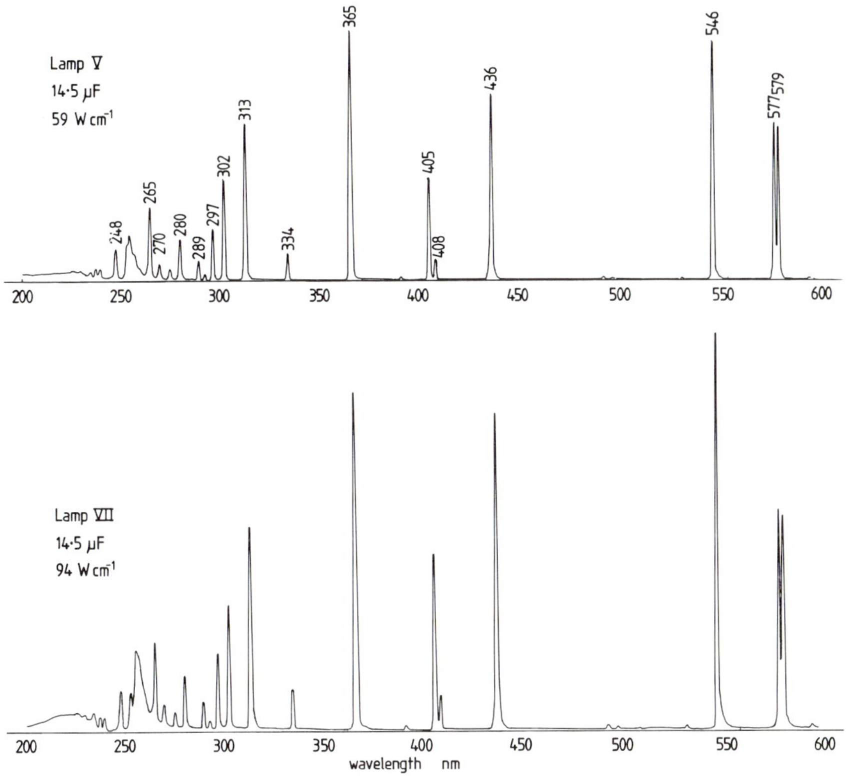

Fig. 8.25. Lamps V and VII. Constant wattage ballast. Spectral output between 200 and 600 nm.

not be resolved by the monochromator. Output in a peak was calculated from the area under the peak on the chart recording, rather than from its height. The dispersion of a grating monochromator is independent of wavelength, but there were significant variations in the widths of the peaks recorded, either because a peak represents more than one line, or because of pressure broadening, often resulting in a long wavelength "tail". Pressure broadening (see Section 8.6.7) was found to be more common for lines arising from transitions with low upper state energies, and in particular the visible lines from 7S.

The observant reader may notice some small discrepancies in the data presented. The distributions shown in Fig. 8.23 and in Table 8.4 were measured when the lamps were new, while those in Figs 8.25, 8.26 and 8.27 were obtained after some ageing of the lamps had taken place, producing small differences in output at short wavelengths (see Section 8.8).

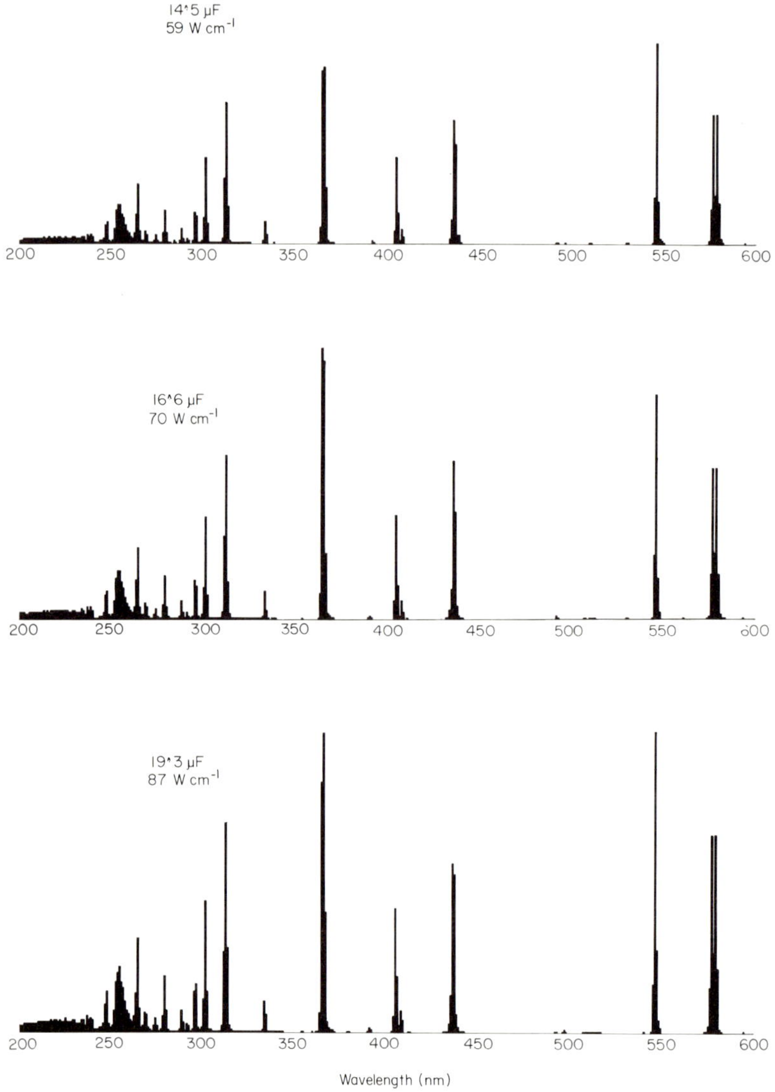

Fig. 8.26. Lamp V. Constant wattage ballast. Spectral output between 200 and 600 nm. Computer-generated 1 nm histograms. Effect of input power.

The data in Table 8.4 indicate that for Lamps V, VI and VII, changing the power input over the range studied has very little effect on the spectral *distribution*, despite a large effect on the total output. This is also shown in Fig. 8.27, where the three cumulative spectra are virtually identical.

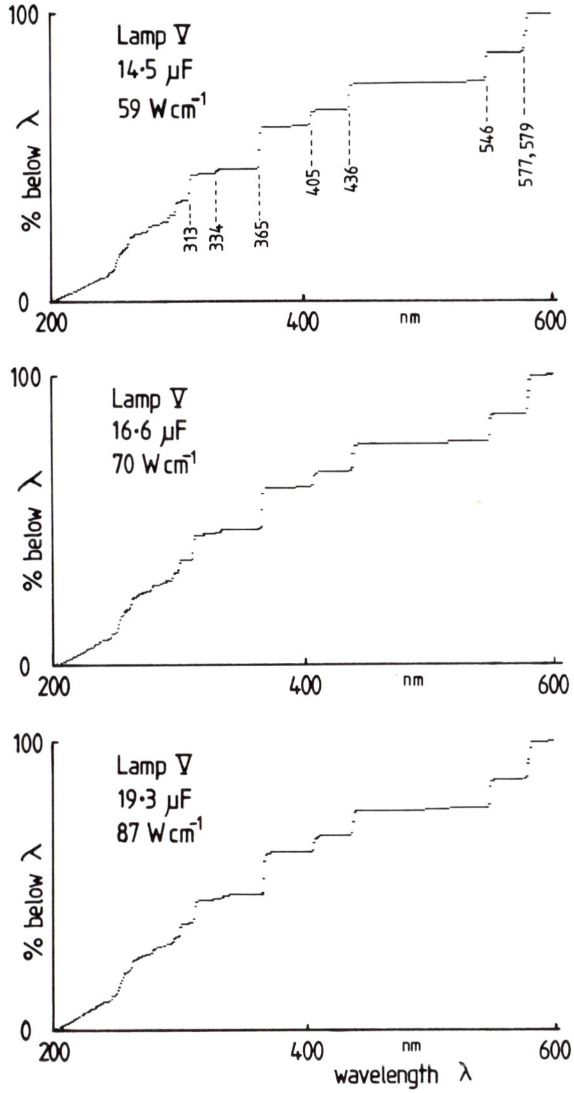

Fig. 8.27. Lamp V. Constant wattage ballast. Cumulative display of spectral output. Effect of input power.

Graphs were drawn of the total UV line output (between 248 and 365 nm) for each of the lamps against the power input. Straight lines were obtained, all with an intercept of about 10 W cm^{-1} on the power axis. At any particular power the total UV line output decreased slightly with

Table 8.4 *Output (in watts per centimetre) in significant lines. Influence of mercury pressure and input power*

	Lamp V						Lamp VI					Lamp VII			
Series capacitance (μF):	12·2	14·5	16·6	18·2	19·3	20·3	12·2	14·5	16·6	18·2	19·3	12·2	14·5	16·6	18·2
Arc power ($W\,cm^{-1}$):†	48	59	70	78	87	92	63	82	95	107	118	71	94	112	126
Wavelength (nm)															
248·3	0·38	0·50	0·60	0·72	0·79	0·88	0·43	0·59	0·74	0·85	0·96	0·48	0·64	0·79	0·91
253–265‡	2·05	2·80	3·50	3·97	4·39	4·71	3·22	4·59	5·58	6·40	7·34	4·43	6·07	7·38	8·32
265·3	0·80	1·05	1·27	1·46	1·54	1·66	0·90	1·15	1·37	1·54	1·71	0·94	1·17	1·40	1·58
269·9	0·16	0·21	0·28	0·31	0·33	0·36	0·19	0·25	0·31	0·35	0·41	0·21	0·27	0·37	0·40
275·3	0·11	0·13	0·16	0·19	0·21	0·24	0·14	0·18	0·23	0·24	0·27	0·16	0·18	0·25	0·30
280·3	0·40	0·55	0·68	0·80	0·88	0·96	0·49	0·66	0·82	0·94	1·07	0·54	0·73	0·89	1·06
289·4	0·18	0·23	0·24	0·32	0·35	0·37	0·27	0·29	0·36	0·43	0·48	0·25	0·35	0·42	0·49
292·5	0·06	0·07	0·10	0·12	0·14	0·14	0·07	0·11	0·12	0·14	0·16	0·07	0·11	0·13	0·16
296·7	0·47	0·61	0·74	0·83	0·92	0·98	0·59	0·77	0·94	1·07	1·21	0·67	0·91	1·10	1·27
302·3	0·98	1·31	1·58	1·79	1·97	2·10	1·16	1·50	1·81	2·00	2·27	1·25	1·61	1·94	2·24
313	1·76	2·27	2·73	3·03	3·32	3·57	2·10	2·65	3·19	3·61	4·04	2·29	2·97	3·55	4·08
334·1	0·26	0·29	0·36	0·39	0·45	0·49	0·30	0·37	0·47	0·53	0·60	0·35	0·44	0·55	0·63
365	2·70	3·50	4·17	4·67	5·10	5·43	3·34	4·24	5·18	5·77	6·41	3·71	4·80	5·85	6·58
Σ UV	10·31	13·52	16·41	18·60	20·39	21·89	13·20	17·35	21·12	23·87	26·93	15·35	20·25	24·62	28·02
404·7	0·85	1·09	1·29	1·46	1·62	1·71	1·17	1·52	1·87	2·10	2·35	1·40	1·89	2·32	2·61
407·8	0·15	0·21	0·24	0·27	0·30	0·34	0·21	0·27	0·34	0·40	0·46	0·26	0·35	0·43	0·50
435·8	1·64	2·10	2·51	2·81	3·07	3·29	2·29	2·96	3·65	4·12	4·53	2·73	3·68	4·48	5·08
546·1	1·66	2·11	2·50	2·81	3·06	3·27	2·34	3·05	3·69	4·10	4·47	2·78	3·71	4·51	5·00
577·0	1·02	1·36	1·66	1·91	2·12	2·30	1·36	1·81	2·23	2·56	2·84	1·49	2·01	2·50	2·86
579·0	0·99	1·31	1·60	1·84	2·07	2·21	1·30	1·74	2·16	2·46	2·73	1·44	1·94	2·41	2·75
Σ Visible	6·31	8·18	9·80	11·10	12·24	13·12	8·67	11·35	13·94	15·74	17·38	10·10	13·58	16·65	18·80
Ratio UV : Visible	1·63	1·65	1·67	1·68	1·67	1·67	1·52	1·53	1·52	1·52	1·55	1·52	1·49	1·48	1·49

† Note that a correction has been made for electrode losses so that this figure represents the power consumed in the positive column. The cathode + anode falls were assumed to be 15 V for the purposes of this correction.

‡ The power emitted in the pressure broadened $6^3P_1 \rightarrow 6^1S_0$ transition. Some authors consider this to be part of the continuum.

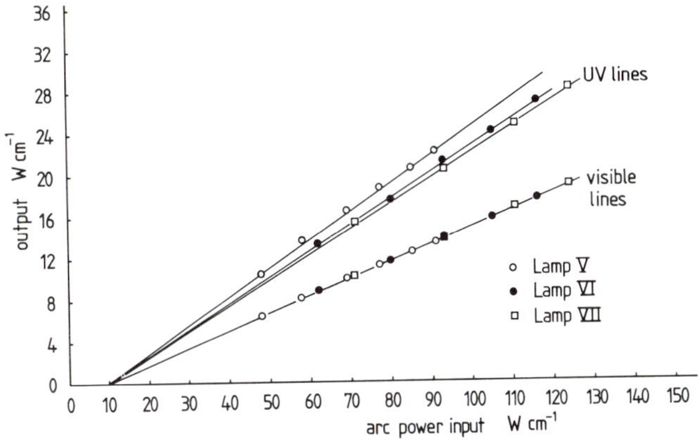

Fig. 8.28. Lamps V, VI and VII. Constant wattage ballast. UV and visible line output as a function of input power.

increasing mercury fill. Similar graphs were drawn for the total visible lines, and here it was found that the results for all three lamps fell on the same straight line, with the same intercept of about 10 W cm^{-1}. The graphs are shown in Fig. 8.28. These may be compared with results obtained with the "Light Bug" (see Section 4.4), which measures radiation in the region between about 250 and 400 nm. Again straight lines were obtained, with the same intercept on the power axis (Fig. 8.29).

The 10 W cm^{-1} intercept is of great interest, and corresponds to the conductive loss from the arc to the lamp envelope, which Elenbaas (1951,

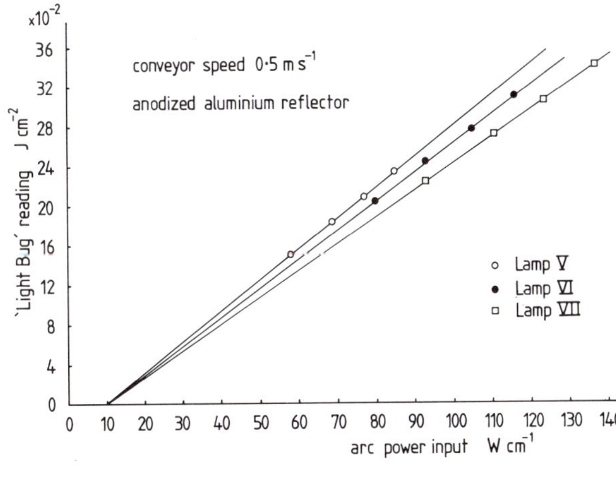

Fig. 8.29. Lamps V, VI and VII. Constant wattage ballast. "Light Bug" reading as a function of input power.

p. 20) found to be largely independent of mercury pressure, lamp bore and power input. The constancy of the conductive loss causes the radiative efficiency for both the UV and visible parts of the spectrum to increase with power input (see Section 8.7).

Whereas variation of power input had little effect on the spectral distributions measured for Lamps V, VI and VII, this was not the case with Lamps II, III and IV. Operation of Lamp II, with either the constant wattage ballast or the reactor ballast, showed that with this lamp, variation of the lamp current and power input caused quite large changes in spectral distribution. It was also found that for a particular *mean* current, or *mean* power input, the spectral distribution measured with the constant wattage ballast was significantly different from that measured with the reactor ballast. However, Fig. 8.12 shows that for a particular *mean* power input, the constant wattage ballast provides a much higher *peak* power input than the reactor ballast, for this rather low pressure lamp. It was found that there was an excellent correlation between the spectral distribution and the peak power input to the lamp.

A spectral distribution for Lamp II operating from the reactor ballast is shown in Fig. 8.30. Spectral distribution is largely determined by T_{eff}. The

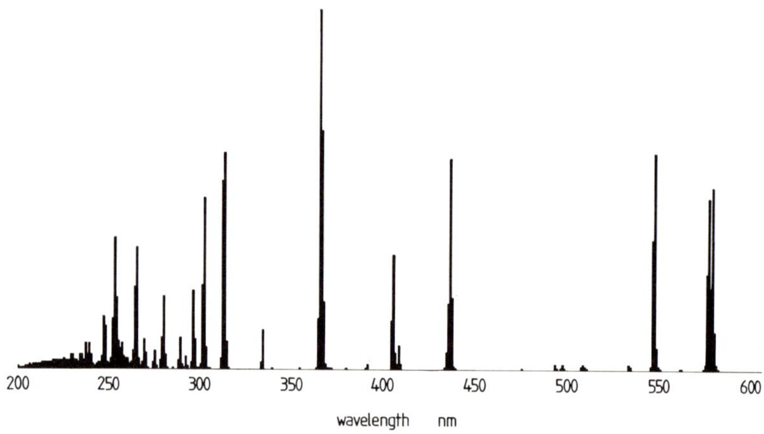

Fig. 8.30. Lamp II. Reactor ballast. Spectral output between 200 and 600 nm. Computer-generated 1 nm histogram. Arc input power 98 W cm^{-1}.

calculation of T_{eff} from line intensities poses several difficulties. Even if a line can be chosen for which G_λ can be confidently taken as unity, the choice of values for R_{eff} and N_0 is somewhat arbitary. However, a useful indication of *comparative* values of plasma temperature for different discharges may be easily obtained by comparing outputs in two peaks which are derived from transitions with significantly different upper state energies. Two convenient peaks are those at 302·3 nm (upper state energy 9·56 eV) and at 404·7 nm (upper state energy 7·73 eV). The ratio of the output at 302·3 nm to that at 404·7 nm should increase with plasma temperature. This simple method has a shortcoming in that if the value of G_λ is different for the two lines, the ratio may vary with pressure as well as with plasma temperature. However, it will be seen that this method does give sensible and instructive results. Gallo (1967) has suggested a somewhat similar procedure, comparing the sum of the 577·0 and 579·0 nm lines with the 546·1 nm line.

Values of the 302·3 : 404·7 nm ratio for Lamp II for both reactor ballast and constant wattage ballast are shown in Fig. 8.31 as a function of peak

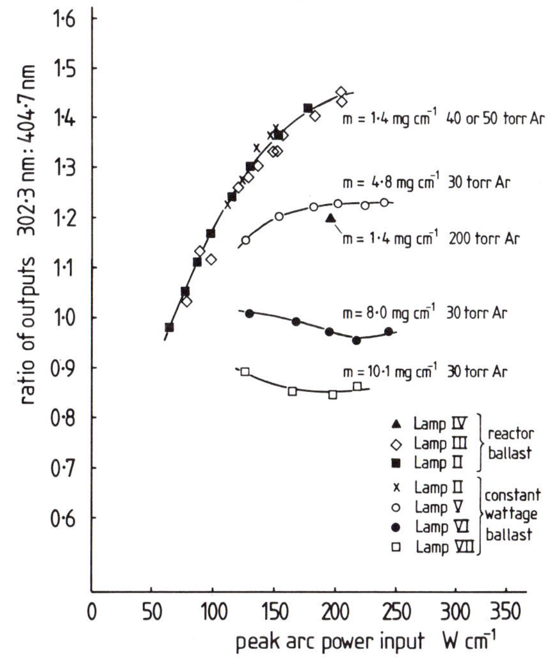

Fig. 8.31. Relationship between 302·3 nm : 404·7 nm ratio and peak arc power input. Effects of mercury pressure and argon pressure.

arc power input. Values are also shown for measurements with the other lamps. It appears that an increase in power for Lamps V, VI and VII brings about an increase in R_{eff} rather than an increase in T_{eff}, while the reverse is true for Lamps II and III, with the lower mercury pressure. It can be seen that increasing the argon pressure reduces the plasma temperature slightly.

A detailed comparison of the spectral outputs of Lamps III and IV showed that for the same input power Lamp IV with the high argon pressure had about 5% less output in the region 200–600 nm. However, the outputs in some lines actually increased, and it appears that a high argon pressure may modify the radial temperature distribution to give a larger value of R_{eff} and a lower value of T_{eff}.

When a lamp is operated from an a.c. supply with a capacitative ballast and a rather low mercury pressure, the light output waveform consists of quite narrow pulses which appear at twice the supply frequency. With the constant wattage ballast, Lamp II gives a light pulse with a half peak width of only about 2 ms, which corresponds roughly to the half peak width of the power input pulse (Fig. 8.12). It is therefore not surprising that as substantial light appears during only a small part of the half cycle, its spectral distribution is determined by the value of the power during the power input pulse rather than the value averaged over the entire half cycle. An extreme example of this general phenomenon is the pulsed xenon lamp, which has a spectral distribution quite different from that of a continuously operating xenon lamp with the same time averaged power input.

The efficiencies with which the more significant lines are generated are compared for Lamps V and VII in Table 8.5. This table also details the transitions which give rise to each of the lines and the associated upper state energy. It can be seen that there is a relationship between the efficiency ratio and the upper state energy, and this relationship is shown in graphical form in Fig. 8.32.

Reducing the mercury pressure increases the plasma temperature, but on the other hand reduces the number of atoms present in the arc. Increasing the plasma temperature increases the population of all excited states, but the form of the Boltzmann distribution law dictates that the higher excited states will experience a higher proportional increase in

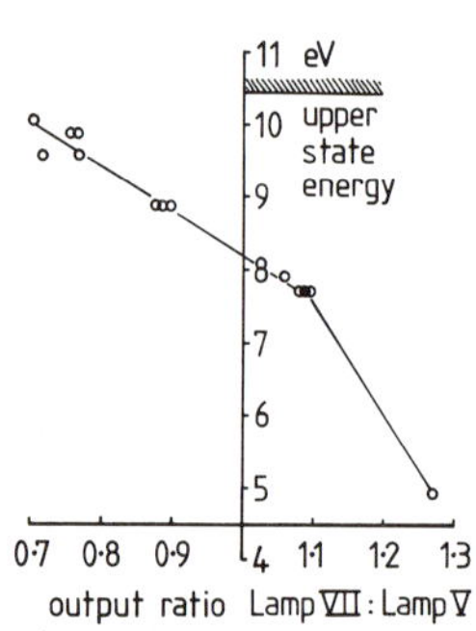

Fig. 8.32. Graphical presentation of data in Table 8.5.

Table 8.5 Comparison of Lamps V and VII. Ratios of efficiency of output for the more significant spectral peaks. Influence of upper state energy

Wavelength (nm)	Transition(s)	Upper state energy (eV)	Lamp VII : Lamp V
248·3	$8^1D_2 \rightarrow 6^3P_1$ $8^3D_2 \rightarrow 6^3P_1$ $8^3D_1 \rightarrow 6^3P_1$	9·879 9·883 9·882	0·76
253–260	$6^3P_1 \rightarrow 6^1S_0$	4·888	1·27
265·3	$7^1D_2 \rightarrow 6^3P_1$ $7^3D_2 \rightarrow 6^3P_1$ $7^3D_1 \rightarrow 6^3P_1$	9·557 9·563 9·560	0·72
269·9	$9^3D_3 \rightarrow 6^3P_2$	10·056	0·71
280·3	$8^3D_3 \rightarrow 6^3P_2$	9·888	0·77
296·7	$6^3D_1 \rightarrow 6^3P_0$	8·847	0·89
302·3	$7^3D_3 \rightarrow 6^3P_2$ $7^3D_2 \rightarrow 6^3P_2$ $7^3D_1 \rightarrow 6^3P_2$	9·565 9·563 9·560	0·77
313	$6^3D_2 \rightarrow 6^3P_1$ $6^3D_1 \rightarrow 6^3P_1$	8·854 8·847	0·90
365	$6^1D_2 \rightarrow 6^3P_2$ $6^3D_3 \rightarrow 6^3P_2$ $6^3D_2 \rightarrow 6^3P_2$ $6^3D_1 \rightarrow 6^3P_2$	8·847 8·859 8·854 8·847	0·90
404·7	$7^3S_1 \rightarrow 6^3P_0$	7·733	1·09
407·8	$7^1S_0 \rightarrow 6^3P_1$	7·928	1·06
435·8	$7^3S_1 \rightarrow 6^3P_1$	7·733	1·08
546·1	$7^3S_1 \rightarrow 6^3P_2$	7·733	1·10
577·0	$6^3D_2 \rightarrow 6^1P_1$	8·854	0·88
579·0	$6^1D_2 \rightarrow 6^1P_1$ $6^3D_1 \rightarrow 6^1P_1$	8·847 8·847	0·88

Transitions taken from the American Institute of Physics Handbook.

population than the lower excited states, and therefore the spectral distribution will shift in favour of lines given by transitions with higher upper state energies.

8.6.5 Spectral changes during warm-up

The decrease in plasma temperature with increasing pressure is in agreement with the form of Fig. 8.1. It is expected that the plasma temperature will pass through a maximum value during the lamp warm-up period. This can be neatly demonstrated by following the development of different wavelengths during the warm-up period. Figure 8.33 compares the development of the 302·3 nm peak with that of the 404·7 nm peak for Lamp V operated with 14·5 μF series capacitance. For each peak the percentage of

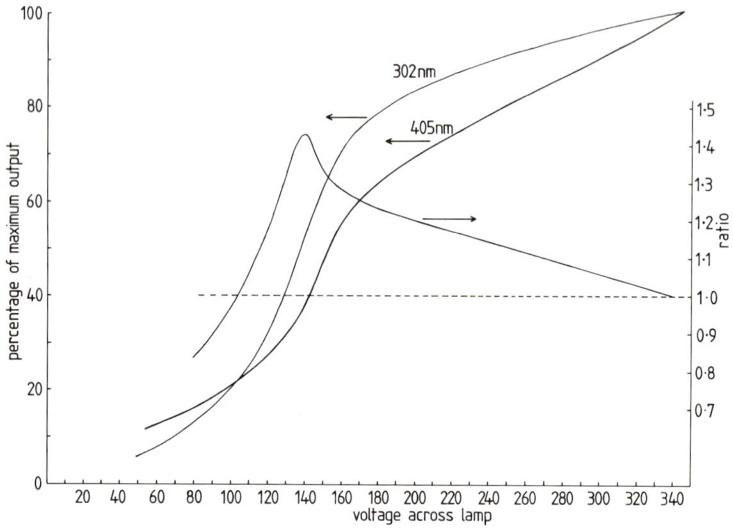

Fig. 8.33. Lamp V. Constant wattage ballast. Development of 302·3 nm and 404·7 nm output during warm-up. Percentage of maximum output plotted against lamp voltage.

maximum output is plotted against the voltage developed across the lamp. The figure also shows how the ratio of the two peaks changes during the warm-up period. It can be seen that the ratio passes through a maximum value when about 140 V (Avometer reading) has developed across the lamp, and it is not unreasonable to assume that at this voltage the plasma temperature passes through a maximum. If one refers back to Fig. 8.9 one can see that the situation of maximum plasma temperature corresponds to a value of $(P^{1/2}/(P - 4{\cdot}5P^{1/4})^{1/3})\, m^{7/12} d^{-3/2}$ of about 0·84 $W^{1/6}\, mg^{7/12}\, cm^{-9/4}$. At this point in the warm-up the value of P is about 20·2 W cm^{-1}. It is now possible to calculate m, and it appears that at the point of maximum plasma temperature there is about 0·99 mg of evaporated mercury per centimetre of arc length.

Elenbaas (1951, p. 46) has given an expression to relate the pressure developed in a mercury lamp to the amount of evaporated mercury, m (in milligrams per centimetre of arc length), and the tube diameter, d:

$$\text{Pressure} = 1{\cdot}25\, md^{-2} \qquad \text{atm.}$$

Use of this expression indicates that the pressure which corresponds to the maximum plasma temperature is about 0·38 atm. The value predicted from expression (8.3) is 0·40 atm, in quite pleasing agreement. It is of interest that Kenty and Karash (1941), using X-rays to measure the axial temperature of a mercury discharge as a function of pressure for a constant power input of 30·7 W cm^{-1}, found a maximum value at 0·40 atm.

8.6.6 Spectral fluctuations during the a.c. cycle

The spectral distribution data which has been presented so far is for the light output averaged over the whole a.c. cycle. When a lamp is operated from an a.c. supply, the plasma temperature varies during each half cycle (Lowke and Zollweg, 1975). Therefore both the total radiative output and the spectral distribution also vary. When the lamp voltage and current of a low pressure mercury lamp pass through zero, the light output is totally extinguished. In contrast, with the medium pressure mercury lamp there is a residual intensity. The ratio of maximum to minimum output varies from line to line, and can be related to the transition involved (Vaidya and Dandawate, 1966). Patterson (1972) has described a method of measuring the spectral distribution of light from a lamp at different points in the half cycle, and has suggested that measuring maximum to minimum output ratios is a valuable source of spectroscopic information.

Figure 8.34 shows waveforms for the light output of Lamp VII at different wavelengths chosen to give a progression of upper state energies (see Table 8.5). It can be seen that there is a *general* similarity between the shapes of the light output waveforms and the shape of the power input waveform (Fig. 8.6). The peak light output corresponds in phase to the peak power input. The exact shape of the light output waveform, and in particular the ratio of maximum to minimum output, varies with wavelength, and there is a clear dependence on the upper state energy for the transition involved.

8.6.7 Pressure broadening of spectral lines

Whereas the lines radiated by the low pressure mercury lamp are very sharp, those radiated by the medium pressure mercury lamp are broadened, the width increasing with pressure. There may be a shifting of lines to shorter, or more commonly longer wavelengths, and the peak profile may become asymmetric, with a long wavelength tail.

The width of a spectral line is influenced by a number of distinguishable effects (Cundall and Gilbert, 1970, p. 15):

(1) *Natural broadening* This is a very small effect which results from the Heisenberg uncertainty principle. There is an uncertainty ΔE in

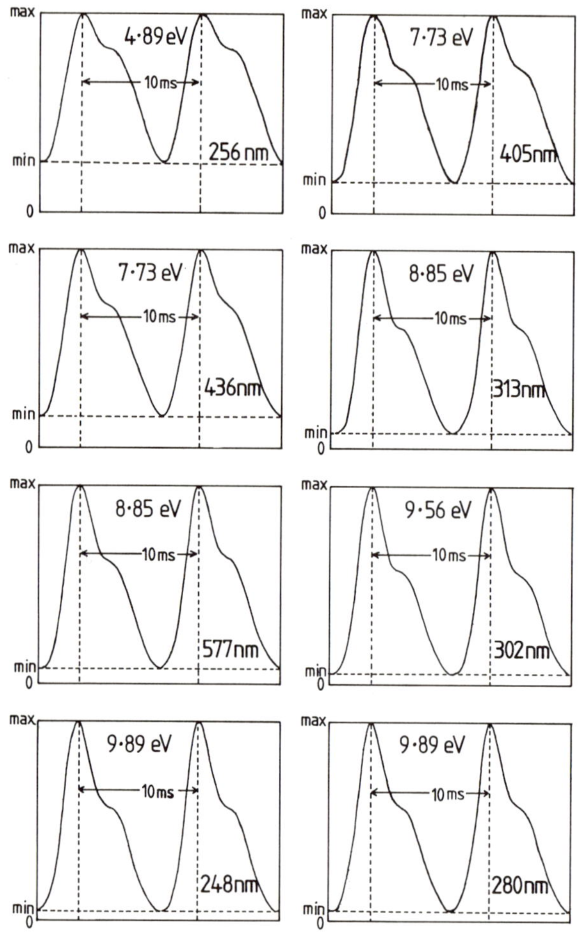

Fig. 8.34. Lamp VII. Constant wattage ballast. Light output waveforms. Maximum: minimum ratio is related to upper state energy of relevant transition.

the energy of an electronic state which is related to the lifetime of that state (τ) by

$$\Delta E \times \tau = h/2\pi.$$

This contributes only 3×10^{-7} nm to the width of the mercury 253·7 nm line.

(2) *Doppler broadening* This arises from the motion of atoms relative to the direction of light propagation, and is an example of the

familiar Doppler effect. The contribution of Doppler broadening is small. For the mercury 253·7 nm line it is 2×10^{-4} nm.

(3) *Pressure broadening* The electronic energy levels of an atom are influenced by the proximity of other particles. The other particle may be an electron or an ion (Stark effect), a foreign atom or molecule (Lorentz broadening), or an atom of the same type (Holzmark broadening). Collisions taking place while an atom is radiating make by far the greatest contribution to the width of spectral lines emitted by medium and high pressure discharges. The various effects are combined in the general term "pressure broadening".

The theory of pressure broadening is very complex (Thorne, 1974, p. 262). In simple terms, if an excited atom undergoes an elastic collision, the proximity of another atom causes shifts in the energy levels of the excited atom, the shifts depending on the distance r between the partners in the collision, as shown schematically in Fig. 8.35. When the pressure is high, elastic collisions are frequent, and there will be a distribution of values of r, resulting in a distribution of wavelengths for a particular transition.

The effect is most pronounced for transitions involving the ground state, as is demonstrated in Fig. 8.35.

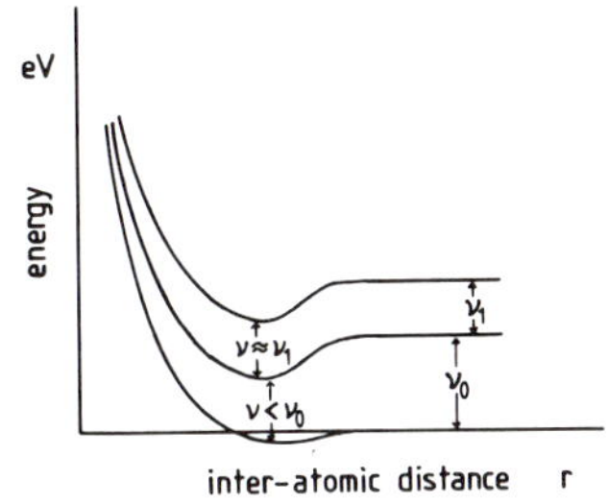

Fig. 8.35. Energy levels during elastic collisions. (From Elenbaas, 1951.)

If the spectra shown in Figs 8.23 and 8.24 had been measured with a very high resolution spectral radiometer, it is likely that radiation at exactly 253·7 nm would have been virtually absent as a result of self-absorption (Coolidge, 1944). Some authors, and particularly those who have studied the UV Standard lamp (see Section 8.6.9) prefer to treat the broadened resonance peak as continuum rather than line radiation.

8.6.8 Sources of continuum radiation

The output from a medium pressure mercury lamp consists predominantly of lines. There is, however, a significant continuum, so that there is some radiation at all wavelengths. The height of the continuum relative to the lines increases with increasing pressure, and if the pressure is increased to several hundred atmospheres, the lines disappear into the continuum. When spectral changes are measured during the warm-up of a lamp, the time taken for the various lines to reach half their maximum output varies somewhat from line to line, and the continuum takes longer than any of the lines.

Recombination of positive ions and electrons releases energy which is emitted as continuum radiation. It can easily be seen by reference to the Saha equation (Section 8.5) that this continuum is proportional to the mercury pressure and increases exponentially with the plasma temperature. The line radiation, which is governed by the Boltzmann distribution law, is also proportional to the mercury pressure, and likewise increases exponentially with the plasma temperature. The result is that the ratio of line to continuum radiation resulting from recombination is independent of the mercury pressure. In fact measured continuum:line ratios increase with pressure and therefore there must be another component in the continuum. It has been suggested (Elenbaas, 1947) that an excited atom can lose energy when in collision with a normal atom and that this energy loss will be emitted as a continuum. This component will be proportional to the square of the mercury pressure, and is sometimes called "molecular radiation". Zollweg *et al.* (1977) have found that electron–atom *Bremsstrahlung* contributes to the continuum, a result of the large electron scattering cross section of mercury.

The continuum radiation from the medium pressure mercury lamp is weak in the visible part of the spectrum, but quite significant in the UV and infra-red. In the infra-red, part of the continuum is radiated by the hot tube and the incandescent electrodes. Figure 8.36 shows the spectral output of the Thorn MBL/D lamp. The radiation axis is logarithmic in order to demonstrate clearly the relative magnitudes of the continuum and line components of the radiative output of the lamp.

8.6.9 The UV Standard

The value, for calibration purposes, of lamps with very well defined and reproducible spectral output characteristics was mentioned in Section 4.3.3.1. Krefft *et al.* (1937) devised a medium pressure mercury discharge

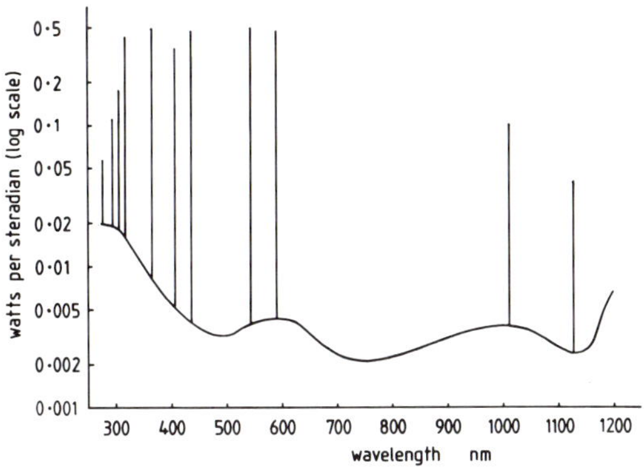

Fig. 8.36. Spectral output (5 nm bandwidth) of Thorn MBL/D 125 W mercury lamp. Adapted from data published in Thorn Lighting Comprehensive Catalogue.

lamp so constructed and operated that it would give a highly reproducible spectral output. This lamp is known as the *UV Standard* and has been available commercially.

It is a 250 W lamp with a fused silica tube, a 1·8 cm internal diameter, and an arc length of 10·5 cm. The ends of the tube are covered so that radiation is emitted only from the central 8 cm of the tube. The lamp is operated from a d.c. supply using a variable series resistor to adjust the power consumption to exactly 250 W. The arc voltage is 130 V, and the mercury dose 3·7 mg cm^{-1}. The 8 cm of exposed arc dissipate 175 W, of which 80 W are lost to the tube by thermal conduction. Of the remaining 95 W, 72% is transmitted through the tube, so that the total amount of measured radiation from the arc is 68 W. The distribution of this 68 W has been very thoroughly investigated by a number of workers including Fischer (1937), Rössler (1939), Coolidge (1944), Zschaeck (1964) and Kasse and Bischoff (1977). The lamp has been described by Elenbaas (1951, p. 111, 1972, p. 165) and Schulze and Bensel (1977).

The UV Standard operates with rather a low mercury pressure and rather a low current, in comparison with the conventional 80 W cm^{-1} medium pressure mercury lamp. However, its spectral distribution, which has been thoroughly characterized both for line radiation *and* continuum radiation, is sufficiently similar to that of the conventional UV curing lamp that it has been used as the basis of mathematical models of the UV curing process (Rubin, 1976; Holman and Rubin, 1978).

Table 8.6 shows the line and continuum output in the various spectral regions for the UV Standard. The figures are taken from Philips Lighting Division's leaflet (LC 48.9–11 E 7111) on their commercial version of the UV Standard.

Table 8.6 Spectral distribution for UV Standard lamp

Region	Wavelengths (nm)	Radiant flux (W) Lines	Continuous	Total
UVC	200–280	4·2	7·1	11·3
UVB	280–315	9·7	1·1	10·8
UVA	315–400	8·2	2·0	10·2
Visible	400–750	21·5	2·6	24·1
Infra-red	750–2000	7·4	3·9	11·3
Total	200–2000	51·0	16·7	67·7

8.7 Non-radiative Losses

Not all the power drawn from the mains supply is converted into radiation. About 10% will be lost as heat in the ballast. The electrode losses may be significant in the case of a low voltage–high current lamp, or negligible in the case of a long high voltage lamp operating at low current. Of the power P supplied to the positive column of the discharge (in watts per centimetre of arc length), part is radiated through the tube wall and the remainder results in the heating of the tube, either by absorption of radiation or by non-radiative loss from the arc. The heating of the tube is an essential part of the operation of the lamp, as it provides a tube temperature sufficient to maintain the required mercury pressure. The hot tube itself loses energy by radiation in the visible and infra-red regions of the spectrum, and by conduction and convection.

There are several possible mechanisms which can be suggested for non-radiative losses from the arc which result in the heating of the tube wall. These are (1) the migration of ions and electrons from the arc with recombination at the wall, (2) convection currents, and (3) thermal conduction. In contrast with the situation with the low pressure lamp (1) is quite insignificant for the medium pressure discharge, and the concentration of current carriers in the dark annular space is very low (Elenbaas, 1951, p. 19). The presence of convection currents in medium pressure mercury lamps has been apparent for many years. When a lamp is operated vertically, hot gas in the arc moves upwards, while colder gas in the dark annular space moves downwards. Sometimes a constriction of the arc is observed near the lower electrode, resulting from the inrush of cold gas into the arc (Kenty, 1938). Figure 8.37 shows a photograph of the arc of

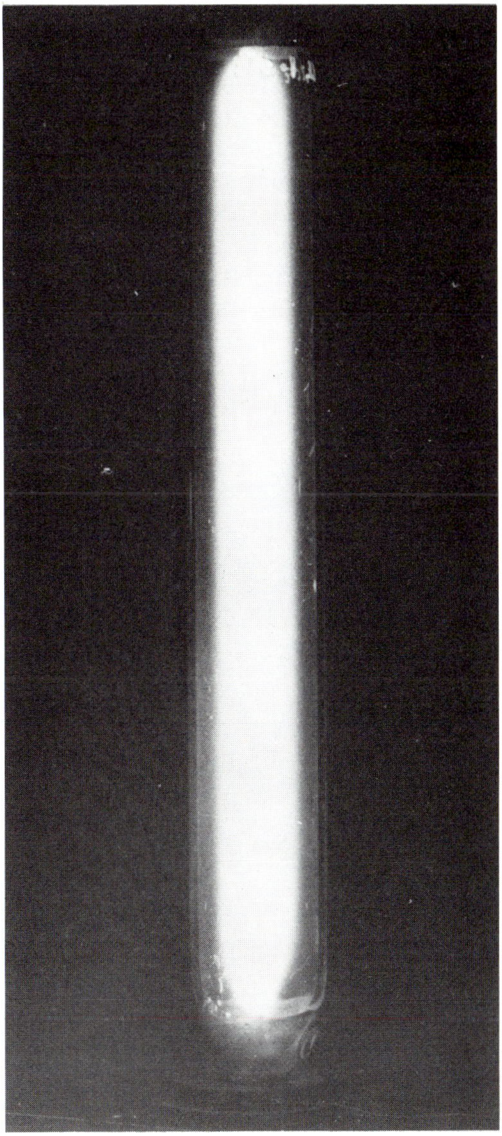

Fig. 8.37. Photograph of arc of Lamp X. Vertical operation. Constant wattage ballast. 14·5 μF series capacitance. Lamp voltage 480 V; lamp current 4·65 A (Avometer readings). Observe the constriction of the arc near the lower electrode.

Lamp X, which demonstrates this constriction. As long as the flow is laminar, very little energy loss takes place due to convection because, except at the ends, gas flows from one volume element into another at the same temperature. The flow pattern which has been described is known as single cell convection. If the mercury fill is increased beyond a critical level, a more complex convection flow pattern results, known as multicelled convection (Fohl, 1975) with a number of short cells with radial inflow and outflow at the ends of the cells. Further increase in mercury fill leads to a loss of laminar flow and the onset of turbulence. The likelihood of energy-dissipating convective flows depends on the mercury fill and the diameter of the tube. The critical parameter is m^2d^{-1} and for the medium pressure mercury lamp convective loss becomes significant when m^2d^{-1} exceeds a value of 20 mg^2 cm^{-3} (Elenbaas, 1951, p. 79).

Convection currents and magnetic effects in horizontal lamps cause the displacement of the arc upwards. Figure 8.38 shows a photograph of the

Fig. 8.38. Photograph of arc of Lamp X. Horizontal operation. Observe the upward displacement of the arc.

arc of Lamp X, operated horizontally. It can be seen that there is a marked displacement from the geometrical axis of the lamp. Figure 8.38 should be compared with Fig. 8.21. Lamp I, with a much lower mercury pressure, shows no discernable displacement of the arc. The critical parameter for convective loss in horizontal discharges is again m^2d^{-1}, but for a particular value of m^2d^{-1} the convective loss is greater for a horizontally operated discharge (Elenbaas, 1951, p. 81).

Several mathematical analyses of convection currents in lamps have been made (e.g. Elenbaas, 1972, Appendix A) and convective flows have been studied experimentally by for instance introducing incandescent particles into the lamp (Kenty, 1938). Conditions necessary for the onset of turbulence, in terms of mercury fill and tube diameter, have been studied experimentally (Elenbaas, 1951, p. 83). However, while it is relatively easy to measure the *total* non-radiative loss, it is less easy to distinguish experimentally between convective and conductive loss.

Elenbaas assumed that the conductive loss remained constant at $10\,\mathrm{W\,cm^{-1}}$, and that the increase in non-radiative loss with increasing mercury pressure was totally due to convection loss. This approach has been criticized by Zollweg (1978), who has suggested that the increased non-radiative loss results from increased conductive loss and increased self-absorption of radiation in the dark annular space. However, whether or not convection currents contribute to energy loss, if they cause turbulence they can lead to instability of the arc.

The heat loss to the walls per centimetre of length is given by

$$\boxed{H = -\,\kappa(\mathrm{d}T/\mathrm{d}r)_R 2\pi R,}$$

where κ is the thermal conductivity of the gas at the wall temperature, R is the radius of the tube, and $(\mathrm{d}T/\mathrm{d}r)_R$ is the radial temperature gradient at the tube wall (Waymouth, 1971, p. 152). The thermal conductivity of gases is almost independent of pressure. Waymouth has shown that *if* the radial temperature profile approximates to a parabola, i.e.

$$T_r = (T_0 - T_\mathrm{w})\,(1 - (r/R)^2) + T_\mathrm{w},$$

where T_r is the temperature at radius r, T_0 is the axial temperature and T_w is the wall temperature, then the radial temperature gradient at the wall is

$$(\mathrm{d}T/\mathrm{d}r)_R = -\,2(T_0 - T_\mathrm{w})/R$$

and therefore

$$\boxed{H = 4\kappa\pi(T_0 - T_\mathrm{w}).}$$

$T_0 - T_\mathrm{w}$ will vary very little from one lamp to another, and therefore H, the conductive loss in watts per centimetre, is almost a constant, independent of tube radius, power input and mercury pressure. The calculated value of H, based on a value for the thermal conductivity of mercury at $1000\,\mathrm{K}$, is $9{\cdot}4\,\mathrm{W\,cm^{-1}}$ (Elenbaas, 1972, p. 138). This value corresponds closely to the value of the intercept seen in Figs 8.28 and 8.29.

As the thermal conduction loss depends on the thermal conductivity of the gas it is somewhat dependent on the nature and pressure of the starting gas used because, with the exception of xenon, the inert gases have higher thermal conductivities than mercury. The most common starting gas, argon, has only a small effect on the conductive loss (Elenbaas, 1951, p. 90).

Elenbaas (1951, p. 21) measured the total radiation from lamps of different mercury fill and tube diameter, as a function of input power, and found that the total power radiated, P_{rad}, in watts per centimetre, could be related to the power input P, also in watts per centimetre, by the equation

$$P_{rad} = 0{\cdot}72\ (P - 10).$$

Note that P_{rad} does not include radiation from the tube itself or the electrodes. The factor 0·72 represents the fraction of radiated power from the arc which escapes the tube and reaches the measuring instrument. Some photons are absorbed in the dark annulus, and some are absorbed by the silica (mainly below 200 nm and above 4 μm in the infra-red). The form of the above expression is in keeping with the results shown in Figs 8.28 and 8.29. As the conductive loss is independent of P, the radiative efficiency P_{rad}/P increases with P.

Elenbaas later decided (1965b) that an even better fit with experimental results would be obtained if the conductive loss was slightly dependent on the power input, and derived the expression

$$P_{rad} = 0{\cdot}75\ (P - 4{\cdot}5P^{1/4}).$$

It seems likely that for the normal range of operation of a medium pressure mercury lamp this empirical expression allows adequately both for conductive loss, and convective loss, if any. However, analyses by Nelson (1961), Lowke and Zollweg (1975) and Lowke *et al.* (1975) have suggested that higher conductive losses will be found with very narrow tubes.

8.8 Ozone-free Lamps

The medium pressure mercury lamp has an output which is rich in short wavelength radiation. Oxygen in the atmosphere absorbs radiation below about 224 nm to produce ozone, which is corrosive and highly toxic. Ozone production by a lamp can be eliminated by using "ozone-free" silica in the manufacture of the lamp. Figure 8.39 shows spectral distributions in the region 200–300 nm for Lamps VIII and IX, both operated under identical conditions from a constant wattage ballast. It can be seen from Table 8.1 that these lamps differ only in that Lamp VIII has a tube made from

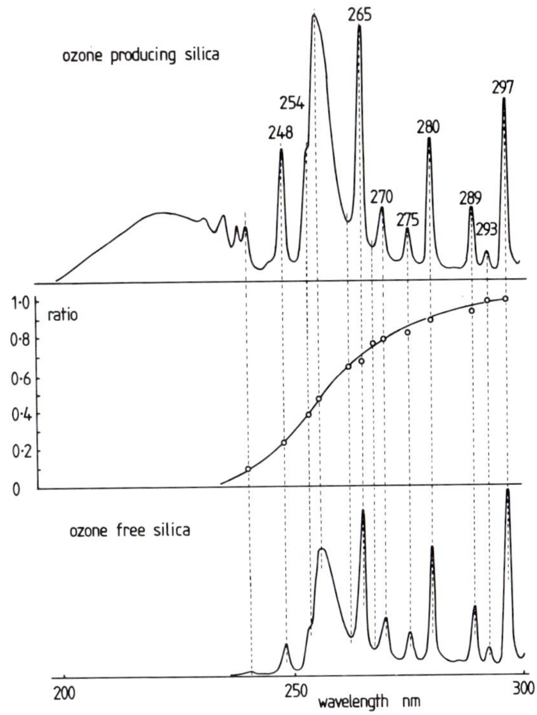

Fig. 8.39. Comparison of the spectral outputs of Lamp VIII (ozone producing) and Lamp IX (ozone free).

ozone-producing silica and Lamp IX has a tube made from ozone-free silica. Figure 8.39 shows that at 297 nm the outputs of the two lamps are identical. However, as the wavelength is decreased there is a progressive depletion in the output of the ozone-free lamp, relative to that of the ozone-producing lamp. The extra absorption in the doped silica reduces the total output below 300 nm by about half. Bassemir and Bean (1974), Bean (1975) and Carlick (1976) have published similar results for Hanovia lamps. The reader may wonder why there is so much depletion at wavelengths well above those responsible for the formation of ozone. The reason may lie in the way in which the spectral transmission of ozone-free silica varies with temperature. Figure 6.7 shows transmission curves for 20 °C (i.e. ambient temperature) and 900 °C (around the upper limit of tube operating temperature) for ozone-free silica. These curves suggest that if the wavelength cut-off is sufficiently high at ambient temperature to prevent ozone formation while the lamp is warming up, loss throughout the

UVC region will be significant when the lamp has reached its operating temperature.

The loss of short wavelength radiation with ozone-free lamps causes some loss in efficiency in lamps used for UV curing.

8.9 Lamp Life

The life of a tungsten filament lamp is normally limited by catastrophic failure. The filament breaks and the lamp is then incapable of producing any more light. With discharge lamps catastrophic failure is uncommon. There is a gradual loss of output until the stage is reached when the lamp can no longer perform its function. The lifetime of a discharge lamp is therefore a rather subjective concept, and depends on the particular application. It depends on the wavelengths which are needed, and how much excess capacity is built into the system. If the lamp can barely perform its function when new, its lifetime will be short, whereas if when new it provides 20 or 30% more output than the minimum required, its effective lifetime will be much longer. Quoted lives for lamps must therefore be interpreted with care, and in particular lives for *lumen* maintenance are not necessarily relevant when *actinic* radiation is required.

The deterioration processes which take place in the medium pressure mercury lamp are very largely associated with the loss of materials from the electrodes. Tungsten and other materials sputtered or evaporated from the electrodes coat the inside wall of the tube and reduce its transmission. Loss of transmission of the tube may result in local overheating and devitrification of the silica, although this should not be a serious problem if the lamp is adequately cooled. Sputtering of electrode materials takes place mainly during the warm-up period. Lamp deterioration therefore depends on the frequency of the switching operation (Eby and Levin, 1979). Sputtering during warm-up can be significantly reduced by increasing the argon pressure (Martt *et al.*, 1960).

When silica is chosen for the envelopes of discharge lamps it is important that it should have a very low hydroxyl content, otherwise water will enter the tube and react with the electrodes, giving a coating of undesirable tungsten oxide on the electrodes, and hydrogen, which will raise the ignition voltage. The silica should also have a low level of metallic impurities, otherwise the intense radiation from the arc may cause solarization.

The UV output of a medium pressure mercury lamp deteriorates more

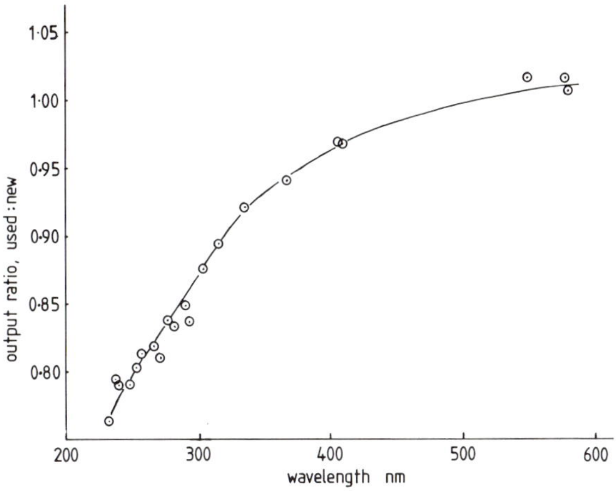

Fig. 8.40. Lamp IX. Constant wattage ballast. Loss of UV output during lamp ageing. Central 10 cm of lamp examined.

rapidly than the visible output. Figure 8.40 shows the ratio of two spectral distributions measured for Lamp IX. The first spectral distribution was measured when the lamp was new, and the second when the lamp was about half way through its useful life. It can be seen that there is little loss at 546, 577 and 579 nm, but as the wavelength is decreased towards 200 nm the ratio of outputs in each spectral line reduces smoothly.

References

Ayotte, A. D. and Hale, R. R. (1967). *Illum. Eng.* **62,** 221–228.

Bassemir, R. W. and Bean, A. J. (1974). Paper presented at 26th Annual Meeting of TAGA, St Paul, Minn., 13–15 May 1974; *TAGA Proc.* **1974,** 133–146.

Bean, A. J. (1975). SME Tech. Pap. (Ser.) FC75-310.

Bleekrode, R., Koedam, M. and Rehder, L. (1975). *Philips Tech. Rev.* **35,** 308–315.

Bouwknegt, A., Nienhuis, H., Schipper, D. J. and Tielemans, P. A. W. (1975). *Philips Tech. Rev.* **35,** 356–359.

Burns, J. A. and Nelson, E. H. (1964). *GEC J. Sci. Technol.* **31,** 3–10.

Carlick, D. J. (1976). *In* "Encyclopedia of Polymer Science and Technology", Suppl. 1, pp. 367–378. Wiley–Interscience, New York.

Coolidge, A. S. (1944). *J. Opt. Soc. Amer.* **34,** 291–301.

Corliss, C. H. and Bozman, W. R. (1962). "NBS Monograph No. 53". US Government Printing Office, Washington D.C.

Cundall, R. B. and Gilbert, A. (1970). "Photochemistry". Nelson, London.

DuRon, B. (1976). SME Tech. Pap. (Ser.) FC76-482.

Eby, J. E. and Levin, R. E. (1979). *In* "Applied Optics and Optical Engineering", Vol. VII (Eds. R. R. Shannon and J. C. Wyant), pp. 1–46. Academic Press, New York.
Elenbaas, W. (1947). *Philips Res. Rep.* **2,** 442–453.
Elenbaas, W. (1951). "The High Pressure Mercury Discharge". North Holland, Amsterdam.
Elenbaas, W. (1965a). "High Pressure Mercury Vapour Lamps and their Applications", Philips Technical Library. Centrex Publishing Company, Eindhoven.
Elenbaas, W. (1965b). *Philips Res. Rep.* **20,** 213–225.
Elenbaas, W. (1972). "Light Sources", Philips Technical Library. Macmillan, London.
Fischer, E. (1937). *Z. Tech. Phys.* **18,** 336–342.
Fohl, T. (1975). *J. Illum. Eng. Soc.* **4,** 265–270.
Gallo, C. F. (1966). *Appl. Opt.* **5,** 1285–1291.
Gallo, C. F. (1967). *Appl. Opt.* **6,** 1563–1565.
Hehenkamp, T. (1965). *In* "High Pressure Mercury Vapour Lamps and their Applications" (Ed. W. Elenbaas), pp. 71–106. Philips Technical Library. Centrex Publishing Co., Eindhoven.
Holman, R. J. and Rubin, H. (1978). *J. Oil Colour Chem. Assoc.* **61,** 189–194.
Kasse, H. and Bischoff, K. (1977). *Optik* (*Stuttgart*) **48,** 451–458.
Keeffe, W. M. (1980). *IEE Proc. A* **127,** 181–189.
Kenty, C. (1938). *J. Appl. Phys.* **9,** 53–66.
Kenty, C. and Karash, W. J. (1941). *Phys. Rev.* **60,** 66 (abstract).
Krefft, H., Rössler, F. and Rüttenauer, A. (1937). *Z. tech. Phys.* **18,** 20–25.
Lienhard, O. E. (1975). SME Tech. Pap. (Ser.) FC75–335.
Lowke, J. J. and Zollweg, R. J. (1975). *J. Illum. Eng. Soc.* **4,** 253–259.
Lowke, J. J., Zollweg, R. J. and Liebermann, R. W. (1975). *J. Appl. Phys.* **46,** 650–660.
Majewski, D. W. (1978). *IES Light. Rev.* (*Australia*) **40,** 155–159.
Martt, E. C., Gottschalk, K. and Green, A. C. (1960). *Illum. Eng.* **55,** 260–268.
Mosburg, E. R., Jr and Wilke, M. D. (1978). *J. Quant. Spectrosc. Radiat. Transfer* **19,** 69–81.
Nelson, E. H. (1961). *GEC J.* (*GB*) **28,** 165–171.
Patterson, P. L. (1972). *J. Opt. Soc. Amer.* **62,** 627–633.
Plante, E. R. and Sessoms, A. B. (1973). *J. Res. Nat. Bur. Stand. A* **77,** 237–242.
Rautenberg, T. H., Jr and Johnson, P. D. (1964). *Appl. Opt.* **3,** 487–492.
Rössler, F. (1939). *Ann. Phys.* (*Leipzig*) **34,** 1–22.
Rubin, H. (1976). *TAGA Proc.* **1976,** 279–301.
Schooley, W. A. (1975). SME Tech. Pap. (Ser.) FC75-334.
Schulze, R. and Bensel, I. (1977). *In* "Ultraviolette Strahlen" (Ed. J. Kiefer), pp. 215–310. Walter de Gruyter, Berlin.
Smithells, C. J. (1976). "Metals Reference Book", 5th edn. Butterworth, London.
Thorne, A. P. (1974). "Spectrophysics". Chapman and Hall, London.
Till, W. S. and Unglert, M. C. (1960). *Illum. Eng.* **55,** 269–281.
Vaidya, W. M. and Dandawate, V. D. (1966). *J. Opt. Soc. Amer.* **56,** 1693–1697.
van de Weijer, M. H. A. (1965). *In* "High Pressure Mercury Vapour Lamps and their Applications" (Ed. W. Elenbaas), pp. 107–157. Philips Technical Library. Centrex Publishing Co., Eindhoven.
Waymouth, J. F. (1971). "Electric Discharge Lamps". MIT Press, Cambridge, Mass.

Zollweg, R. J. (1978). *J. Appl. Phys.* **49,** 1077–1091.
Zollweg, R. J. (1979). *J. Illum. Eng. Soc.* **8,** 126–131.
Zollweg, R. J., Lowke, J. J. and Liebermann, R. W. (1975). *J. Appl. Phys.* **46,** 3828–3838.
Zollweg, R. J., Liebermann, R. W. and Burnham, R. (1977). *Bull. Amer. Phys. Soc.* **22,** 196 (abstract); *J. Quant. Spectrosc. Radiat. Transfer* **18,** 133–141.
Zschaeck, H. (1964). *Exp. Tech. Phys.* **12,** 373–391.

9 Metal Halide Lamps

9.1 Shortcomings of the Mercury Spectrum

Although the medium pressure mercury lamp is an efficient converter of electrical energy into radiant energy, there are regions of the spectrum where the output is very low indeed. This is of particular importance when the medium pressure mercury lamp is used for illumination, as the output in the visible consists only of a few strong lines, with very little between 436 and 546 nm, and beyond 579 nm. The radiation at 436 nm and below contributes little to the total lumen output, and the lack of radiation in the red part of the spectrum above 600 nm gives the mercury light its characteristic blue-green colour.

There are three logical approaches to the improvement of the lumen efficiency and/or the colour balance of the medium pressure mercury lamp:

(1) The first involves the use of fluorescent powders to convert short wavelength radiation into more useful longer wavelength radiation. Lamps for illumination consist of a short silica discharge tube housed inside a larger glass bulb, and it is convenient to coat the inside of the glass bulb with the fluorescent material.

(2) In some lamps a tungsten filament is used as a ballast. As the filament provides a continuum output with a peak at about 800 nm, its red bias counteracts the blue bias of the mercury discharge and helps to give a better colour balance.

(3) The third approach involves the addition of other metallic elements to the discharge, selected to give spectral lines where enhancement is required (Sarychev, 1967; Schirmer *et al.*, 1969; Sucov, 1970). This is the approach adopted in the metal halide lamp.

9.2 Principles

There are considerable problems involved in adding other metallic elements to the mercury fill of a discharge lamp. No other metallic element has a volatility which matches that of mercury (see Fig. 7.1), and few have sufficient vapour pressure at discharge tube temperatures to be viable propositions as additives to modify the spectral output (Waymouth, 1971a, p. 209). Another problem is that many metals, and in particular the alkali and alkali earth elements, react vigorously with silica at discharge tube temperatures. Cadmium (with strong lines at 509 and 644 nm) and zinc (468, 472, 481 and 636 nm) had limited commercial success but were found to lower the lumen efficiency of the mercury discharge, and also attacked the silica (Dalrieu and Taxil, 1964).

These problems are elegantly overcome by introducing the desired element as a halide. Most metallic elements can be introduced in this way (Waymouth, 1971a, p. 211). Halides are usually, but not always, more volatile than the parent metal. Notable exceptions are the halides of lithium and sodium. The vapour pressure curves for a variety of metallic iodides are shown in Fig. 9.1. Other data have been given by Work (1981). The halides, apart from being more volatile, are less likely to attack the silica tube than the parent metal. Iodides are favoured as they are normally the most volatile of the halides and the least chemically aggressive (Vermeulen, 1974). There is some use of chlorides and bromides, in conjunction with iodides, but fluorides are too reactive (Reiling, 1966; Drop *et al.*, 1975).

The concept of the metal halide lamp is not new. Indeed the patent literature dates back to 1900. However, early attempts to construct metal halide lamps were unsuccessful because available techniques did not allow the high loadings required to keep the halides in the vapour phase. Also, metal halides are prone to being hygroscopic and it was not possible to eliminate all traces of water. The presence of water leads to the formation of hydrogen, which has extremely adverse effects on the operation of the lamp. Interest in the concept was revived in the early 1960s, following the successful introduction of the incandescent tungsten halogen lamp.

In the high temperature of the plasma the metal halides break down to the constituent elements. The vapour pressure of the metal is higher at the centre of the discharge than it would be if the metal were added to the lamp in elemental form (Beijer *et al.*, 1968). The halogen atom does not contribute to the spectral output, as its non-metastable excitation levels are very much higher than those of the metal (Rautenberg and Johnson, 1964; Gallo, 1967; Waymouth, 1971b).

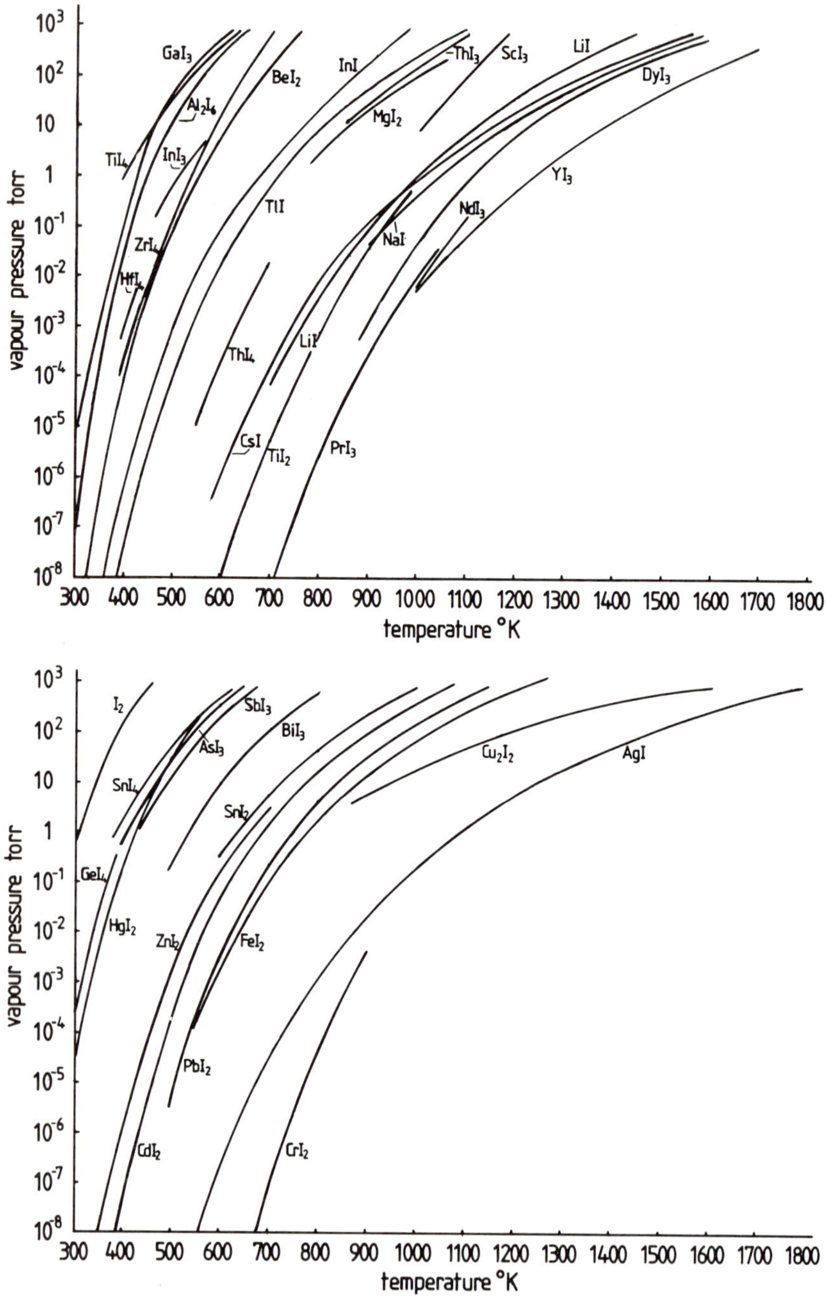

Fig. 9.1. Vapour pressure curves for metallic iodides.

The metal halide must fulfil certain requirements:

(1) The metal must have a strong output in the spectral region requiring enhancement.
(2) It must have a fairly low vapour pressure at ambient temperature, otherwise the lamp will be difficult to start.
(3) It must be stable at normal wall temperatures.
(4) It must have a fairly high vapour pressure at normal wall temperatures.
(5) It must be unstable at the temperature of the discharge.

It is convenient that the majority of metallic elements in the periodic table form iodides which fulfil requirements (2)–(5).

The data shown in Fig. 9.1 were obtained by use of the major compilations of Stull (1947) and Smithells (1976), supplemented by a search of *Chemical Abstracts* back to 1946. The data are by no means comprehensive but do enable the reader to draw some general conclusions about the relationship between the volatility of a metallic iodide and the position of the metal in the periodic table. There is, however, a complicating factor. Lamps for illumination purposes often make use of combinations of iodides, e.g. $NaI + ScI_3$, and there is evidence of possible complex formation, e.g. $NaScI_4$, to give an iodide more volatile than either of the parent iodides (Hirayama *et al.*, 1977). This may be a common phenomenon when combinations of iodides are used.

9.3 Construction and Operation

Metal halide lamps are similar in general construction and operation to conventional medium pressure mercury lamps, and much of the description of the medium pressure mercury lamp in Chapter 8 is relevant also to the metal halide lamp. However, there are important differences of detail in the construction of the lamps and in the ballast requirements. Great care is required in the fabrication of metal halide lamps, and it is particularly important to keep the halide as dry as possible (Tchang, 1974; Vermeulen, 1974).

9.3.1 *Electrodes*

Alkaline earth oxides cannot be used as emissive materials for the electrodes because the halogen attacks the oxide and removes the metal as a halide (Bouwknegt *et al.*, 1975). For this reason the electrodes used are of pure tungsten, or use thorium oxide or rare earth oxides such as those of yttrium, dysprosium, gadolinium and terbium (Zollweg *et al.*, 1970; Datta, 1975) as emissive materials. Pure tungsten may exhibit an increase in work function in a halide atmosphere. Although the presence of halides places some restraints on electrode design, the operation of the tungsten halogen cycle helps to return sputtered or evaporated tungsten to the electrodes (Vermeulen, 1974). When thorium is used an emitter, the thorium iodide regenerative cycle returns thorium to the cathode tip and so helps to sustain cathode activity (Keeffe, 1980). Some authors (Heller *et al.*, 1969; Bouwknegt *et al.*, 1975) have stressed the need to prevent halides condensing on the electrodes, as this would affect the ignition of the lamp.

9.3.2 *Gas fill*

The fill of a metal halide lamp normally consists of sufficient mercury to provide, when fully vaporized, a mercury pressure in the medium pressure range (e.g. 1–5 atm), a starting gas, usually argon, or a mixture of argon and neon, at a few torr pressure, as well as the metal halide(s). The amount of the metal halide additive is normally much less than the amount of mercury, and even so it may be present in excess (i.e. some will remain unvaporized). The partial pressure of the metal additive in the discharge will normally be less than 5% of that of the mercury. However, most additive metals have both ionization and average excitation energies significantly lower than those of mercury, with the result that the total spectrum may nevertheless be dominated by the lines of the additive metal (Delrieu and Taxil, 1964; Rautenberg and Johnson, 1964).

The very small amounts of metal halide required pose problems in supplying the exact dose. The metal halide is not necessarily added as such, but may be incorporated as separate elements, or the metal may be supplied as an amalgam with the mercury. Taxil *et al.* (1977) have suggested adding iodine in a gaseous organic form, as methyl or ethyl iodide.

The metal halide lamp has so far been treated as a modified medium pressure lamp. The reader may be forgiven for wondering why the mercury is present at all. There are several reasons:

(1) Mercury condenses at ambient temperature and therefore permits easy starting, but provides a high pressure at the normal running temperature, and therefore a high impedance and a high loading (Reiling, 1964).

(2) The high mercury atom density reduces the rate of migration of the metal and iodine atoms to the wall.
(3) Mercury reacts with excess iodine to give HgI_2, so reducing the concentration of molecular iodine, which, for reasons which will become apparent, is deleterious to the operation of the lamp.

Actually, not all metal halide lamps are based on the mercury discharge. Some are based on xenon (Gallo, 1971; Schäfer, 1979). Although the great majority of metal halide lamps are based on mercury, often the spectral lines of the additive metal(s) so dominate the output that it is no longer realistic to think of them as simply filling gaps in the mercury spectrum.

9.3.3 Circuit requirements

Metal halide lamps often have the disadvantage that they will not operate satisfactorily from a ballast designed for a similarly rated medium pressure mercury lamp of the same dimensions. When the lamp is cold, there will be, in the vapour phase, a small pressure of either molecular iodine or HgI_2. Both are electron-attracting molecules, i.e.

$$I_2 + 2e \longrightarrow 2I^-$$

$$HgI_2 + 2e \longrightarrow Hg + 2I^-$$

and therefore interfere with the avalanche effect, so that a higher starting voltage is required. In a properly designed lamp there should be almost no molecular iodine present. HgI_2 has a vapour pressure of only 2×10^{-5} torr at 25 °C, and adds only about 50–75 V to the starting voltage for a short lamp (Waymouth, 1971b). However, if there is any water present. HI can be formed, and as this has a boiling point of −37 °C, it can provide an electronegative gas at ambient temperature (Koury and Waymouth, 1966).

Figures 8.5, 8.8 and 8.11 show that there is a just perceptible re-ignition voltage spike in the lamp voltage waveform of the medium pressure mercury lamp, just after current zero. This spike is more pronounced in the case of metal halide lamps. The re-ignition voltage passes through a maximum during warm-up when the lamp has reached a temperature of about 200 °C (Franke *et al.*, 1967; Waymouth, 1971a, p. 259). This is a result of the increasing vapour pressure of HgI_2. Above 200 °C, the increasing vapour density and heat capacity of the mercury gas, coupled with the decreasing stability of the iodine ion, cause a reduction in the re-ignition voltage. It is sometimes found that a lamp will start satisfactorily, but will go out after a few seconds operation, only to start again after cooling down (Waymouth, 1971a, p. 263; Gardner, 1975).

During normal operation, the re-ignition voltage depends on the quality of the electrodes (Lake and Kramel, 1967), and on the ionization energy of the additive metal (Zollweg *et al.*, 1975). It can be controlled by the addition of trace amounts of caesium halide (Gungle *et al.*, 1970). It is reduced by increasing the lamp current (Franke *et al.*, 1967).

The ballast used with a metal halide lamp must provide a high enough voltage for ignition, and for the re-ignition spikes during the warm-up period. It is often desirable to use a high voltage pulse ignitor. Suitable ballast types have been described by Cohen (1965).

As the metal additive usually has an ionization energy very much lower than that of mercury it often plays a dominant role in the production of current carriers (Taxil, 1974), particularly as the plasma temperature is reduced. The mobility of the current carriers, however, is determined by the mercury pressure, and so it is the mercury pressure which largely determines the resistivity of the plasma. However, if an additive element has a significant arc-constricting or arc-fattening effect, it will affect the impedance and voltage of the tube by altering the effective cross section of the arc (Waymouth, 1971a, pp. 216–222).

9.4 Arc Constriction

The conditions for a wall-stabilized arc were described in Section 8.5. In the case of some metal halide discharges these conditions are not met and a "constricted" arc results. This occurs when the power radiated by a volume element exceeds the electrical power input, and d^2T/dr^2 becomes positive. Whether or not this happens depends on the energy level diagram for the additive metal. It is most common for a metal which radiates many lines, and for which many of the excited states lie at energy levels below half the ionization level (Lowke, 1970). An analysis by Waymouth (1971a, p. 295) has shown that if the "average" excitation level is less than $0{\cdot}585V_i$, where V_i is the ionization energy, a constricted arc will result. Arc constriction is an undesirable phenomenon because it causes an unstable arc which is particularly susceptible to convection currents and magnetic influences. If the lamp is operated in a horizontal mode, the arc moves towards the top of the tube, overheating and possibly softening the silica (Drop *et al.*, 1975). The lower part of the tube overcools, and condensation of the metal halide may occur. Convection currents also affect non-constricted arcs, but to a lesser extent. Various techniques have been proposed to overcome the problem, e.g. (a) the creation of a weak magnetic field to offset the effect of the convection current (Kenty, 1938; Drop *et al.*, 1974); (b) changing the geometry of the tube to an arched shape, so that the arched arc follows the axis of the tube (Koury *et al.*, 1975; Fromm *et al.*, 1979); (c) addition of "arc-fattening" halides. While some metal halides cause arc constriction, others, and particularly the alkali metal halides, have the opposite effect (Waymouth, 1971a, p. 218). The alkali metals have very low ionization energies (that of caesium is the lowest of any element, 3·89 eV), and are therefore able to contribute current carriers, and therefore power consumption into the lower temperature regions.

9.5 Spectral Output

The plasma of the metal halide lamp is formed under conditions which allow an approximation to local thermal equilibrium (Gallo, 1966), and therefore the line spectrum is influenced by the same thermodynamic factors as in the medium pressure mercury lamp. However, instead of only two atomic species being present (argon and mercury), there are at least four and possibly more. There are also molecular species. The metal additives will almost certainly have both ionization and excitation energies significantly lower than those of mercury. Although the partial pressures of metal additives will be much lower than that of the mercury, a very much higher fraction of the metal atoms will be in excited states. This has the effect of reducing the plasma temperature (Rautenberg and Johnson, 1964; Gallo, 1967), and depressing the output in the mercury lines accordingly. When a lamp is warming up, the spectrum is dominated at first by mercury radiation because the mercury evaporates more rapidly than the halide. However, as the halide evaporates, and the metal is fed into the plasma, the metal spectrum grows at the expense of the mercury spectrum and the plasma temperature drops.

Reiling (1964) has given an excellent set of spectral distributions for a variety of metal halide lamps. As there is little ionization of mercury, there is little recombination of mercury ions and electrons. The result of the depression of the mercury lines and recombination spectrum is a lower output in the region below 250 nm and metal halide lamps can normally be considered ozone free, even when undoped silica is used.

Vermeulen (1974) has divided the spectra of metal halide lamps into three general types. First, there are the spectra which are dominated by a few strong lines. Examples are given by sodium, thallium, indium and gallium. The second type is given by metals which have a large number of weak lines which merge together to form a pseudo-continuum. Examples are scandium, thorium and dysprosium. The third type is a genuine continuum which is formed by *molecular* radiation from a metal halide which is not totally dissociated into its constituent atoms at a radial position where the plasma temperature is hot enough to excite the molecule. The prime example is tin, which forms a stable monohalide (Chalmers *et al.*, 1975; Drop *et al.*, 1975). This is not the only source of continuum radiation. Waymouth (1971a, p. 228) has suggested that the background continuum present in the spectra of many metal halide lamps is due to radiative dissociation of molecules from an excited state. Molecular species not only give rise to band and continuum radiation, but also, in the cooler regions, may absorb radiation, and further modify the spectrum (Slagg, 1967; Ashurkov and Sarychev, 1975; Sarychev *et al.*, 1979).

A considerable problem with metal halide lamps is that of axial segregation of the components of the discharge. If the components are not uniformly mixed along the length of the discharge, the spectral distribution will vary, e.g. there may be predominantly mercury radiation at one end and predominantly additive metal radiation at the other. This is most prevalent in discharge tubes operated in the vertical mode, and is caused by the coupling of radial diffusion and vertical convection (Fischer, 1976).

It was shown in Chapter 8 that for a lamp operated from an a.c. supply, the output in a particular spectral line responds to the supply cycle, and that the maximum to minimum output ratio depends on the upper state energy for the transition involved. As upper state energy levels for additive metals are almost invariably lower than those for mercury, the ratio is reduced (Reiling, 1964; Frier and Henderson, 1973; Rothwell and Keeffe, 1980).

9.6 Non-radiative Losses

This is a difficult area, and it is not easy to draw any general conclusions. Reiling (1964) and Jack and Koedam (1974) have suggested that the electrode losses will be much the same as in a conventional medium pressure mercury lamp with the same electrode construction. However, the radial heat transfer from the arc to the tube is very much complicated by the presence of molecular species, which not only have thermal conductivity characteristics quite different from those of mercury, but also are involved in association and dissociation reactions which are exothermic or endothermic.

Dobrusskin (1971) has given a table of energy balances for a variety of metal halide lamps for illumination purposes, showing that the radiation efficiency is considerably higher than for a mercury lamp, and also that heat lost by conduction and convection is much less. This is not unreasonable if little of the radiation from the discharge is absorbed by the envelope. Several authors have applied the Elenbaas technique of plotting total energy radiated, per centimetre of arc, against the energy input per centimetre of arc. Whereas Elenbaas found that for the medium pressure mercury lamp $P_{rad} = 0{\cdot}72(P - 10)$ (see Section 8·7), he found (Elenbaas, 1972, p. 188) that for a metal halide lamp containing thallium, indium and sodium iodides $P_{rad} = (P - 20)$. This relationship was also given by Beijer *et al.* (1968). The higher intercept of 20 W cm^{-1} found with the halide lamp was ascribed to energy lost by dissociation of the iodides in the discharge and subsequent recombination of the atoms near the tube wall. Heller *et al.* (1969) and Vermeulen (1974) also state that the conductive energy loss is double that for a mercury lamp. The gradient of unity is of great interest, because it implies that all the radiation from the arc escapes from the tube. However, Jack and Koedam (1974) examined a lamp with the same mix of iodides (i.e. Tl, In and Na) but found a gradient of only 0·65. For a tin halide discharge it was even lower (0·60), while for a dysprosium iodide discharge it was much higher (0·90).

There are considerable discrepancies between published data for energy balances, and Zollweg (1974) has stressed the difficulties both of measurement and interpretation of experimental results.

9.7 Applications of Metal Halide Lamps

Metal halide lamps were initially developed primarily for illumination purposes, to provide a high lumen efficiency and a good colour balance. The majority of metal halide lamps manufactured are used for illumination, and the majority of published papers deal with this aspect. However, there are photochemical processes for which the spectral output of the mercury discharge is not ideal, and metal halide lamps have been developed and used for these processes. The areas of greatest interest are the graphic arts, UV curing (Beeson, 1976; Priest and Eckel, 1976) and industrial synthesis. In this section some of the more significant lamp types will be classified according to the additive metal used.

9.7.1 Thallium

Thallium was one of the first metals examined as an additive for lamps for illumination, because it has a very strong line at 535·0 nm. This is close to the wavelength for peak eye sensitivity (555 nm). The thallium halide lamp has been extensively studied (Larson *et al.*, 1963; Delrieu and Taxil, 1964; Gallo, 1966, 1967, 1970; Ayotte and Hale, 1967; Sarychev *et al.*, 1967; Otto, 1974).

Perhaps the most interesting photochemical application of the thallium halide lamp is for the photonitrosation process (see Section 1.4.1.4). Nitrosyl chloride has an absorption spectrum which extends to 600 nm, and the economics of the process dictate that the longest effective wavelength should be used. Very high powered thallium halide lamps have been developed for this process, with ratings up to 60 kW (Fischer, 1978; Kanoh and Shinjoh, 1981).

9.7.2 Gallium

Undoubtedly the most commonly used additive for photochemical purposes is gallium. Many graphic arts materials, and particularly those based on diazo systems, are sensitive to radiation in the region between 400 and 420 nm, and there are numerous lamps available commercially for photoprinting, platemaking, etc., which contain gallium iodide. Gallium provides two strong lines at 403·3 and 417·2 nm, and is usually the first choice if intense blue light is needed (Ayotte and Hale, 1967; Beeson and Furmidge, 1972). It is sometimes combined with lead (Philips, 1967; Gion *et al.*, 1974; Dobrusskin *et al.*, 1975). Figure 9.2 shows the spectral output of a Philips HPM 15 lamp, which contains gallium and lead iodides. The optimization of gallium iodide lamps has been discussed by Prikupets *et al.* (1972), who were able to devise a lamp with 30% conversion of electrical power into radiation between 400 and 420 nm. Figure 9.3 shows the spectral output of a short arc gallium iodide lamp operating at a high mercury pressure, a high axial potential gradient and a high power input. The mercury lines are strongly depressed, and the 403·3 and 417·2 nm gallium lines are broadened and the peaks are shifted to longer wavelengths, with signs of reversal at 403·3 and 417·2 nm.

Gallium iodide lamps have interesting applications in UV curing, based on the ability of the long wavelength blue radiation to penetrate into thick films. Nakabayashi (1975) recommended a gallium iodide lamp followed by a conventional medium pressure mercury lamp for the curing of opaque white coatings. A gallium iodide lamp has little short wavelength radiation,

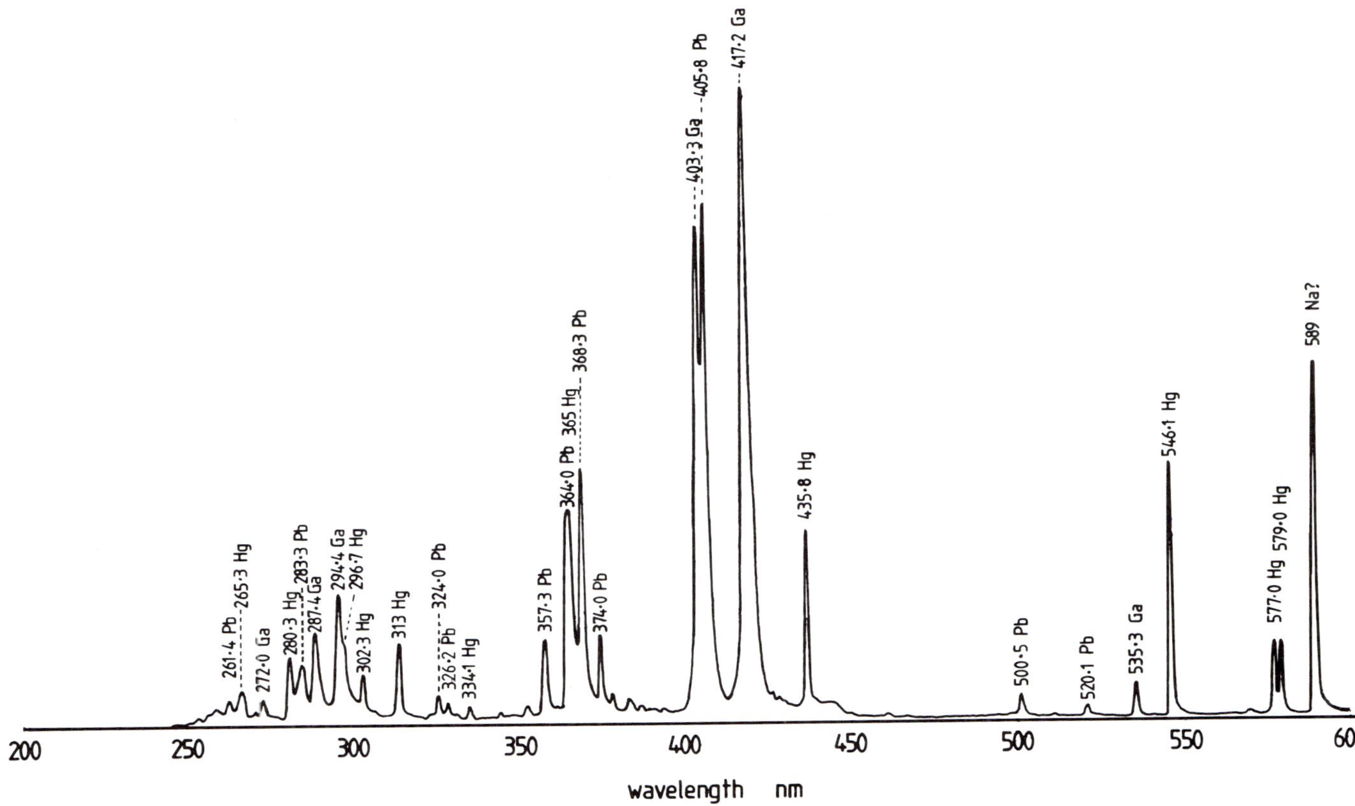

Fig. 9.2. Spectral output of Philips HPM 15 lamp.

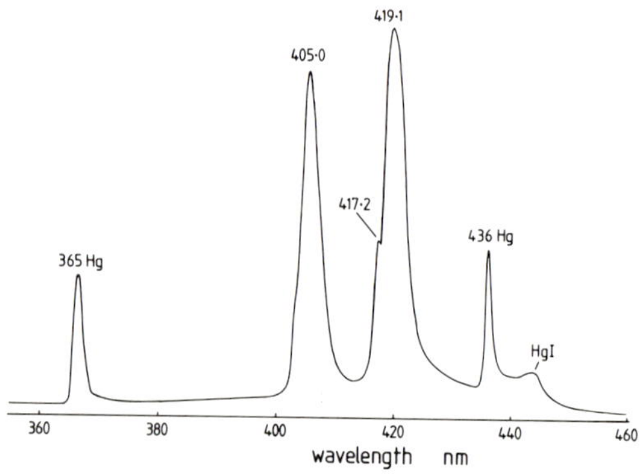

Fig. 9.3. Spectral output of Original Hanau short arc gallium iodide lamp (50 mm arc, 28 mm diameter tube). Lamp voltage 200 V. Lamp current 8 A. Reactor ballast. Note the asymmetric broadening of the two strong Ga lines. The peaks are shifted to longer wavelengths and there is clear evidence of self-absorption.

and so surface cure is poor when free radical systems are used. This can be turned to advantage in a technique for the control of surface gloss (see Section 10.6).

The spectral output of an electrodeless lamp doped with gallium halide is shown in Fig. 10.10.

9.7.3 *Lead*

Lead provides strong lines at 364·0, 368·3 and 405·8 nm. Sarychev *et al.* (1979) described a lead iodide lamp which provided an enhancement of 1·5 times in the region between 355 and 380 nm, and three times in the region 400–420 nm, relative to a conventional mercury lamp. Philips (1967) gave a table of enhancements for a lead iodide lamp for various wavelength regions. The greatest enhancement, 2·73 times, was found for the region 375–425 nm. Addition of gallium iodide to the lead iodide discharge allowed a further large enhancement to 5·08 times. However, in the region 350–400 nm, the lead iodide by itself provided a larger enhancement, 1·75 times, than the lead iodide/gallium iodide combination (1·20 times). A spectral output for a lead iodide lamp is shown in Fig. 9.4. This lamp contains no mercury, and is used for the treatment of psoriasis.

9.7.4 *Iron*

Iron has been found to be a very useful additive when enhancement in the region between 350 and 450 nm is required. Iron gives a complex multiline

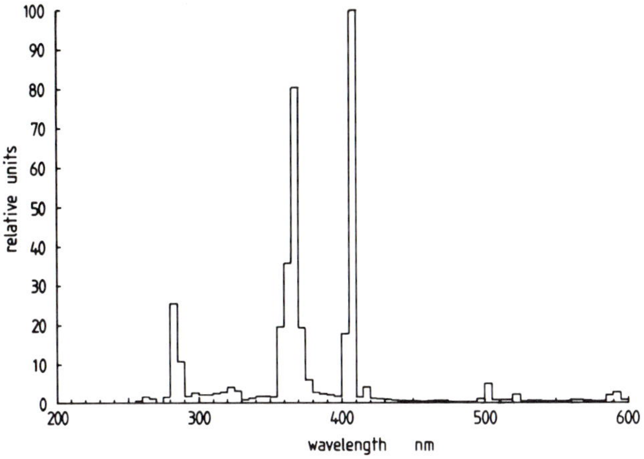

Fig. 9.4. Spectral output of Original Hanau lead iodide lamp. (Data provided by W. C. Heraeus GmbH, Department Original Hanau, and reproduced with permission.)

spectrum which is particularly rich in this region. Gardner (1975) and Gardner *et al.* (1975) have described lamps for UV curing containing iron iodide. Lamps of various lengths up to 96·5 cm were described, operating with a current of 6·7 A, a potential gradient of about 20 V cm^{-1} and a power input of 120 W cm^{-1}. The output in the region between 350 and 400 nm was found to be 3·5 times that measured for a conventional

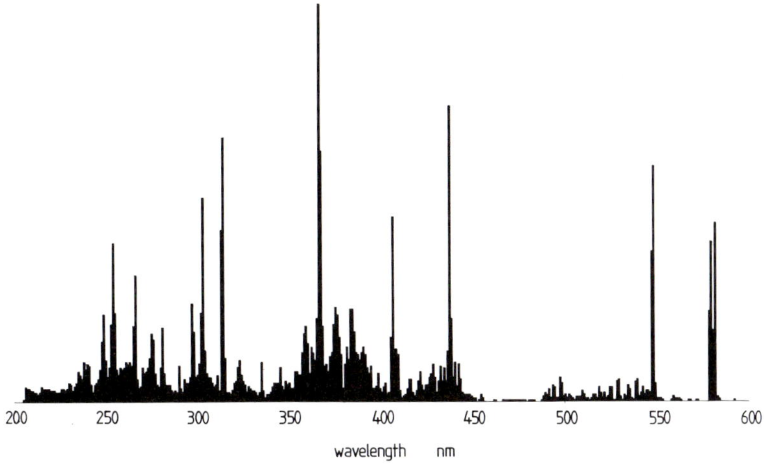

Fig. 9.5. Spectral output of iron iodide lamp. The ratio of iron to mercury radiation was found to be dependent on the cooling of the lamp.

80 W cm^{-1} medium pressure mercury lamp. Cure rates for UV curing inks and particle board fillers were enhanced by a factor of two or three. The use of iron iodide lamps for the curing of coatings pigmented with TiO_2 has been described by Hulme *et al* (1976).

Figure 9.5 shows the spectral output of a lamp similar to Lamp III (Table 8.1) but containing a small quantity of iron iodide. A multitude of rather weak iron lines fills the gaps between the mercury lines and greatly increases the total output between 350 and 450 nm. This spectral distribution may be compared with that for an electrodeless iron halide lamp shown in Fig. 10.7, where the ratio of iron to mercury line output is quite a lot higher.

9.7.5 Antimony

Schäfer (1979) has described an interesting antimony iodide lamp designed for the UV curing of inks and coatings. It is of interest for two reasons. First, it does not use mercury as a buffer gas, but rather neon or xenon. Secondly, its output, unlike that of almost every other additive lamp, is concentrated in the UVC and UVB, between 200 and 320 nm. The spectral output is shown in Fig. 9.6. High intensity antimony lamps are also used for water sterilization.

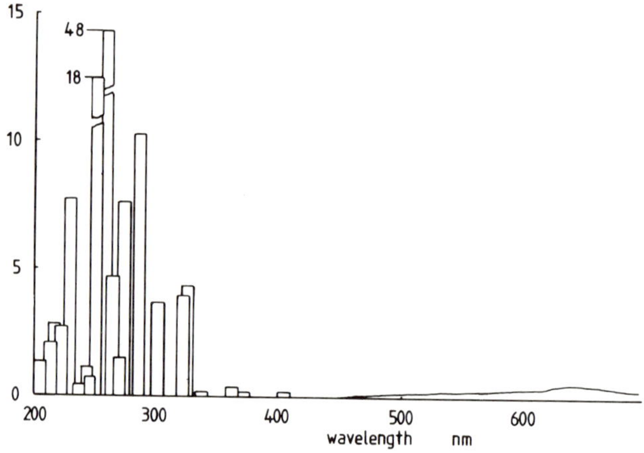

Fig. 9.6. Spectral output of antimony iodide lamp. (Taken from British patent 1,552,334, W.C. Heraeus GmbH, Department Original Hanau, and reproduced with permission.)

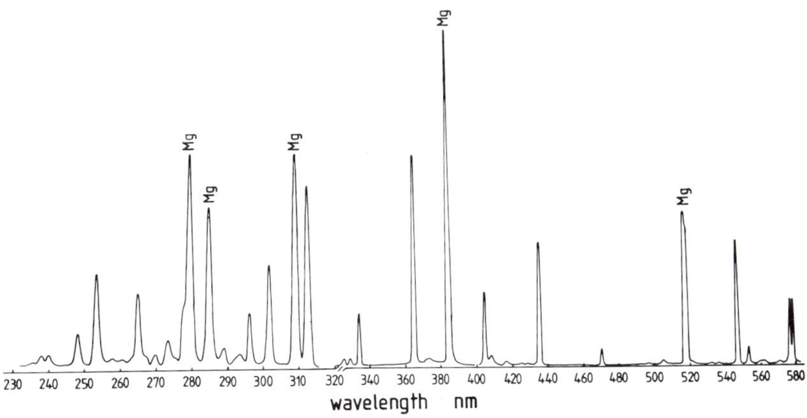

Fig. 9.7. Spectral output of Thorn MBIL magnesium iodide lamp.

9.7.6 *Magnesium*

The Thorn MBIL 1800W UV curing lamp utilizes magnesium iodide. A detailed description has been given by Beeson and Furmidge (1976). It is similar in construction to a 1600 W metal halide lamp designed for the floodlighting of sports stadia. The arc length is 19 cm and the bore of the tube 2·0 cm. It is designed to run at 130 ± 15 V and a maximum current of 14 A. The low voltage allows operation from a 220–240 V mains supply without the need for a step-up transformer. However, a high voltage (3–4 kV) pulse ignitor is required. A reactor ballast is used, employing components originally designed for high pressure sodium street lamps, and therefore readily available at low cost. This lamp is used as the basis of the Thorn/Beasley French "Uvicure" system, in which an array of lamps parallel to each other, with a flat reflector, is placed above a conveyor system. The lamps are parallel to the direction of movement of the conveyor. The "Uvicure" system is described in British Patent 1,440,077 (1976) and by Hankins (1974) and Bridge and Poulter (1975).

The Thorn lamp is designed for a full power rating of 100 W cm^{-1} and a useful life of 1000 h is claimed. The spectral distribution of the radiation emitted is shown in Fig. 9.7. The extra lines of magnesium are clearly visible.

References

Ashurkov, S. G. and Sarychev, G. S. (1975). *Svetotekhnika* (5), 10–11.
Ayotte, A. D. and Hale, R. R. (1967). *Illum. Eng.* **62,** 221–228.

Beeson, E. J. G. (1976). *Radiat. Curing* **3** (2), 17–19.
Beeson, E. J. G. and Furmidge, K. F. (1972). *Light. Res. Technol.* **4,** 250–253.
Beeson, E. J. G. and Furmidge, K. F. (1976). *Professional Printer* **20** (1), 4–6.
Beijer, L. B., Jacobs, C. A. J. and Tol, T. (1968). *Philips Tech. Rev.* **29,** 353–362.
Bouwknegt, A., Nienhuis, H., Schipper, D. J. and Tielemans, P. A. W. (1975). *Philips Tech. Rev.* **35,** 356–359.
Bridge, N. K. and Poulter, S. R. C. (1975). "Advances in Printing Sciences and Technology: Proceedings of the 13th International Conference of Printing Research Institutes", 179–187.
Chalmers, A. G., Wharmby, D. O. and Whittaker, F. L. (1975). *Light. Res. Technol.* **7,** 11–18.
Cohen, S. (1965). *Illum. Eng.* **60,** 660–666.
Datta, R. K. (1975). US Patent 3,919,581 (General Electric Co.).
Delrieu, P. and Taxil, A. (1964). *Bull. Soc. Franc. Electriciens* **5,** 80–93.
Dobrusskin, A. (1971). *Light. Res. Technol.* **3,** 125–132.
Dobrusskin, A., Leyendecker, H., and Schmid, G. (1975). *Int. Kongr. Reprogr. Inf., 4th* (*Fachref*), 9–14; *Chem. Abstr.* **84,** 52070.
Drop, P. C., De Groot, J. J., Jack, A. G. and Rouweler, G. C. J. (1974). *Light. Res. Technol.*, **6,** 212–216.
Drop, P. C., Fischer, E., Oostvogels, F. and Wesselink, G. A. (1975). *Philips Tech. Rev.* **35,** 347–353.
Elenbaas, W. (1972). "Light Sources", Philips Technical Library. Macmillan, London.
Fischer, E. (1976). *J. Appl. Phys.* **47,** 2954–2960.
Fischer, M. (1978). *Angew. Chem. Int. Ed. Engl.* **17,** 16–26.
Franke, A., Gungle, W. C., Ring, J. F. and Waymouth, J. F. (1967). *Illum. Eng.* **62,** 204–213.
Frier, J. P. and Henderson, A. J. (1973). *J. Illum. Eng. Soc.* **3,** 83–86.
Fromm, D. C., Seehawer, J. and Wagner, W. J. (1979). *Light. Res. Technol.* **11,** 1–8.
Gallo, C. F. (1966). *Appl. Opt.* **5,** 1285–1291.
Gallo, C. F. (1967). *Appl. Opt.* **6,** 1563–1565.
Gallo, C. F. (1970). *Appl. Opt.* **9,** 2711–2716.
Gallo, C. F. (1971). *Appl. Opt.* **10,** 2517–2520.
Gardner, P. J. (1975). SME Tech. Pap. (Ser.) FC75-333.
Gardner, P. J., Morris, J. C., Watson, W. R., Silver, H. G. and Scholz, J. A. (1975). *J. Illum. Eng. Soc.* **5,** 45–49.
Gion, H., Okumura, H. and Nakamura, N. (1974). *Natl. Tech. Rep. Matsushita Electr. Ind.* **30,** 332–339.
Gungle, W. C., Koury, F. and Waymouth, J. F. (1970). US Patent 3,514,659 (Sylvania Electric Products Inc.).
Hankins, W. C. (1974). *Offset Printing* (70), 34–36.
Heller, J., Amlong, U. J., Erbs, H., Heymel, G. and Netzel, H. (1969). *Elektrie* **23,** 170–172.
Hirayama, C., Castle, P. M., Liu, C. S. and Zollweg, R. J. (1977). *J. Illum. Eng. Soc.* **6,** 209–214.
Hulme, B. E., Hird, M. J. and Marron, J. J. (1976). SME Tech. Pap. (Ser.) FC76-492.
Jack, A. G. and Koedam, M. (1974). *J. Illum. Eng. Soc.* **3,** 323–329.
Kanoh, T. and Shinjoh, K. (1981). *J. Illum. Eng. Soc.* **10,** 140–144.

Keeffe, W. M. (1980). *IEE Proc. A* **127,** 181–189.
Kenty, C. (1938). *J. Appl. Phys.* **9,** 53–66.
Koury, F., Gungle, W. C. and Waymouth, J. F. (1975). *J. Illum. Eng. Soc.* **4,** 106–110.
Koury, F. and Waymouth, J. F. (1966). US Patent 3,250,940 (Sylvania Electric Products Inc.).
Lake, W. H. and Kramel, L. F. (1967). *Illum. Eng.* **62,** 631–637.
Larson, D. A., Fraser, H. D., Cushing, W. V. and Unglert, M. C. (1963). *Illum. Eng.* **50,** 434–439.
Lowke, J. J. (1970). *J. Appl. Phys.* **41,** 2588–2600.
Nakabayashi, S. (1975). SME Tech. Pap (Ser.) FC75-322.
Otto, J.-L. (1974). *Rev. Gen. Electr.* (*France*) **83,** 583–587.
Philips NV Gloeilampenfabrieken (1967). Fr. Patent 1,467,694.
Priest, D. K. and Eckel, R. A. (1976). *Radiat. Curing* **3** (2), 2,4–7.
Prikupets, L. B., Bukhanov, Yu. A and Sarychev, G. S. (1972). *Svetotekhnika* (5), 5–7; *Chem. Abstr.* **82,** 24234.
Rautenberg, T. H., Jr and Johnson, P. D. (1964) *Appl. Opt.* **3,** 487–492.
Reiling, G. H. (1964). *J. Opt. Soc. Amer.* **54,** 532–540.
Reiling, G. H. (1966). US Patent 3,234,421 (General Electric Co.).
Rothwell, H. L. and Keeffe, W. M. (1980). *J. Illum. Eng. Soc.* **10,** 40–46.
Sarychev, G. S. (1967). *Svetotekhnika* (1), 15–16.
Sarychev, G. S., Gavrilkina, G. N., Ermakov, A. I. and Bukhanov, Yu. A. (1967). *Svetotekhnika* (6), 1–4; *Chem Abstr.* **69,** 55371.
Sarychev, G. S., Gavrilkina, G. N. and Ashurkov, S. G. (1979). *Svetotekhnika* (9), 5–8.
Schäfer, J. (1979). Brit. Patent 1,552,334 (Original Hanau Quarzlampen GmbH).
Schirmer, H., Friedrich, F. J. and Seehawer, J. (1969). *Tech. Wiss. Abhandl. Osram Gesell.* **10,** 25–32.
Smithells, C. J. (1976). "Metals Reference Book", 5th edn. Butterworth, London.
Stull, D. R. (1947). *Ind. Eng. Chem.* **39,** 540–550.
Slagg, N. (1967). *Illum. Eng.* **62,** 621–625.
Sucov, E. W. (1970). *J. Opt. Soc. Amer.* **60,** 777–781.
Taxil, A. (1974). *Rev. Gen. Electr.* (*France*) **83,** 577–582.
Taxil, A. M. V., Divoux, J. C. and Demaeyer, J. H. (1977). US Patent 4,021,696 (ITT Industries Inc.).
Tchang, L. I. (1974). *Rev. Gen. Electr.* (*France*) **83,** 595–598.
Vermeulen, J. (1974). *Rev. Gen. Electr.* (*France*) **83,** 589–594.
Waymouth, J. F. (1971a). "Electric Discharge Lamps". MIT Presss, Cambridge, Mass.
Waymouth, J. F. (1971b). *Proc. IEEE* **59,** 629–633.
Work, D. E. (1981). *Light. Res. Technol.* **13,** 143–152.
Zollweg, R. J., Ennulat, D. F., Slagg, N. and Ward, P. C. (1970). US Patent 3,530,327 (Westinghouse Electric Corp.).
Zollweg, R. J. (1974). Contribution to discussion following paper by Jack and Koedam (1974).
Zollweg, R. J., Liu, C. S., Hirayama, C. and McNall, J. W. (1975). *J. Illum. Eng. Soc.* **4,** 249–253.

10 Electrodeless Lamps

10.1 Current Coupling

The medium pressure mercury lamps described in Chapter 8 are exclusively those employing electrodes as current coupling devices. Although the majority of lamps used for chemical applications are of this type, a significant development in recent years has been that of the electrodeless lamp manufactured by Fusion Systems Corporation of Rockville, Maryland, USA. By a combination of genuine advantages and an energetic marketing policy, Fusion Systems Corporation has captured a significant fraction of the UV drier market. Instead of using electrodes to feed energy into the discharge, the discharge tube absorbs microwave energy fed via waveguides into a microwave chamber in which the tube is housed.

The general concept of electrodeless lamps is not new. They have found applications in spectroscopy, photochemistry and analytical chemistry (see Gleason and Pertel, 1971, and references therein). However, they have been almost exclusively low powered, low pressure lamps, designed to produce light of high spectral purity. The application of the concept to high powered medium pressure lamps is of great interest, and deserves a detailed description.

The FSC lamp system is covered by US Patents 3,872,349, 3,911,318 and 4,042,850 and UK Patent 1,482,950, and has been described in a number of papers published by FSC personnel (Eastlund *et al.*, 1975; Eastlund, 1976; Greenblatt, 1976; Levine, 1977; de Saussure, 1978; Spero, 1978; Spero and Matthews, 1979).

10.2 Construction and Operation

The FSC lamp system is modular and consists of two parts, an irradiator and a power supply. The irradiator is shown in Figs 10.1 and 10.2. It contains a microwave chamber formed by an anodized aluminium reflector of semi-elliptical cross section, with flat ends, and a fine wire mesh screen. The lamp itself is a closed tube of transparent vitreous silica, of length 24·4 cm, varying in internal diameter from 8 mm near the ends to 6 mm at the centre. The lamp is located so that its axis lies at the focus of the ellipse, and it acts as a dissipative load.

Microwave energy is generated by two 1500 W magnetrons and is fed through waveguides into the chamber via rectangular slots cut in the back of the reflector. The microwave frequency used is 2450 MHz. The magnetrons and waveguides are cooled by a filtered air flow, and this air is also passed through small circular holes cut in the back of the reflector, and over the lamp.

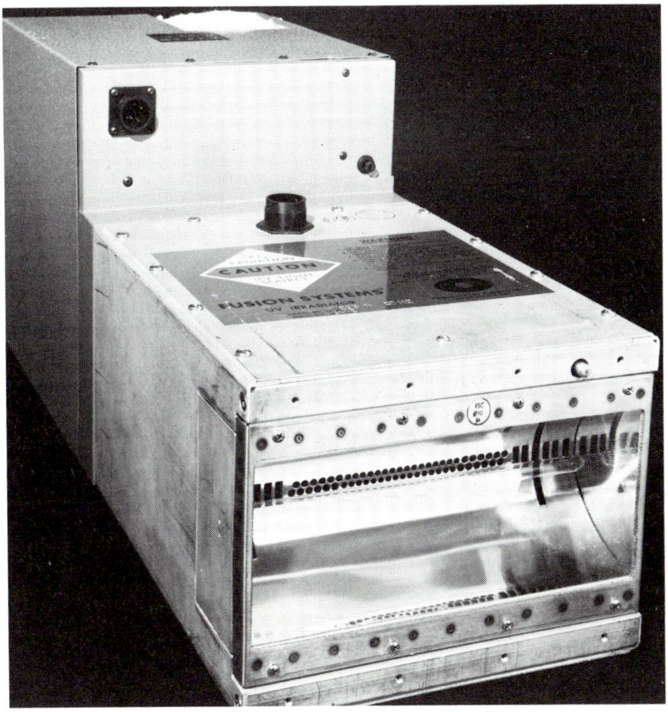

Fig. 10.1. View of irradiator showing lamp and reflector.

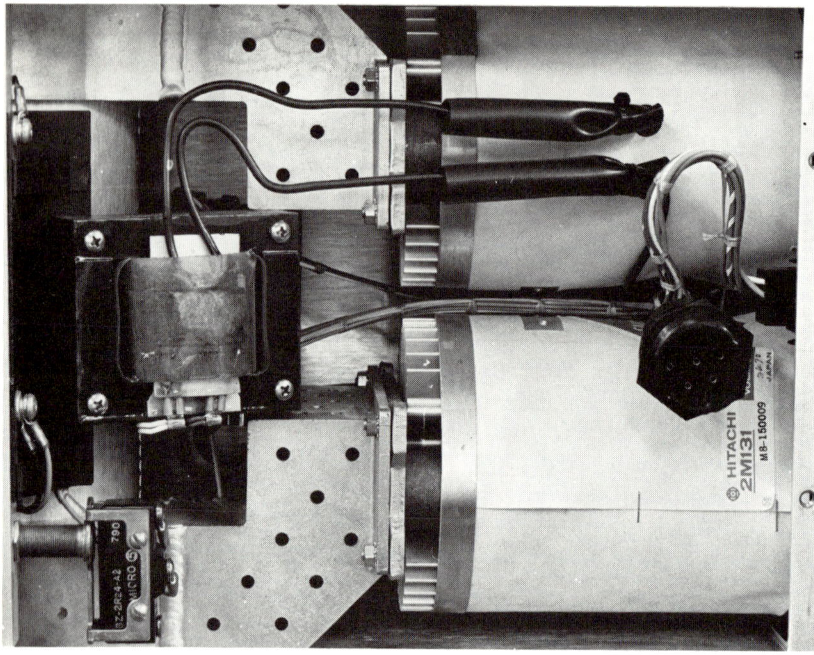

Fig. 10.2. View of irradiator showing magnetrons and waveguides.

A description of the principles of operation is given in US Patent 4,042,850. The system is designed to avoid resonances or cavity-like effects, as these would set up standing waves which would result in the concentration of microwave energy in local regions of the tube and a non-uniform light output. In order to avoid resonances, the coupling slots are offset both axially and azimuthally, and the magnetrons are matched pairs which differ in frequency by 15 MHz.

The power supply contains two full wavelength voltage doubler circuits which supply a regulated source of high voltage power to each magnetron. The system is well equipped with fail-safe devices. For instance, a microwave detector close to the irradiator will automatically switch off the system if the leakage exceeds 5 mW cm^{-2}.

When the system is switched on, the magnetron filaments warm up first. This takes about 10 s. The high voltage is then supplied to the magnetrons. A probe in one of the waveguides picks up microwave energy and this is used to activate a small low pressure mercury lamp situated behind the reflector. The 254 nm radiation passes through holes in the back of the reflector and strikes mercury droplets in the main tube, causing the emission of photoelectrons. These electrons accelerate in the microwave field and an avalanche occurs with a rapid increase in the number of current carriers. The electrons absorb energy from the field and the mercury rapidly evaporates as the lamp warms up. The time between ignition and full light output is only 2–3 s, very much less than that for a medium pressure electrode lamp operated from a conventional ballast.

The lamp fill is normally mercury and an inert gas, but considerable modification of the spectral distribution is possible if suitable metal halides are added to the fill.

10.3 Advantages of the Electrodeless Lamp

Although the electrodeless lamp system is rather more complicated than the medium pressure electrode lamp operated from a conventional ballast, it does have some very significant attractions:

(1) The lamp warms up very quickly from cold; in about 12 s as compared with 20 s to 3 min for a conventional electrode lamp. If the magnetron filaments are kept heated in a standby mode, the lamp can be extinguished and then re-ignited almost instantaneously whenever radiation is required. Therefore, in the standby mode, the lamp has an *instant-on* capability, which may be of considerable practical advantage in certain applications.

(2) The lamp has no electrodes and has a much narrower and lighter tube than a conventional electrode lamp. This results in a much lower heat capacity, and a reduction in the amount of infra-red radiated by the materials of the lamp (as opposed to the discharge). When the lamp is extinguished, the residual radiation of energy is at a very low level, and insufficient to cause overheating of heat-sensitive materials which are stationary in front of the lamp. This means that the lamp has a virtually *instant-off* capability, which eliminates the need for shutters and simplifies the construction of the reflector module.

(3) The lack of electrodes eliminates the lamp deterioration processes associated with electrodes (see Section 8.9), and very long lifetimes are claimed—about three times those of conventional electrode lamps.

(4) The lack of any possible interaction with electrodes allows a wide choice of additives (i.e. metal halides) to enhance parts of the spectrum for particular applications. The high power density in the

lamp and the lack of cold spots behind electrodes probably makes it easier to incorporate the less volatile metal halides, and maximize the enhancement obtained.

(5) There are no electrode losses.

(6) The discharge fills virtually the entire length of the tube, and therefore its length is only slightly less than the overall length of the module. With an electrode lamp, the inter-electrode distance is inevitably several centimetres less than the overall length of the lamp because of the space taken by the electrode seals and connections. Unlike the electrode lamp, the electrodeless lamp provides the possibility of forming a modular array of lamps end to end giving an uninterrupted radiation pattern.

10.4 Disadvantages of the Electrodeless Lamp

There are a number of disadvantages which must be set against the attractions stated above:

(1) The lamp is available in one size only, and therefore to cover a wide sheet or web it is necessary to butt a number of irradiators end to end. This is not a disadvantage of the electrodeless lamp *per se*, only of what is commercially available at the time of writing. However, a shorter, lower powered unit has been announced (Anon., 1981), and also an intriguing range of spherical lamps designed as deep UV sources for microlithography (Ury *et al.*, 1982).

(2) Because the irradiator module contains not only the lamp and reflector but also the magnetrons, waveguides, etc., it is quite large—26·7 cm × 20·3 cm × 35·6 cm. The size is even greater if an integral blower is mounted over the magnetrons. This may make it difficult to fit in situations where space is limited.

(3) There is some attenuation of the radiation emitted by the lamp in the wire mesh screen. Microscopic examination shows that the open area is about 90% of the whole. Attenuation by the wire mesh screen will be discussed further in Section 10.5.1.

(4) The two magnetrons deliver 3000 W to the lamp. However, the magnetrons generate microwave energy with rather low efficiency, with the result that the total system requires a power input of 5500 W. This represents a power loss greatly in excess of the ballast and electrode losses experienced with electrode lamps (Chapter 8).

(5) Although the lamps should have very long lifetimes, and they are

guaranteed for 3000 h, the magnetrons have a limited lifetime, and require replacement at regular intervals (about every 5000 h), so the cost benefit of long lamp life is largely offset by the cost of magnetron replacement.

10.5 Spectral Output

A claim frequently made in the papers which have been published about the FSC electrodeless lamp is that the UV output is very much greater than that of a conventional 80 W cm^{-1} electrode lamp of the type used for UV curing. For instance, Spero and Matthews (1979) have given the following analysis of the power input and the power output of the two types of lamp:

	Electrode lamp (W in^{-1})	Electrodeless lamp (W in^{-1})
Total power input	240	550
Total lamp output	200	300
UV	34	109·5
Visible	56	83·6
Infrared	110	50·0
Convected	—	56·9

1 W in^{-1} = 0·394 W cm^{-1}.

The claim that the electrodeless lamp gives over three times as much UV radiation as the conventional electrode lamp is of great interest, as it does appear to be inconsistent with the generally accepted concepts of mercury discharges as described in Chapter 8. Electrode losses in conventional lamps are quite small, and may represent an insignificant fraction of the total power input for a long tube using a moderate current.

The non-radiative losses of the conventional electrode lamp are fairly well defined (see Section 8.7). The non-radiative losses for the electrodeless lamp are less easy to predict, but it seems unlikely that they will be less than those for the electrode lamp. The absorption by the tube should depend on the thickness and quality of the silica used, and therefore there should be no significant difference in this respect. Levine (1977) has pointed out that the electrodeless discharge fills the tube cross section and that there should be much less self-absorption of radiation in cooler gas regions than is the case with the conventional arc. In fact the plasma cannot completely fill the tube and there must be some insulating layer of cooler gas to prevent the silica melting. However, visual and photographic examination of the lamp shows that the discharge does appear to occupy almost the complete cross section.* On the other hand, the mercury pressure in the electrodeless lamp is quite a lot higher than is normal in the electrode lamp. Levine quotes a range of 5–20 atm. Therefore, although the dark annulus may be narrow, it will still contain quite a large number of atoms. Figure 10.3 shows the spectral distribution of the output of the FSC mercury lamp in the region between 246 and 272 nm. The self-absorption of the resonance radiation and the extreme pressure broadening are clearly visible.

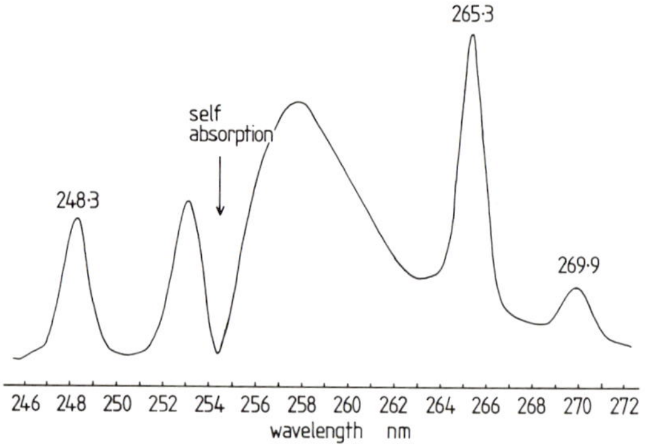

Fig. 10.3. Spectral distribution in 246–272 nm region for FSC H lamp (reflector blackened). Compare Figs 8.23 and 8.24.

The narrowness of the insulating layer implies that the radial temperature gradient at the tube wall may be very high, and therefore that the conductive heat loss, $H = -\kappa(\mathrm{d}T/\mathrm{d}r)_R 2\pi R$ (Waymouth, 1971, p. 152), may also be high, despite the low value of R. Convective losses are also difficult to predict. However, it is known that the combination of a high mercury pressure and a narrow bore lamp is conducive to turbulent gas flows and significant convective loss (Fohl, 1975).

The figures quoted by Spero and Matthews for the conventional electrode lamp suggest that the total visible output is quite a lot higher than the total UV output. The data presented in Chapter 8 indicates that on the contrary the *total* UV output is considerably greater than the total visible output. This has also been found for the UV Standard lamp (Table 8.6) and by workers who have investigated total energy balances for mercury arcs (e.g. Rautenberg and Johnson, 1964; Jack and Koedam, 1974). The probable reason for the discrepancy is that the figure of 34 W in^{-1} is taken

† Recent measurements by Ury *et al.* (1982) of the radiant radial power distribution for a spherical lamp 0·8 in (20·32 mm) in diameter have shown that there is a minimum at the centre of the sphere, and a peak about half way between the centre and the inside wall of the lamp.

from Canrad Hanovia (US) data for the 200 W in^{-1} electrode lamp and is in fact only the sum of the outputs in the most significant *lines* between 248 and 365 nm. Tables 8.6 shows that about one-third of the UV output of the UV Standard (incidentally a rather low pressure lamp), is continuum, and therefore it is clear that the figure of 34 W in^{-1} is a gross underestimate of the *total* UV output of the 200 W in^{-1} electrode lamp. The figure of 109·5 W in^{-1} given for the UV output of the electrodeless lamp is rather higher than the figure of 97 W in^{-1} quoted in some previous papers (e.g. Levine, 1977). This figure is the sum of the *total* outputs in 10 nm intervals between 200 and 400 nm.

The figures given by Spero and Matthews for the infra-red and convective heat outputs of the two types of lamp are worth some comment. If one uses Elenbaas's expression, $P_{rad} = 0{\cdot}75(P - 4{\cdot}5P^{1/4})$ (see Section 8.7), one can see that loss to the tube is about 76 W in^{-1} in the case of a 200 W in^{-1} lamp, and about 105 W in^{-1} for a 300 W in^{-1} lamp. The tube will lose energy both by infra-red radiation, mainly above 1·9 μm (Rautenburg and Johnson, 1964), and by convection. The ratio of the types of loss will depend on the type of cooling employed. For instance, if without forced cooling a tube operates at 800 °C, and then is subjected to an air flow which reduces its temperature to 600 °C, the infra-red output from the tube will be reduced to only about 50% of its original value, and the remaining 50% will be lost as convected heat. There is obviously a minimum temperature at which the tube can operate, and as the electrode lamp has a higher surface area it will inevitably generate more infra-red radiation. However, the sum of infra-red and convected heat should be higher for the electrodeless lamp at 300 W in^{-1} than for a conventional electrode lamp at 200 W in^{-1}. Also a great amount of convected heat is generated by the magnetrons used to power the electrodeless lamp.

In order to clarify the situation I have made as direct as possible a comparison between the spectral outputs of FSC electrodeless lamps and those of the conventional electrode lamps described in Chapter 8. Four electrodeless lamp types, comprising the standard mercury (H type; six samples) and three additive lamps (coded A, D and M by FSC; one sample of each) were examined.

10.5.1 "Light Bug" measurements

The FSC irradiator module was mounted over a conveyor belt moving at 0·5 m s^{-1}, so that the lamp was horizontal and oriented perpendicular to the direction of travel of the conveyor. "Light Bug" measurements were made. Details of the instrument are given in Section 4.4. The results are given in Table 10.1, and may be compared with the data shown in Fig. 8.29. The reflector materials and dimensions were similar for the two types of lamp (electrode and electrodeless) and therefore the results give a good general indication of the UV output of each lamp in the region between 250 and 400 nm. Note that the H lamp gives an output equivalent to that of an electrode lamp operating at between about 90 and 110 W cm^{-1}, depending on the mercury pressure. Note also the very high values given by the additive lamps. The reasons for these very high values will be explained later.

Table 10.1 Light Bug measurements: FSC electrodeless lamps

Lamp type	Light Bug reading (J cm^{-2})
Mercury H (six tubes)	0·270, 0·266, 0·259 0·263, 0·265, 0·265
Additive A	0·354
Additive D	0·459
Additive M	0·317

The Light Bug provided a method of estimating the true attenuation caused by the wire mesh screen. The reflector was removed from the irradiator module, and was used to replace a similarly shaped reflector in an electrode lamp module. Light Bug measurements with and without the screen in position showed that the attenuation, due to the screen, of the integrated irradiance received by the instrument was 14%. The attenuation will clearly depend on the angle at which light strikes the screen. Light striking the screen at a normal to the plane of the mesh was found to be attenuated by 11%.

10.5.2 Measurement of spectral distribution

The measurement of the spectral distribution of radiation from electrodeless lamps poses some difficulties because the reflector functions not only as a reflector, but also as part of the microwave chamber. It is preferable to examine only radiation emitted directly from the lamp, and to eliminate reflected radiation. It was feared that blackening the entire reflector might prejudice the operation of the lamp. To overcome this problem a 5 cm square area of the reflector was blackened by spraying carbon black from a specially filled aerosol. The carbon black film remained unchanged after many measurements, but was easily wiped off with no residual effect on the reflector surface. The sprayed area was directly behind the lamp, and offset from the centre so that the nearest edge was 2 cm from the microwave slits. Measurements were made with the lamp horizontal. A collimated beam was taken from a 2·8 cm length of the lamp to a monochromator, as shown in Fig. 10.4. The scanning of the spectrum and the manipulation of the detector response data was controlled by a small computer, as described in Section 4.3.3.3.

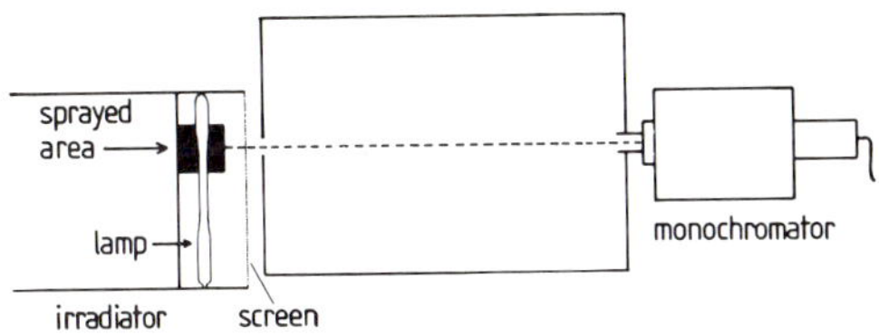

Fig. 10.4. Measurement of spectral distribution.

The computer was programmed to produce 1 nm interval histograms in tabular form. The units, as produced by the computer, were arbitrary, and a calibration procedure was needed. The Light Bug method, as described in Section 4.2, was used, and it was gratifying that all four lamp types, despite their very different spectral distributions, gave calibration factors within 2% of each other. Figure 10.5 shows a 1 nm interval histogram in graphical form for one of the H tubes (No. 4H13996). Also shown is a histogram for an electrode lamp with an arc input power of 115 W cm^{-1}, measured under identical conditions. Histograms for the three additive lamps are shown in Figs 10.6, 10.7 and 10.8. Twenty nanometre histograms were derived, and these are shown in Table 10.2. The data for the electrode lamp are included for comparison.

The light output waveforms for the mercury lamp H at a variety of wavelengths were displayed on an oscilloscope and the results are shown in Fig. 10.9. The waveform is repeated every half cycle; as in Fig. 8.34, the ratio of maximum to minimum light output varies with wavelength and can be related to the upper state energy of the electronic transition involved.

10.5.3 Interpretation

Table 10.2 shows that the total output between 200 and 600 nm of the electrodeless lamp with the standard mercury fill is about the same as the electrode lamp with the same power input. However, the distribution of radiation is markedly different, and the histograms shown in Fig. 10.5 show the extremes of spectral distribution found in the medium pressure range. The electrodeless lamp gives significantly less UV radiation, and

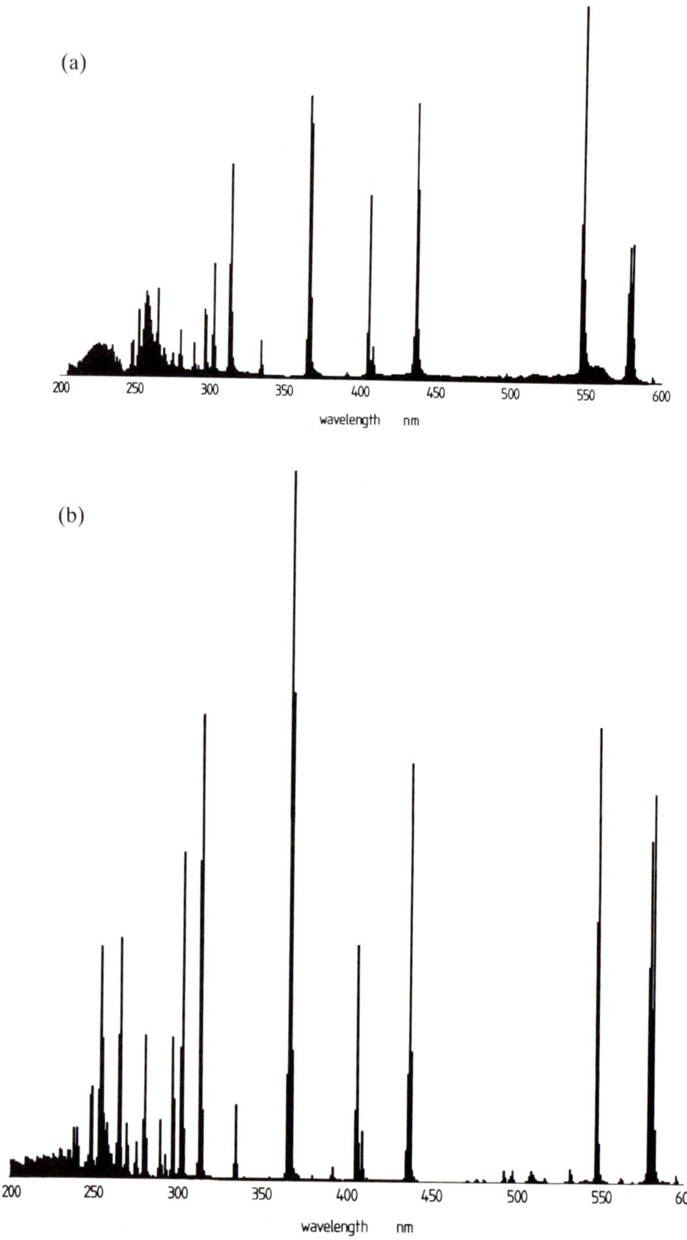

Fig. 10.5. One nanometre histograms: (a) electrodeless mercury lamp, tube 4H13996; (b) electrode lamp, type III, 115 W cm^{-1}. Both lamps operated horizontally.

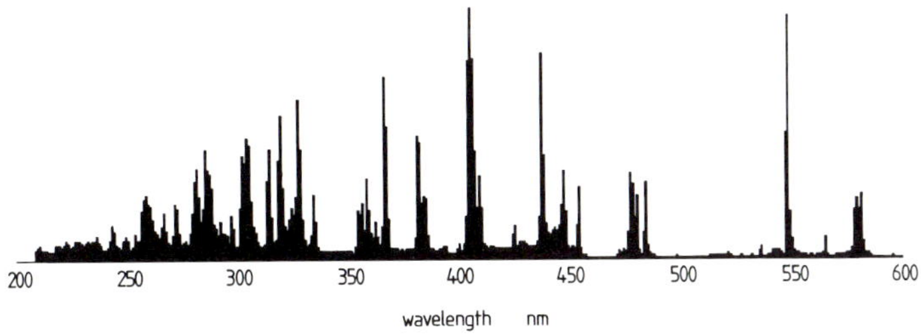

Fig. 10.6. Spectral output of FSC "A" lamp. One nanometre histogram.

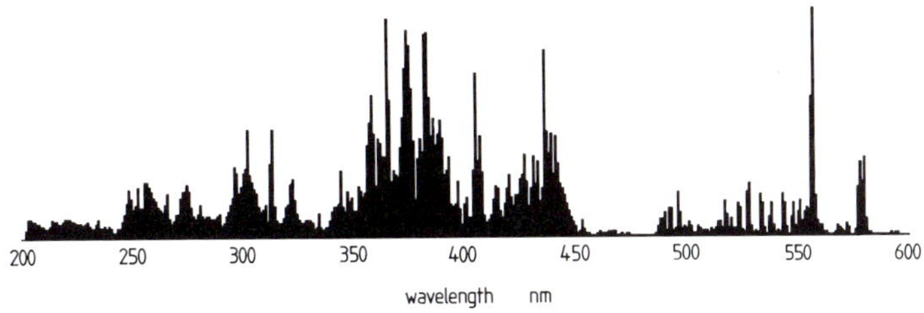

Fig. 10.7. Spectral output of FSC "D" lamp. One nanometre histogram.

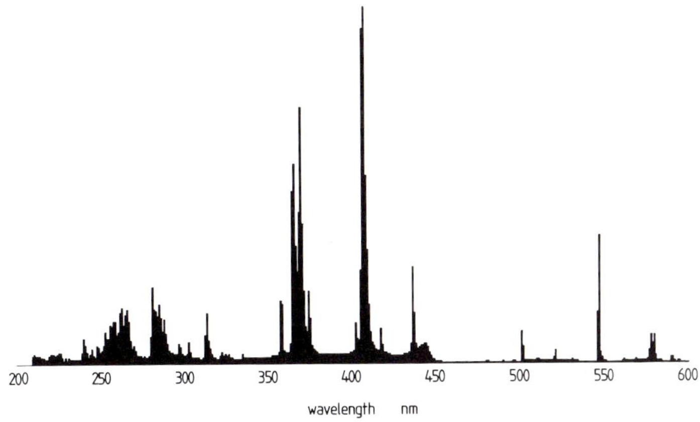

Fig. 10.8. Spectral output of FSC "M" lamp. One nanometre histogram.

significantly more visible radiation than the electrode lamp. This is a result of the much higher mercury pressure in the electrodeless lamp, and the greater number of atoms which take part in the plasma. The lower plasma temperature enhances the output of the visible lines which have 7S upper states, while depressing the UV lines which are emitted from higher states. There is also a much higher continuum in the visible part of the spectrum.

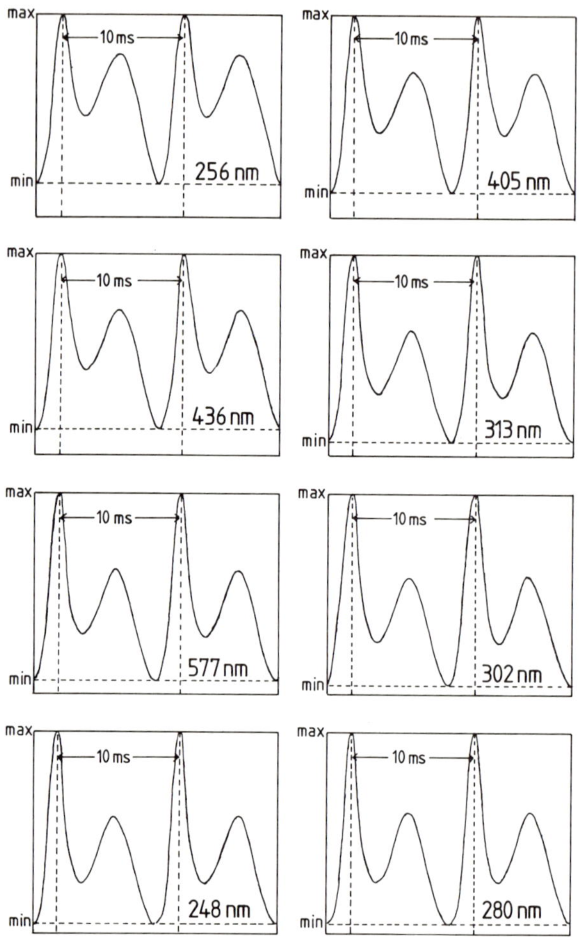

Fig. 10.9. Light waveforms for different wavelengths. H lamp (tube 3H4625). 50 Hz.

The spectra of the additive lamps show that the additive elements have very significant effects on the spectral distributions. The mercury lines are all depressed. Analysis of the extra lines introduced by the additives shows that the A lamp contains tin, the D lamp iron and the M lamp lead. All these elements have fairly volatile iodides. The spectrum of the D lamp may be compared with the spectrum of the iron iodide electrode lamp published by Gardner *et al.* (1975), and with the spectrum of the iron iodide electrode lamp shown in Fig. 9.5. All three additive lamps have less radiation below 280 nm than the standard mercury lamp, but more in

Table 10.2 Twenty nanometre histograms for electrodeless lamps. Horizontal operation. Output in walls per centimetre. The data for an electrode lamp (arc power input 115 W cm^{-1}) are included for comparison

Wavelength interval (nm)	FSC H lamp 4H13996	Electrode lamp III	FSC D lamp D304	FSC M lamp M48	FSC A lamp A100
200–220	1·65	1·83	1·51	0·59	0·77
220–240	3·86	2·63	1·45	1·04	1·71
240–260	4·41	5·35	3·10	2·74	2·80
260–280	4·10	3·96	3·09	3·09	3·10
280–300	2·62	2·95	2·85	3·74	4·48
300–320	4·99	7·50	4·34	1·60	6·02
320–340	0·99	0·71	1·91	0·93	3·84
340–360	0·39	0·19	5·07	1·67	2·19
360–380	5·29	7·78	10·32	9·55	3·03
380–400	0·49	0·26	8·42	1·31	2·80
400–420	3·14	2·31	4·48	8·87	5·44
420–440	4·89	3·90	6·45	2·19	3·18
440–460	0·95	0·03	2·62	1·06	2·53
460–480	0·76	0·08	0·32	0·22	1·73
480–500	0·85	0·24	1·25	0·27	0·81
500–520	0·99	0·27	1·07	0·71	0·33
520–540	1·11	0·21	1·99	0·48	0·39
540–560	6·74	4·13	3·00	1·47	2·57
560–580	4·28	5·98	1·56	0·96	1·56
580–600	0·84	0·52	0·35	0·41	0·36
Sum: 200–400	28·79	33·16	42·06	26·26	30·74
Sum: 400–600	24·55	17·67	23·09	16·64	18·90
Sum: 200–600	53·34	50·83	65·15	42·90	49·64

certain higher wavelength regions. The D lamp is of particular interest because of its large output in the important region between 360 and 420 nm. We have found that this lamp is particularly effective for the curing of thick pigmented films, especially those with titanium dioxide. Reference to the spectral sensitivity curve of the Light Bug, shown in Fig. 4.26, indicates that peak response is at about 350 nm. This explains why the additive lamps give higher Light Bug readings than the standard mercury lamp.

10.6 Applications

FSC electrodeless lamps are now being extensively used for the UV curing of inks and coatings in a wide variety of applications, and in particular where the short lamp length is adequate for the area to be irradiated. They have often been the first choice for the curing of prints on containers such as two-piece cans, plastic pots and bottles, and tubes, especially when on-mandrel irradiation is used (see de Saussure, 1978, and Section 14.8.5), although excellent electrode lamps are now becoming available for this application. The main attractions of the electrodeless lamp system have been described in Section 10.3. A further attraction is the small volume of the discharge, which leads to a high radiant emittance and the possibility of focusing the radiation from the lamp into a narrow beam on the substrate. The narrow tube also reduces the possibility of "lamp re-entry depletion" (Section 14.8.2). Although the magnetrons are wasteful of energy, the heat dissipated is blown on to the coating being cured, and often has a beneficial effect on the curing process.

The main reasons for the commercial success of the electrodeless lamp system have been that (a) it introduced a $120\,W\,cm^{-1}$ lamp when $80\,W\,cm^{-1}$ was still the norm for electrode lamps, and it is often possible to replace two $80\,W\,cm^{-1}$ lamps by a single $120\,W\,cm^{-1}$ lamp; (b) it is a well engineered modular system available "off the shelf" which can be easily fitted by manufacturers of printing and coating machinery, if there is space available for the rather bulky irradiator module; (c) it is safe to use with most heat-sensitive substrates without the need for a shutter; (d) its rapid strike/restrike capability, and the lack of influence of switching frequency on lamp life, makes it an attractive proposition in applications where there is batch processing, or continuous processing with frequent stoppages; (e) the range of additive lamps available allows some interesting and novel processes. For instance, one lamp (the V lamp, containing gallium) has a strong output in the blue part of the visible spectrum, but little short wavelength UV (Fig. 10.10). Free radical curing materials remain wet on the surface because of oxygen inhibition, but cure underneath (Matthews,

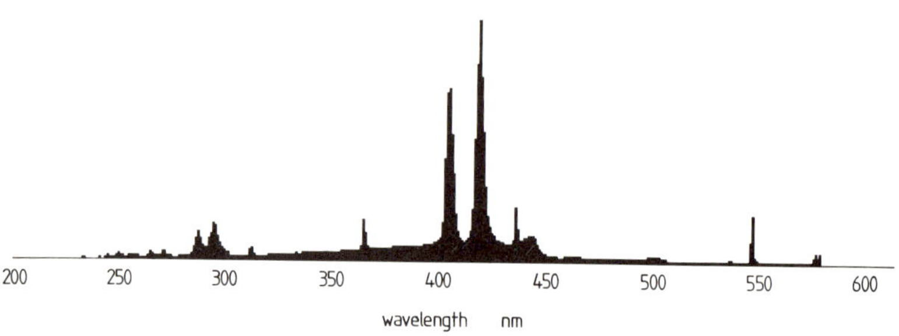

Fig. 10.10. Spectral output of FSC "V" lamp. One nanometre histogram. Note that the scale of the vertical axis is 2·5 times that in Figs 10.6, 10.7 and 10.8. Compare with Fig. 9.3.

1980). If matting agents are incorporated in the formulation, the shrinkage which takes place below the surface forces the matting agents to the surface. The surface is then cured by a standard mercury lamp to give the desired matt finish.

References

Anon. (1981). *UV/EB News* **1** (6), 4.

de Saussure, E. C. (1978). SME Tech. Pap. (Ser.) FC78-522.

Eastlund, B. J., Ury, M. G. and Wood, C. H. (1975). SME Tech. Pap. (Ser.) FC75-336.

Eastlund, B. J. (1976). SME Tech. Pap. (Ser.) FC76-481.

Fohl, T. (1975). *J. Illum, Eng. Soc.* **4,** 265–270.

Gardner, P. J., Morris, J. C., Watson, W. R., Silver, H. G. and Scholz, J. A. (1975). *J. Illum. Eng. Soc.* **5,** 45–49.

Gleason, W. S. and Pertel, R. (1971). *Rev. Sci. Instr.* **42,** 1638–1643.

Greenblatt, M. (1976). SME Tech. Pap. (Ser.) FC76-514.

Jack, A. G. and Koedam, M. (1974). *J. Illum. Eng. Soc.* **3,** 323–329.

Levine, L. S. (1977). *Int. J. Radiat. Phys. Chem.* **9,** 819–829.

Matthews, J. C. (1980). *Fifth Radiation Curing Conference, Boston, 23–25 May*, AFP-SME, 189–207.

Rautenberg, T. H., Jr and Johnson, P D. (1964). *Appl. Opt.* **3,** 487–492.

Spero, D. M. (1978). SME Tech. Pap. (Ser.) FC78-545.

Spero, D. M. and Matthews, J. C. (1979). *J. Radiat. Curing* **6** (8), 6–10.

Ury, M. G., Matthews, J. C. and Wood, C. H. (1982). *Proc. Soc. Photo-Opt. Instr. Eng.* **334,** 241–246.

Waymouth, J. F. (1971). "Electric Discharge Lamps". MIT Press, Cambridge, Mass.

11 Xenon Lamps

11.1 Inert Gas Discharges

The inert gases are clearly of considerable interest as potential fillings for discharge lamps because (a) they exist in atomic rather than molecular form, and therefore there are no vibrational, rotational or dissociative processes to dissipate energy which might otherwise be radiated; (b) they are chemically inert and will therefore not attack the electrodes, or a glass or silica envelope; and (c) they can be used at any desired pressure at ambient temperature—and therefore there is no need for a warm-up period as with mercury or sodium.

On the other hand, the inert gases have complete outer shells and therefore their ionization energies are high compared with those of metals. Furthermore, their lowest excited states (which are metastable) are much higher than those of mercury. Relevant data are shown in the following table:

Element	Ionization energy (eV)	Energy of lowest excited state (eV)
Mercury	10·4	4·67
Xenon	12·1	8·32
Krypton	14·0	9·91
Argon	15·8	11·6
Neon	21·6	16·6
Helium	24·6	19·8

Transitions involving the ground state give line spectra at quite short wavelengths, i.e. in the vacuum UV, while those which occur *between* excited states give lines mainly in the visible and infra-red regions. There are no lines in the xenon spectrum between 147 and 334 nm (Striganov and Sventitskii, 1968). The UV lines are weak because they require high upper state energies.

Xenon has the lowest ionization energy and the lowest thermal conductivity of the inert gases. It is also the most efficient in converting electrical energy into light (Hoyt and McCormick, 1950; Elenbaas, 1972, p. 178). Xenon is the most expensive of the inert gases, and its use in discharge lamps became a commercial proposition in the 1930s, after economical methods had been developed for separating it from liquid air.

Because of the large number of electrons in the outer shell of the xenon atom the energy level diagram is very complex (White, 1934, p. 274), and a large number of lines is possible. Indeed, Striganov and Sventitskii (1968) list no fewer than 589 lines for the neutral xenon atom. Lines involving the ground state will not be transmitted by a glass or silica envelope. The strongest lines are in the infra-red, and therefore, if a xenon lamp is to be a viable source of UV or visible light, it is necessary to operate under conditions which favour processes producing continuum radiation. This implies a high plasma density, and therefore a high current density.

11.2 Continuously Operating Lamps

Figure 11.1 shows the spectral output of a small tubular xenon lamp (3·8 cm arc, 1·1 cm bore) with a fill pressure of 1·7 atm, operated on direct current with a lamp voltage of 30 V and a lamp current of 35 A. Although the lamp has a high watts per centimetre rating, the spectrum is typical of a medium pressure lamp, with a strong line spectrum and a weak continuum. The line spectrum is largely confined to the infra-red, and therefore a lamp of this type is an inefficient source of UV and visible radiation, in contrast to the medium pressure mercury lamp.

A more efficient production of visible radiation can be achieved by increasing the fill pressure, as in the compact source xenon lamp. This is an electrode-confined† arc, usually operated on direct current, and first studied by Schulz (1944, 1947). Examples are shown in Fig. 11.2. Cold fill pressures are typically up to about 15 atm and during operation the

† That is, electrode confined rather than tube confined. Anderson (1951) distinguished between an electrode-stabilized arc (independent of lamp operating position) and a convection-controlled arc (the arc is significantly affected by convection currents; the arc spot on each electrode depends on lamp orientation, and the arc is significantly bowed when the electrode axis is horizontal).

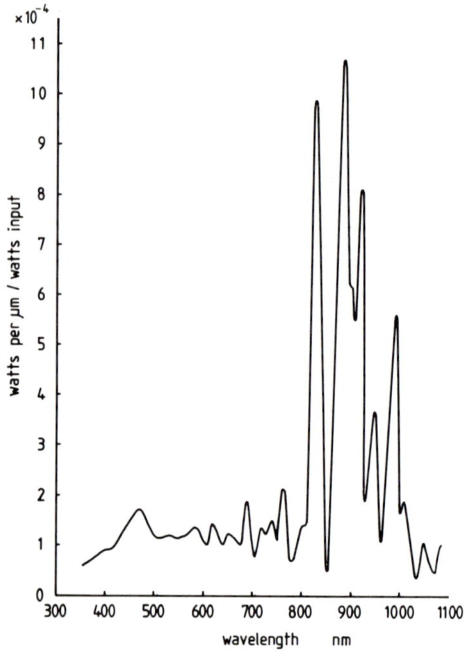

Fig. 11.1. Spectral output of a d.c. operated tubular xenon lamp. Current density $37\,A\,cm^{-2}$. (From Goncz and Newell, 1966.)

pressure may be up to 60 atm. The spherical bulb has the ideal shape for withstanding the pressure. The cathode of the d.c. lamp is small, with a pointed tip, and is usually made from thoriated tungsten. The anode is much larger and usually made from pure tungsten. Power ratings may be as low as 30 W or as high as 30 kW. At the higher power ratings (above 10 kW) the anode is normally water cooled.

Xenon presents two significant problems as a fill for a discharge lamp. There is local thermal equilibrium in the plasma, and therefore the laws of thermodynamics may be applied. In the Saha equation (Section 8.5) the ratio of statistical weights g_i/g_0 is 6 for xenon, whereas for mercury it is only 2. Also, at the predominant electron velocities in discharge lamps, the electron collision cross section for xenon is about 50 times smaller than that for mercury (Elenbaas, 1972, p. 176), which means that for a particular pressure and power input per centimetre the axial field in a xenon lamp is only about one-seventh of that in a mercury lamp. Therefore a high power input per centimetre requires a high current. Xenon lamps, even at quite high pressures, operate at low voltages and high currents. For instance a Thorn XE/D 2 kW compact arc, with a 6 mm interelectrode gap, operates

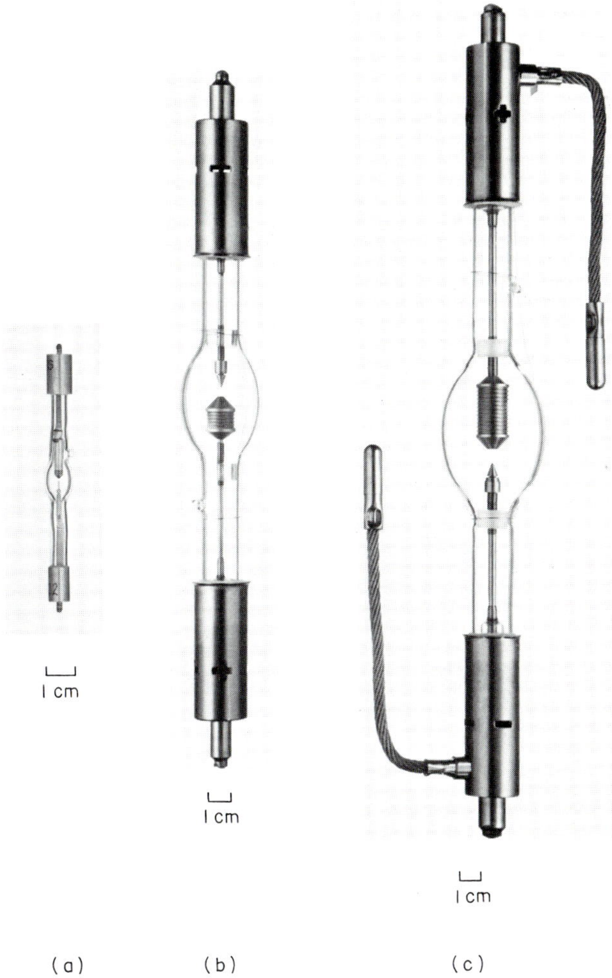

Fig. 11.2. Compact source xenon (CSX) lamps. Photographs by courtesy of Philips Lighting Division. (a) CSX 75W-2 OF (lamp power 75 W); (b) CSX 900W OF (lamp power 900 W); (c) CSX 2500W OF (lamp power 2500 W).

at 23·5 V and 85 A. High powered lamps operate with currents of several hundred amps. High currents make severe demands on the electrodes, cause large electrode losses, and necessitate heavy conductors in the external circuitry to avoid high circuit losses. The second problem is a consequence of the high cold fill pressure. This necessitates very high ignition voltages, and pulse starters delivering tens of thousands of volts are used.

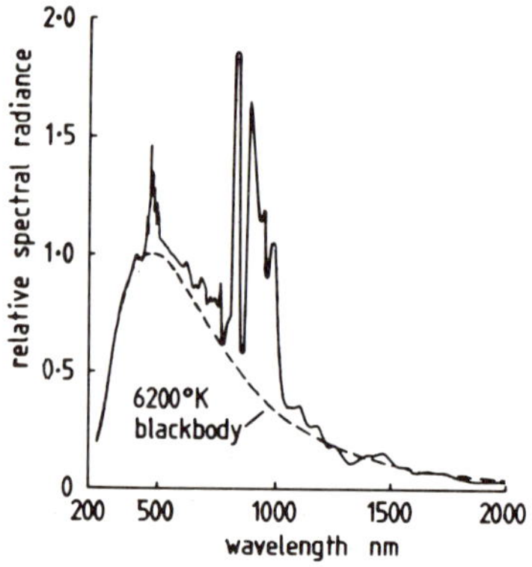

Fig. 11.3. Spectral energy distribution of a 6·5 kW compact source xenon lamp. (From Rehmet, 1980, with permission of Institute of Electrical Engineers.)

The spectral output of a 6·5 kW compact arc is shown in Fig. 11.3. The spectrum is still dominated by lines in the infra-red, but the continuum is enhanced by the higher pressure and higher current density. The continuum is largely due to electron–ion recombination processes (free–bound) and to a lesser degree *Bremsstrahlung* (free–free) processes (Finkelnburg, 1934; Rehmet, 1980). The spectral distribution of the continuum approximates to that of a black body. The colour temperature is in fact much lower than the actual arc temperature, because the emissivity of the arc increases with wavelength. Schirmer (1959) has shown that the relationship between arc temperature (T) and colour temperature (T_c) is given by

$$T = \frac{T_c}{1 - 6{\cdot}25 \times 10^{-5} T_c}.$$

Arc temperatures in xenon discharges are higher than those in mercury discharges because the excited states of the xenon atom are at higher energy levels.

The UV output of the compact source xenon arc has been studied by Baum and Dunkelman (1950), Searle *et al.* (1964) and Klein (1968). The radiance of the arc is greatest near the cathode. The efficiency of the xenon arc for UV production is poor compared with that of a high pressure mercury arc, but the very high radiance of continuum radiation from what is almost a point source makes it a useful illumination source for monochromators. The axial field and the UV output can be enhanced by adding a small quantity of mercury.

The output of the compact source xenon lamp may be boosted for short periods by superimposing on the continuum discharge large pulses from a capacitor discharge. This also alters the spectral distribution, the enhancement in output being much higher in the UV than at longer wavelengths (Hodgson and Keene, 1972; Beck, 1974).

High powered xenon compact arc lamps are very expensive, and it is of interest that lamps which are capable of being dismantled into their component parts have been devised to allow replacement of the cathode and anode head, and cleaning of the envelope (Stearn and Colliver, 1966; Popp and Weninger, 1973).

Most applications of compact source xenon arcs exploit the fact that the colour temperature of the continuum approximates to that of noon sunlight. The diameter of the arc increases with current, and the colour temperature remains fairly constant (Rehmet, 1980). Most lamps are used for illumination, for film projectors, searchlights, lighthouses, etc., but they are also used as solar simulators in applications where the actinic effects of sunlight are of interest. The very high power density in a very small source allows their use as sources in arc image furnaces.

Elenbaas (1949) studied tubular xenon lamps with cold fill pressures between 1 and 20 atm, and compared them with tubular medium pressure mercury lamps. Whereas the total radiation emitted by the mecury discharge is given by

$$P_{rad} = 0{\cdot}72(P - 10)$$

(see Section 8.7), the equivalent expression for the xenon discharge is

$$P_{rad} = 0{\cdot}55(P - 12).$$

The conductive loss from the arc to the tube is almost the same because the thermal conductivity of xenon is almost the same as that of mercury. The factor of 0·55 for xenon is less than the factor 0·72 for mercury because the resonance lines of xenon do not escape the tube.

Tubular xenon lamps operated by a.c. are used for various illumination purposes. Of interest are the long arc xenon (LAX) lamps used for large scale illumination. These lamps are unique among discharge lamps in that no ballast is required. They are connected directly across the a.c. mains, presenting an impedance of 3–4 Ω (Ryan, 1977). They are long lamps (1–2 m) running at low voltage (e.g. 200 V) with a high current (60–80 A).

11.3 Flashed and Pulsed Lamps

Xenon flashlamps have been used for many years for photographic purposes, for increased illumination in dim lighting conditions, and for specialist applications such as high speed photography and stroboscopes (Edgerton, 1970). They were introduced during the 1930s (LaPorte, 1937), and their use during the second world war for night aerial reconnaissance led to the development of large scale installations. The technology was quite well advanced by the late 1940s, and several excellent reviews appeared in the literature during this period (Edgerton, 1946; Carlson and Pritchard, 1947; Aldington and Meadowcroft, 1948). More recent uses, such as for flash photolysis and laser pumping, have encouraged a detailed examination of the physics of flashlamps.

Aspden (1959, p. xv) has commented that "the electronic flashtube is the most complicated artificial light source known. It is also the most flexible". Although much work has been done since 1959 when that statement was made, the xenon flash tube still remains an intriguingly mysterious object, and is flexibility in terms of the power, duration and spectral distribution of the light emitted still excites the imagination.

11.3.1 Construction

The flashlamp consists of a glass or silica envelope with two electrodes. These are usually of tungsten and of fairly massive construction to carry the large currents involved. Some lamps are polarized in so far as the cathode is slightly different from the anode and may be coated or doped with an emissive material. The xenon pressure is normally between 200 and 800 T when the lamp is cold, but is very much higher during the discharge (Guenther, 1970). The envelope is usually a fairly narrow tube, which may be linear, U-shaped or helical. The bore of the tube is kept narrow (compared with that of a medium pressure mercury or metal halide lamp) to confine the discharge and maintain a high current density. Lamps may be cooled by natural convection, by forced air cooling or by water passing through a jacket around the tube. Recent developments in the construction and sealing of lamps have been described by Ryan (1977). A selection of linear xenon flashlamps are shown in Fig. 11.4.

Transparent vitreous silica is a more attractive envelope material than glass because (a) it has better transmission properties, (b) it has a lower coefficient of expansion and therefore withstands thermal shock better, (c) it has a lower electrical conductivity.

11.3.2 Operation

The operation of a xenon flashlamp is basically simple. A capacitor is charged to a certain voltage and then the stored energy ($\frac{1}{2}CV^2$) is allowed to

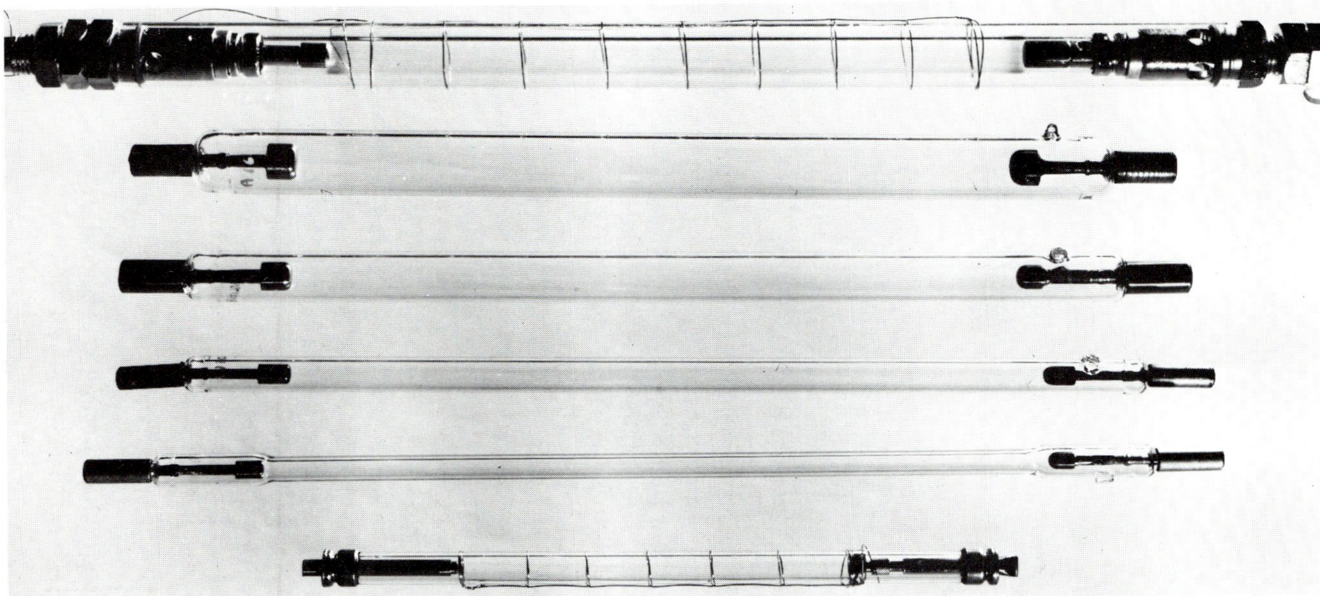

Fig. 11.4. A selection of linear xenon flash tubes. At the top is an EG & G FX225B6 water-cooled lamp (6 mm bore, 152 mm arc). At the bottom is an EG & G FX38C3 air-cooled tube (4 mm bore, 76 mm arc). These tubes are fitted with trigger windings. They have end cap seals. The metal end cap is soft soldered to the silica envelope. The other four lamps are Wingent air-cooled lamps, all of 150 mm arc, and with bores of 3, 5, 8 and 12 mm. These lamps have Vitreosil tubes, and graded seal construction. All the lamps in the figure are polarized. The air-cooled lamps have the cathode on the right. The water-cooled lamp has the cathode on the left.

discharge (usually to earth) through the lamp. During the discharge a fraction of the energy is converted to radiation.

A xenon flash tube may be operated either in a single flash mode with each flash being selectively triggered, or alternatively, flashing may take place at regular intervals, triggering being automatic and usually linked to mains frequency, or to the requirements of a process. If the interval between flashes is insufficient for complete deionization to take place a continuous discharge may occur. This is called "holdover" (Edgerton and Cahlander, 1961; Goncz, 1966).

11.3.3 Triggering the discharge

In most flashlamps the discharge is initiated by an external trigger, which normally consists of a fine wire wrapped around the tube, as shown in Fig. 11.4. This is fed with a high voltage pulse (e.g. 20 kV) which creates very high potential gradients within the gas column, causing local ionization and a reduction in the impedance of the tube which allows the capacitor to discharge itself through the tube.

The voltage to which the capacitor is charged must be above a certain value before discharge can take place. This voltage increases with xenon pressure and with the distance between the electrodes. For a 150 mm interelectrode distance and a pressure of 200 T the minimum voltage for reliable discharge is about 650 V, while for 400 T it is about 820 V. The discharge starts as a thin streamer often confined to a path along the inner surface of the tube (LeCompte and Edgerton, 1956), which expands over a period of microseconds to fill the tube. The rate of expansion is determined by the ratio of the electric field to the pressure (Baker and King, 1975). The larger the tube diameter the longer the period of expansion (Buck *et al.*, 1963). If the tube diameter is large, and the flash duration short, complete filling of the tube cross section may not take place (Heard, 1963a). Various methods of triggering have been discussed by Hook *et al.* (1972).

If the voltage is very high the capacitor may discharge spontaneously without the need for external triggering. Again, the voltage required (the spontaneous breakdown voltage) depends on the length of the lamp and the fill pressure, and is typically in the region of 2500 V for a 150 mm tube. Marshak (1963) found that the presence of a trigger electrode around the tube, even when uncharged, decreased the spontaneous breakdown voltage. Spontaneous discharge is unacceptable for most applications because the exact timing of the flash is normally critical, and also because discharge may occur before the capacitor has reached its full voltage. If operation above the spontaneous breakdown voltage is required, an external triggering electrode cannot be used, and a spark gap or thyratron is more appropriate. These essentially constitute a break in the circuit which is completed on triggering, and can carry the high current surge.

11.3.4 The single mesh circuit

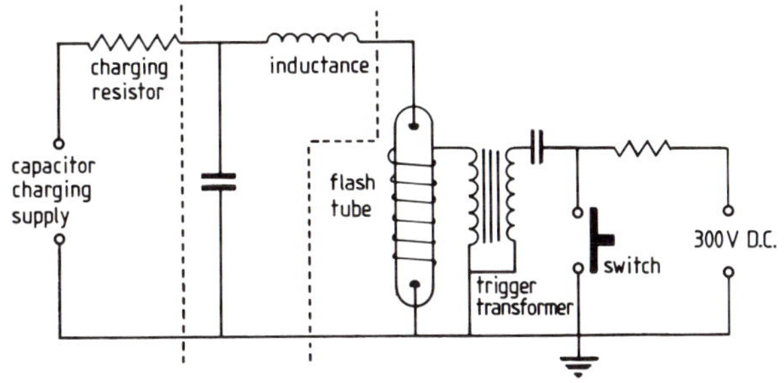

Fig. 11.5. The single mesh flash tube circuit with external triggering. The broken lines delineate the single LC section. In the multiple mesh circuit (see Section 11.3.8) several LC sections are joined in series. An alternative to external triggering is "series injection triggering". In this case the secondary of the trigger transformer is in series with the lamp and forms part of the discharge circuit. Lower voltages are required, and triggering is more reliable. However, the inductance of the transformer secondary will affect the pulse shape and the pulse duration.

Figure 11.5 shows a simple flash tube circuit with an external trigger electrode. This circuit is known as a "single mesh" circuit as it contains a single capacitor inductor combination. The discharge circuit may be represented in its simplest form as follows:

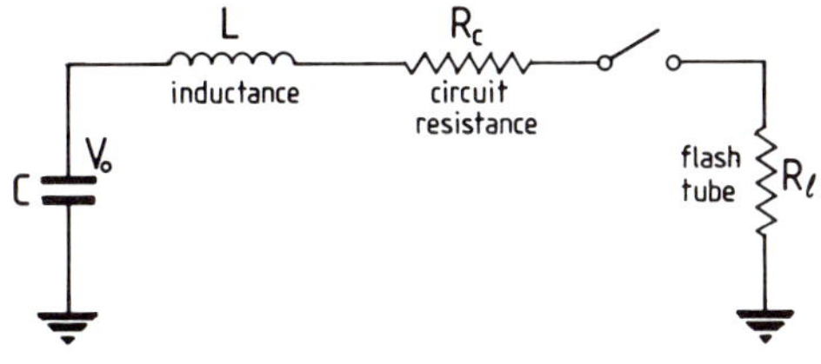

11.3.5 The damping parameter

The successful exploitation of flashlamps requires an understanding of the factors which determine the shape and duration of the current pulse. The variation of current with time is given by the equation

$$L\frac{\mathrm{d}i}{\mathrm{d}t} + V_l + R_c i + \frac{1}{C}\int_0^t i\mathrm{d}t = V_0. \tag{11.1}$$

V_0 is the initial capacitor voltage. V_l is the voltage between the electrodes of the lamp. R_c is the circuit resistance. The solution of this equation requires a knowledge of how the resistance of the lamp varies with the current passing through it. Goncz (1965) found that the resistivity of the plasma in a xenon flashlamp could be related to the current density by the expression

$$\rho = kj^{-1/2},$$

where j is the current density, and k is a constant which is independent of the tube dimensions, but is slightly dependent on the pressure. It follows that the resistance of the tube, R_l, can be related to the current by the expression

$$R_l = \frac{4}{\pi}\frac{l}{d}ki^{-1/2} = Ki^{-1/2},$$

where l is the interelectrode distance and d the bore of the tube. The voltage across the lamp is then

$$V_l = V_e \pm K|i|^{1/2},$$

where V_e is the electrode voltage drop. V_e for most lamps is very small compared with V_l and can be neglected.

Substituting for V_l in (11.1),

$$L\frac{\mathrm{d}i}{\mathrm{d}t} \pm K|i|^{1/2} + R_c i + \frac{1}{C}\int_0^t i\mathrm{d}t = V_0. \tag{11.2}$$

This equation was first solved by Markiewicz and Emmett (1966a), who made the following substitutions and normalizations: $\tau = t/T$, where $T = (LC)^{1/2}$; $I = (Z/V_0)i$, where $Z = (L/C)^{1/2}$; $\alpha = K(V_0 Z)^{1/2}$; $\beta = R_c/Z$. Equation (11.2) then becomes the two-parameter equation

$$\frac{\mathrm{d}I}{\mathrm{d}\tau} \pm [\alpha + \beta|I|^{1/2}]|I|^{1/2} + \int_0^t I\mathrm{d}\tau = 1. \tag{11.3}$$

In a well designed circuit β should be small, otherwise there will be an unacceptable circuit loss. Markiewicz and Emmett solved equation (11.3) for the case when $\beta = 0$. Under such circumstances equation (11.3) reduces to the single-parameter equation

$$\frac{\mathrm{d}I}{\mathrm{d}\tau} \pm \alpha|I|^{1/2} + \int_0^t I\mathrm{d}\tau = 1.$$

The shape of the current pulse depends only on the value of α, which is known as the "*damping parameter*". Markiewicz and Emmett computed the shapes of current pulses for different values of α, and their results are shown in Fig. 11.6. A value of $\alpha = 0{\cdot}77$ gives a "*critically damped*" pulse. Values over 0·77 give an overdamped pulse, while values under 0·77 give an underdamped pulse. Overdamped pulses give a rapid current rise and a slow decay, while underdamped pulses give a slower current rise and current reversal (ringing) during the decay. In general a critically damped pulse has the most desirable shape. The value of α is determined by the lamp dimensions and fill pressure, and by the operating conditions. Therefore the shape of the current pulse can be predicted and, conversely, conditions can be chosen to give a desired pulse shape.

Later authors (Holzrichter and Emmett, 1969; Anan'ev *et al.*, 1974; Dishington *et al.*, 1974; Brown and Nee, 1977) have considered the case where $\beta \neq 0$. As β is increased, the value of α for critical damping decreases. It is of interest that Brown and Nee (1977) repeated the analysis for the case $\beta = 0$, and found critical damping at $\alpha = 0{\cdot}84$ rather than 0·77.

In measuring the relationship between plasma resistivity and current density Goncz examined a wide range of flashlamps ranging in bore from 1·3 to 28 mm, with current densities from 400 to 10 000 A cm^{-2}. Goncz used pulse-forming networks to provide rectangular current pulses, and made measurements of current and voltage on the current plateau, i.e. under conditions where $\mathrm{d}i/\mathrm{d}t = 0$, and where the arc had filled the tube. Goncz found that for xenon at 200–400 T pressure, the constant $k = 1{\cdot}13\,\Omega\,\mathrm{A}^{-2}$. The value of k was 10% higher at 800 T. If we use Goncz's value of $1{\cdot}13\,\Omega\,\mathrm{A}^{-2}$, then

$$R_1 = 1{\cdot}28\left(\frac{l}{d}\right)i^{-1/2}$$

and therefore

$$\alpha = \frac{1{\cdot}28\,l}{d(V_0 Z)^{1/2}} = \frac{1{\cdot}28\,l}{dV_0^{1/2}(L/C)^{1/4}}$$

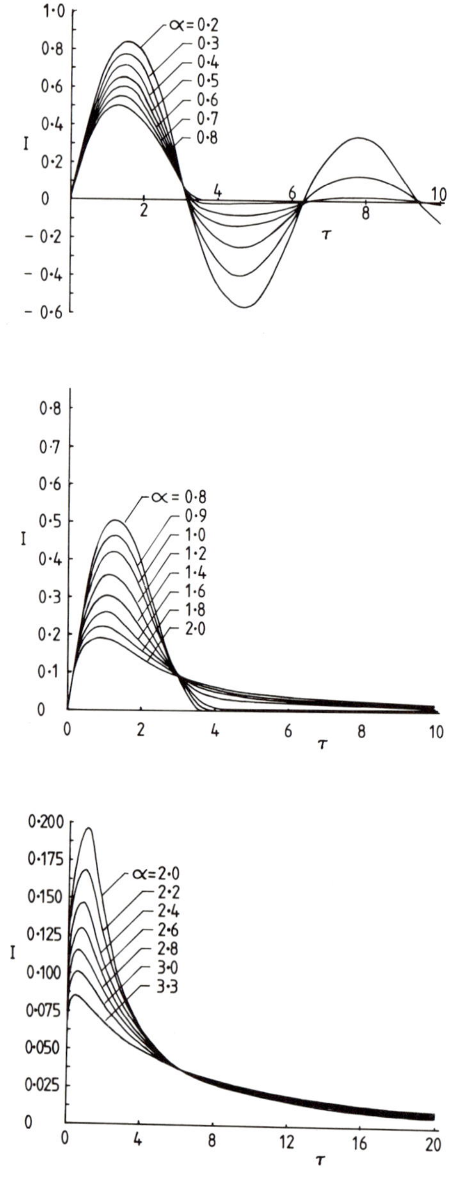

Fig. 11.6. Computed current pulse shapes from analysis of Markiewicz and Emmett (1966a). (Copyright © 1966 IEEE; reproduced with permission.)

11.3.6 Measurement of voltage and current pulses

(See Leonard, 1965.)

A storage oscilloscope is required for the display of these transient phenomena. The voltage at which lamps operate is in general too high for direct application across the terminals of an oscilloscope, and so a voltage-dividing network consisting of two high value resistors in series across the lamp terminals is used. If these are, for instance, in the ratio of 9 : 1, and the oscilloscope is connected across the lesser of the two, the voltage measured by the oscilloscope will be one-tenth of that across the lamp.

There are two methods of measuring current pulses. The more common is to place a low, accurately known resistance in series with the lamp, and measure the voltage drop across it using an oscilloscope. The resistance should be a small fraction of an ohm and should have a high heat capacity so that it does not warm up and change its value during use. The problem with this method is that even if the resistor used has a very small inductance associated with it, the very large rate of change of current during the early part of the pulse will cause an appreciable inductive impedance to be developed and a voltage proportional to the rate of current change. $\mathrm{d}i/\mathrm{d}t$ may be up to $10^9\ \mathrm{A\,s^{-1}}$, and therefore, even if the inductance of the resistor is only $10^{-2}\ \mu\mathrm{H}$, an inductive voltage of the order of 10 V will be generated (Markiewicz and Emmett, 1966b). An alternative method is to place a conducting coil around or near the lamp earth lead and measure the voltage developed across it. This method requires careful design if satisfactory results are to be obtained (Leonard, 1965; Markiewicz and Emmett, 1966b).

Figure 11.7 shows current pulses measured by use of a steel resistor in the earth lead for a lamp of 8 mm bore, 150 mm arc and 200 T pressure. Various paper capacitors were charged to 1500 V and then discharged through the lamp. No inductor was included in the circuit, but the inductance of the leads was measured to be about 3 μH. The damping parameters calculated for the five values of capacitance are as follows: at 33 μF, $\alpha = 1{\cdot}13$; at 66 μF, $\alpha = 1{\cdot}34$; at 99 μF, $\alpha = 1{\cdot}48$; at 132 μF, $\alpha = 1{\cdot}60$; at 165 μF, $\alpha = 1{\cdot}69$. As can be seen from Fig. 11.7, the measured pulse shapes fit quite well with those predicted from calculation. Note that the peak current through the lamp, and the time taken for this to be reached, is only weakly dependent on the capacitance above 66 μF. The main effect of increasing capacitance is to increase the damping parameter and slow the rate of decay. Results similar to those shown in Fig. 11.7 have been given by Edgerton (1946), Aldington and Meadowcroft (1948) and Edgerton *et al.* (1954).

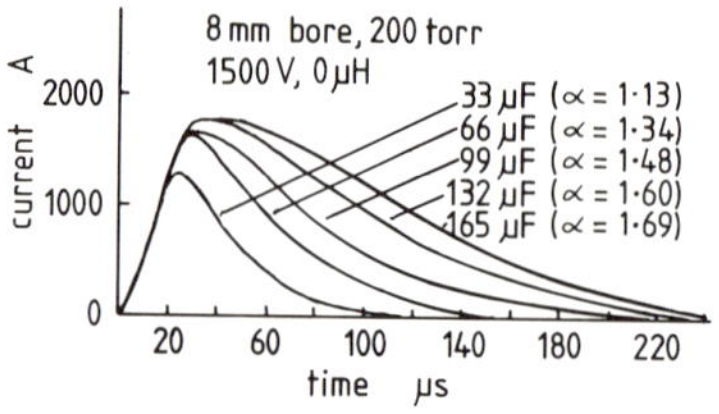

Fig. 11.7. Effect of capacitance on current pulse.

Peak currents, and current densities, are very much higher than those found in medium pressure mercury lamps. Goncz's expression for resistivity shows that for currents of the order of kiloamps, the resistance of a flash tube will typically be between 0·1 and 10 Ω, whereas the resistance of a medium pressure mercury lamp will be normally tens or hundreds of ohms (see Fig. 8.6). The very low resistance of flash tubes during much of the current pulse means that unless the leads are kept as short as possible, and are of heavy gauge, R_c will be comparable with R_l and a substantial energy loss will take place in the leads (Barnes, 1964; Oliver and Barnes, 1969).

Figure 11.8 shows how, for the 8 mm bore lamp, the shape of the current pulse responds to variations in capacitor voltage at constant capacitance.

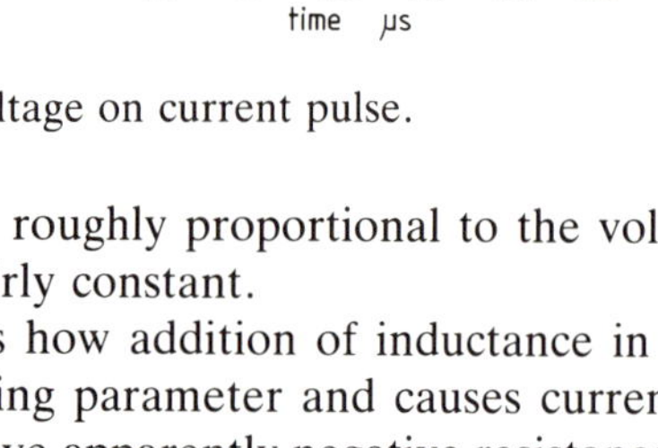

Fig. 11.8. Effect of voltage on current pulse.

The peak current is roughly proportional to the voltage, while the pulse duration remains fairly constant.

Figure 11.9 shows how addition of inductance in series with the lamp decreases the damping parameter and causes current reversal. This may cause the lamp to have apparently negative resistance for part of the pulse (Buck *et al.*, 1963). The current pulse shown in Fig. 11.9 has a much slower current rise time than those for the pulses shown in Figs 11.7 and 11.8. Note that in Fig. 11.6 the peak current appears at a higher value of τ as the damping parameter is reduced. Also, as $t = (LC)^{1/2}\tau$, an increase in L changes the relationship between t and τ, and also causes a decrease in the rate of current rise.

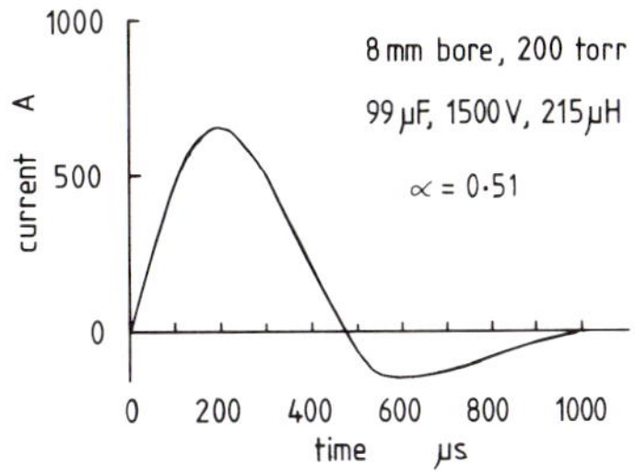

Fig. 11.9. Effect of added inductance on current pulse.

Figures 11.10 and 11.11 show pulses obtained with a 5 mm bore lamp with 400 T of xenon, with capacitance 165 μF and voltage 1800 V. Figure 11.10 shows the current pulse obtained without added inductance. The calculated value of α is 2·46, and it can be seen that this is a very overdamped pulse. There are some oscillations visible near the peak. This phenomenon was also observed by LeCompte and Edgerton (1956) who used a U-shaped tube to photograph arc cross sections during the build-up period, and found that the arc area oscillated. This oscillation could be related to similar oscillations in the current through the tube and in the conductance of the tube.

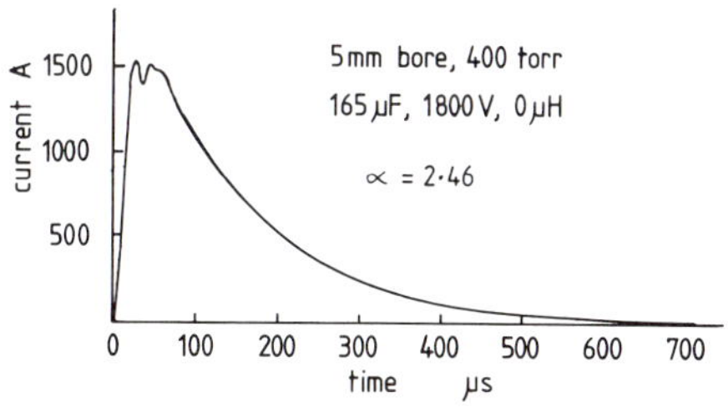

Fig. 11.10. Overdamped pulse showing current oscillation.

Figure 11.11 shows the effect of adding 215 μH series inductance. The calculated damping parameter is reduced to 0·84, predicting a current pulse which is almost critically damped, and the actual measurement is in agreement. Figure 11.11 also shows the voltage pulse and the resistance of the tube calculated from the current and voltage pulses. It is interesting to observe that the voltage across the lamp does not fall smoothly, but drops dramatically near the beginning of the pulse and then rises to a maximum before falling to a low constant value. The voltage falls rapidly during the early part of the pulse because the rapid rise in current induces a voltage drop across the series inductor equal to $L(\mathrm{d}i/\mathrm{d}t)$. Although the voltage on

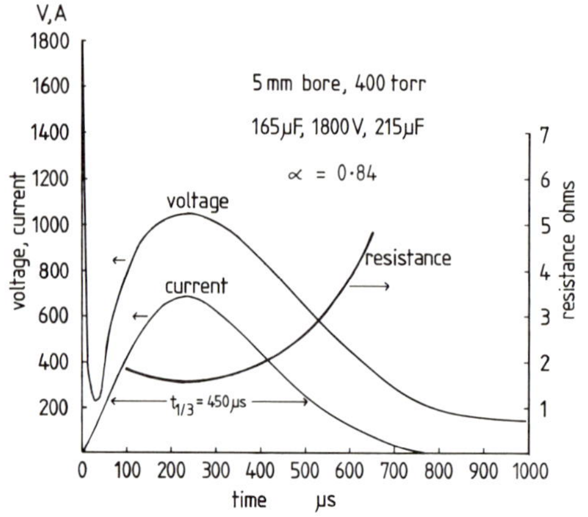

Fig. 11.11. Current, voltage and resistance transients for an (almost) critically damped pulse.

the capacitor may fall smoothly, this voltage is split between the series resistance of the leads (which should be very low), the series inductive impedance, and the resistance of the lamp. Series inductance decreases the voltage across the lamp during the current rise, and increases it during the current fall, and therefore effectively delays the dissipation of energy by the lamp.

The voltage does not fall to zero. Below a certain voltage, the "extinction voltage" (Marshak, 1963), the arc is lost and no further discharge is possible. The extinction voltage is typically between 100 and 200 V. Because the voltage does not fall to zero, not all the energy stored in the capacitor is discharged. However, the fraction remaining is usually very small.

The exact values of current during the period of rapid current rise are open to doubt for reasons already mentioned, and therefore resistance values were not calculated for this part of the pulse. To check Goncz's expression for resistivity, values of R_l calculated were plotted against $i^{-1/2}$, and the result is shown in Fig. 11.12, from which can be obtained the relationships

$$R = 1{\cdot}36\left(\frac{l}{d}\right)i^{-1/2},$$

$$\rho = 1{\cdot}20j^{-1/2},$$

which are in reasonable agreement with Goncz's result.

Several other workers have examined the relationship between lamp

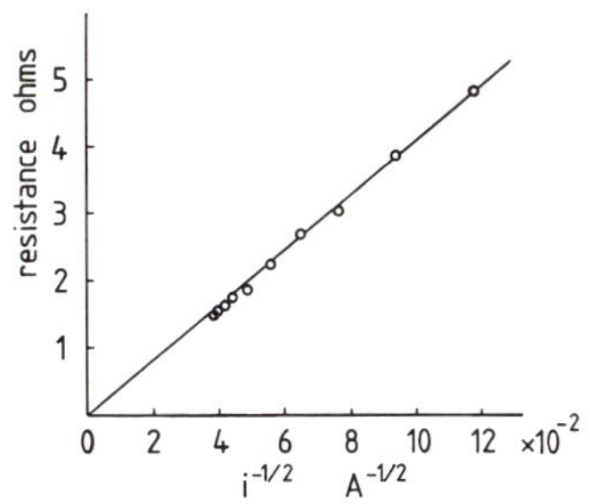

Fig. 11.12. Graph of R_l against $i^{-1/2}$ R_l taken from Fig. 11.11.

resistance and current. Demenik *et al.* (1968) examined lamps of internal bore between 11 and 36 mm, and with 300 T xenon, operating on a single mesh circuit. Resistance values were calculated for the current maximum ($di/dt = 0$). The relationship between resistivity and current density was found to be

$$\rho = (0{\cdot}8 - 0{\cdot}9)\, j^{-(0{\cdot}55 - 0{\cdot}50)}.$$

Demenik *et al.* suggested that the impedance of the tube resulted from electron–ion collisions rather than electron–atom collisions, on the grounds that the electron collision cross section is much greater for ions than for atoms. They deduced theoretically that the resistivity should be proportional to $j^{-6/11}$. Oliver and Barnes (1969) found a $j^{-1/2}$ dependence and measured the coefficient k for xenon, krypton and argon as a function of pressure. They found that for each gas the value of k increased with pressure over the range 250–750 T, and that the highest values of k were found with xenon and the lowest with argon. Their values of k for xenon were somewhat higher than those of Goncz, a result ascribed by Oliver and Barnes to impurities in their xenon. Mavroyannakis (1971) reviewed attempts to derive a relationship between resistivity and current density and deduced theoretically that $\rho = kj^{-3/5}$ for the part of the pulse near the current maximum.

Several authors (Heard, 1963b; Genkin and Mel'nikov, 1966; Anan'ev *et al.*, 1974) have measured lamp resistance as a function of current and have found that for a particular current value the resistance is normally higher during the period of current rise than it is during the period of decay. This hysteresis is due to thermal inertia and a finite deionization time. However, if the pulse is very strong evaporation of silica from the inner wall of the tube will raise the resistivity on the trailing edge of the pulse (Balagurov *et al.*, 1972).

The analysis of Markiewicz and Emmett is perhaps oversimplified, but it does provide a useful starting point, and is quite effective in predicting the shapes of current pulses. Converting a relationship between resistivity and current density into a relationship between resistance and current (or vice versa) requires either the assumption that the arc fills the tube, or an understanding of how the diameter of the discharge develops during the early part of the pulse. Dishington *et al.* (1974) refined the analysis of Markiewicz and Emmett to take account of the behaviour of the lamp during the time the arc is expanding to fill the tube. They found that the diameter to which the arc expands after a particular time is dependent on the energy which has been fed into it. This implies that if a low energy pulse is fed into a wide bore tube the arc may never expand sufficiently to fill the tube. This is important, because it has been suggested that energy fed into the lamp after the arc has filled the tube is more efficiently converted into light than energy fed in during arc expansion (Buck *et al.*, 1963).

11.3.7 Pulse duration

For many applications of xenon flash tubes the duration of the light pulse is of crucial importance. The *total* duration may be difficult to define if the pulse is overdamped or underdamped. It is therefore normal to define the pulse duration as the time interval between the point at which a certain fraction of the peak amplitude is reached during the rise and the point at which the same fraction of the peak amplitude is reached during the decay. Fractions commonly used are $\frac{1}{2}$, l/e, $\frac{1}{3}$ and 1/10.

There are two situations commonly encountered with single mesh circuits. If there is no added inductance in the circuit the situation approximates to that of a capacitor discharging through a resistor. If the resistance, R, were constant, and the initial voltage on the capacitance was V_0, then after time t the voltage would drop to

$$V = V_0 e^{-t/RC}.$$

After RC seconds the voltage would drop to a fraction l/e of its initial value. The power passing through the resistor is V^2/R and therefore the power after time t would be

$$P = \left(\frac{V_0^2}{R}\right) e^{-2t/RC}.$$

The power would therefore drop to a fraction l/e of its peak value after $RC/2$ seconds. The light pulse duration may be expected to be *roughly* similar to the power pulse duration.

In practice, of course, the resistance of a xenon flash tube is not constant, and if R is taken as the minimum resistance, corresponding to the maximum current density, then $RC/2$ will somewhat underestimate the pulse duration. In practice it is found that the pulse duration of an RC discharge lies between $RC/2$ and RC seconds, R being in ohms and C in farads.

It is often desirable to add series inductance in order to obtain a critically damped pulse. If the choice of lamp parameters (length, bore, fill pressure), initial capacitor voltage, capacitance and inductance are such that a damping parameter corresponding to critical damping is obtained, then it can be seen from Fig. 11.5 that the current pulse duration is about 3τ, i.e. $3(LC)^{1/2}$. Markiewicz and Emmett also gave power pulse shapes, and the power pulse duration is again about 3τ for critical damping.

11.3.8 Multiple mesh circuits

For some applications it is desirable to use a fairly lengthy flat topped current pulse. This can be obtained by using a multiple mesh circuit consisting of two or more LC networks (sections) in series. Goncz (1966) gives the pulse duration as

$$T = 2(L_t C_t)^{1/2},$$

where L_t is the total network inductance and C_t the total network impedance. Perlman (1966) found experimentally that for N sections, each containing inductance L and capacitance C, the pulse duration (one-third peak amplitude points) is given by

$$T = 2{\cdot}7N(LC)^{1/2}.$$

It can be seen that as the number of sections is increased the pulse is lengthened. The pulse also becomes more rectangular in shape.

11.3.9 Spectral output

Comparison of Figs 11.1 and 11.3 shows that the increase of pressure and current density increases the ratio of continuum radiation to line radiation. For maximum output in the actinic region of wavelengths it is clear that the line radiation, which is largely in the infra-red, is of little value. As the spectral distribution of the continuum radiation approximates to that of a black body (or a grey body) at a particular temperature, a logical approach to the increase of UV and blue light from a xenon lamp is to increase the

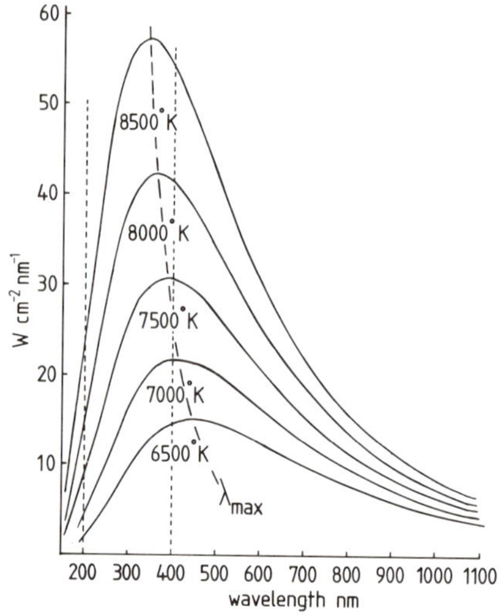

Fig. 11.13. Black body curves.

effective colour temperature of the discharge. Figure 11.13 shows black body curves calculated for a variety of temperatures relevant to xenon discharges. Figure 5.1 shows how the percentage of the total output of a black body radiator which is radiated between 200 and 400 nm depends on the temperature. It can be seen that as the temperature is raised above about 5000 K there is a rapid increase in the UV output.

Despite the transient nature of the pulse discharge the discharge column can be considered to be in local thermal equilibrium and therefore to be characterized by a plasma temperature T. The output in any wavelength region is a function of T and the spectral emissivity. As the spectral emissivity varies with wavelength, the colour temperature T_c differs from T. Another temperature which is relevant is the brightness temperature $T_{b\lambda}$. This is the temperature at which a black body of the same surface area would need to be in order to have the same output as the plasma at a particular wavelength. The main determinant of the spectral output of the xenon flash tube is the current density. As the current density increases, the opacity and emissivity of the discharge increase, so that ultimately the plasma radiates as a black body. The prominent spectral lines reach the black body limit first, followed by the continuum, with opacity at longer wavelengths being achieved at lower values of current density (Emmett *et al.*, 1964; Trenholme and Emmett, 1970). The colour temperature is therefore generally lower than the plasma temperature, and the brightness temperature increases with wavelength (Goncz and Newell, 1966).

Gusinow (1973) has derived theoretically a relationship between plasma temperature T and current density j (in amps per square centimetre) for the rather extreme case of a fully ionized plasma, the surface of which radiates as a black body:

$$T = 603\,(j\sqrt{R})^{4/11},$$

where R is the bore of the tube (in centimetres).

Figures 11.14–11.17 show spectral outputs of two xenon flash tubes measured under conditions chosen to illustrate the salient features of the spectral distribution. The method of measurement involved the use of an integrating radiometer, so that the spectral output measured was for the whole pulse, rather than for an instant within the pulse. It will be shown later that the instantaneous spectral distribution varies during the pulse.

The spectral distributions can be divided into four principal components: (a) continuum radiation which extends over the entire spectral range; (b) line radiation between 220 and 300 nm; (c) line radiation between 420 and 550 nm; (d) line radiation above 750 nm. It can be seen that the relative contribution of each of these components varies very significantly with the operating conditions.

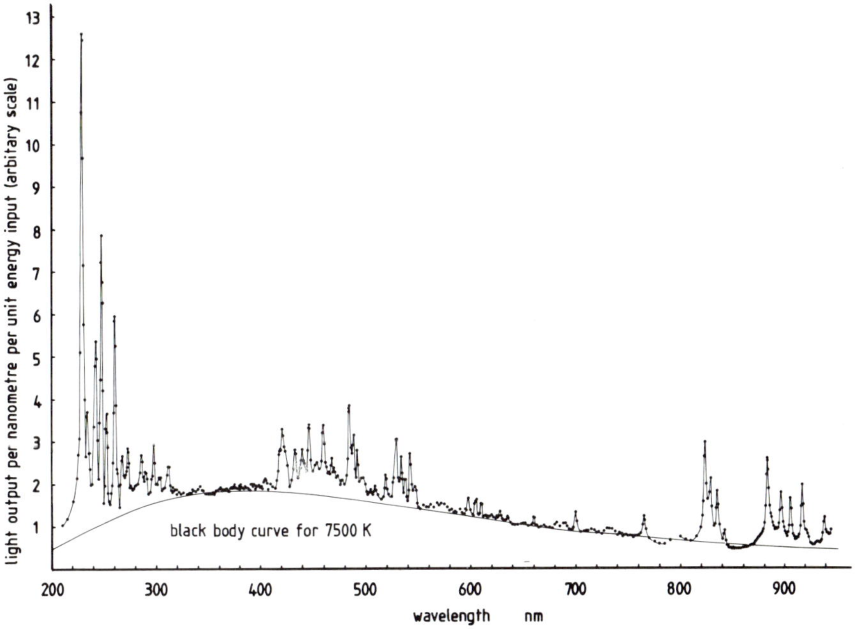

Fig. 11.14. Spectral output of xenon flash tube. 200 T xenon pressure (cold). 150 mm arc, 8 mm bore. 99 μF charged to 1500 V. $\frac{1}{2}CV^2 = 111$ J.

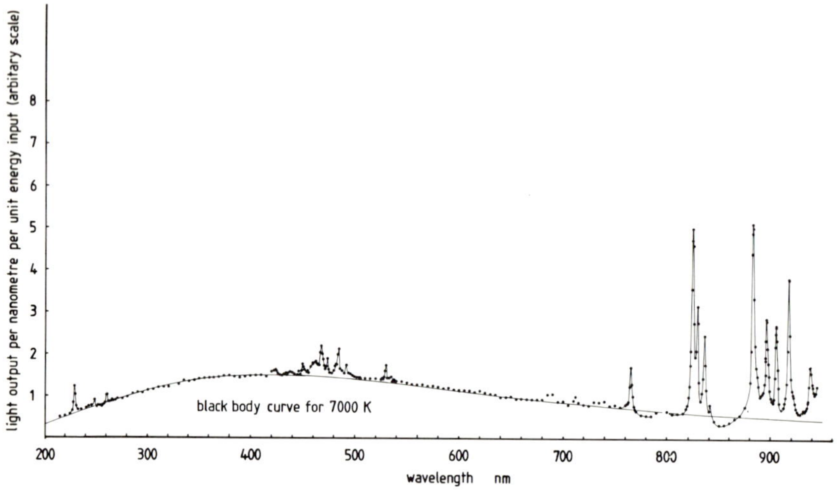

Fig. 11.15. Spectral output of xenon flash tube. 200 T xenon pressure (cold). 150 mm arc, 8 mm bore. 99 μF charged to 1500 V, 215 μH added series inductance. $\frac{1}{2}CV^2 = 111$ J.

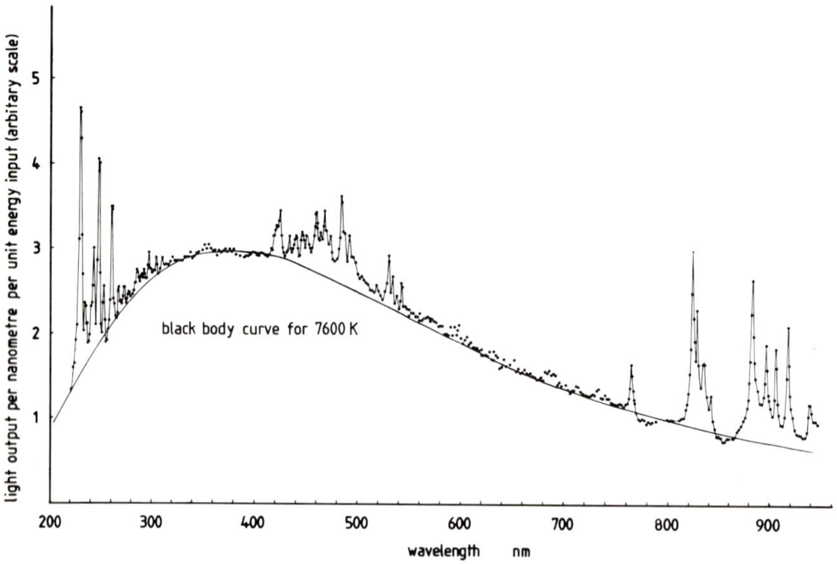

Fig. 11.16. Spectral output of xenon flash tube. 400 T xenon pressure (cold). 150 mm arc, 4 mm bore. 165 μF charged to 1800 V. $\frac{1}{2}CV^2 = 267$ J.

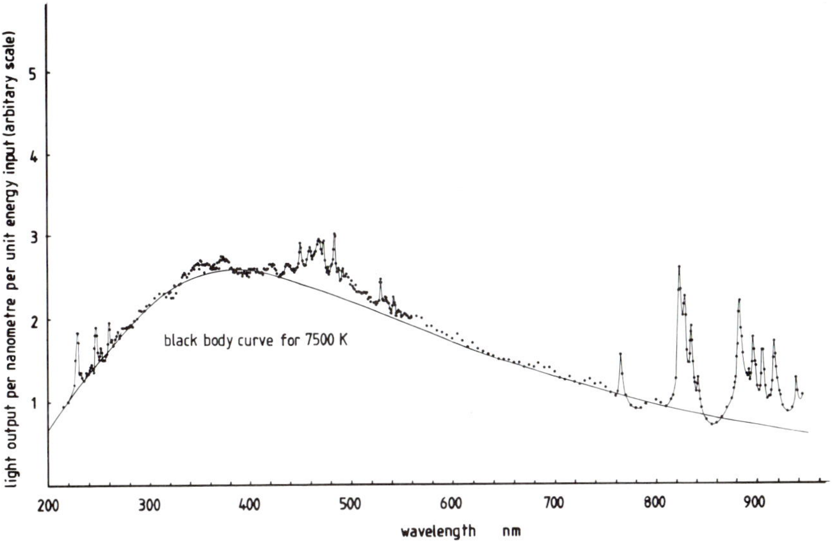

Fig. 11.17. Spectral output of xenon flash tube. 400 T xenon pressure (cold). 150 mm arc, 4 mm bore. 165 μF charged to 1800 V, 215 μH added series inductance. $\frac{1}{2}CV^2 = 267$ J.

The spectral distribution of the continuum radiation approximates to a black body curve. The continuous curves drawn in Figs 11.4–11.17 are the calculated black body curves which best fit the measured distribution. The colour temperature increases with the current density.

The lines between 220 and 300 nm are almost eliminated by the addition of series inductance, which also lengthens the pulse and markedly reduces the peak current density. The most probable explanation of these short wavelength lines is that they are the spectra of Xe(II). Striganov and Sventitskii (1968) list 202 lines for Xe(II) between 200 and 300 nm. By using their quoted intensities a histogram spectrum was drawn based on 1 nm wavelength intervals. The positions of the peaks were found to correspond well to those shown in Figs 11.14 and 11.16. Gusinow (1975) has published spectra for xenon flash tubes for the region 250–300 nm, and the well resolved lines correspond to those tabulated by Striganov and Sventitskii for Xe(II).

The line radiation in the region 420–550 nm is mainly the spectrum of Xe(I), where the upper state energy of the transition involved is quite high (about 11 eV).

The line radiation above 750 nm is also due to Xe(I), but mainly involves lower upper state energies (about 10 eV). Oliver and Barnes (1969) observed that the efficiency with which the near infra-red lines are produced decreases with increasing current density, and they suggested that this could be explained by the rapid saturation of the emissivity at these wavelengths.

The spectral distributions shown in Fig 11.14–11.17 were for flashes obtained under fairly modest operating conditions, well below the maximum limit of operation (i.e. the explosion limit—see Section 11.3.12). Nearer the limit, changes in the silica envelope affect the spectral distribution of the radiation escaping the tube. The inside wall of the tube experiences a rapid temperature rise, with two possible results. Silica evaporating from the wall enters the discharge and the lines of Si(I) appear, either in emission or absorption (Andreev and Gavrilov, 1974a; Andreev *et al.*, 1976). Secondly, the high temperature of the inside layer of silica causes a reversible increase in opacity for wavelengths below 300 nm (Podmoshenskii *et al.*, 1968; Belov *et al.*, 1969; Andreev and Gavrilov, 1974b). This opacity lasts only for a few milliseconds (Balagurov *et al.*, 1972) but may have a significant effect on the spectral distribution.

The spectral distributions shown in Figs 11.14–11.17 are for the entire light pulse. However, if the spectra had been taken at various times, e.g. 10 μs, 20 μs, 30 μs, etc., after the triggering of the pulse, each of these spectra would have been quite different from the integrated spectrum.

It is instructive to measure light pulses for different wavelengths. This was done for conditions giving the spectrum shown in Fig. 11.14. It was found that if a wavelength at which a prominent Xe(II) line appears was examined, the shape of the pulse was as shown in Fig. 11.18. The peak light

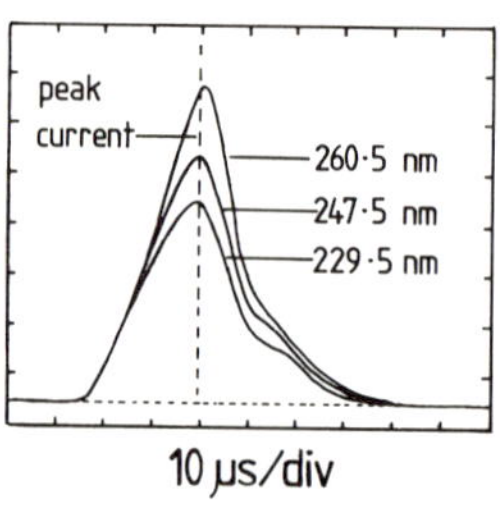

Fig. 11.18. Light pulses for wavelengths corresponding to Xe(II) lines.

corresponds almost exactly to the peak current (and peak power). It has generally been found that Xe(II) lines appear early in the flash (Ignat'ev and Podgaetskii, 1971; Bokova *et al.*, 1974), and may indeed peak before the current peak because of attenuation by reversible opacity of the tube. If a wavelength is chosen for which the output is predominantly continuum, the pulse shape is quite different, as shown in Fig. 11.19. The

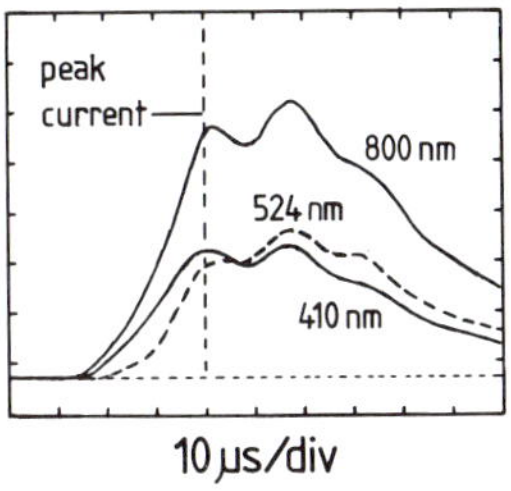

Fig. 11.19. Light pulses for wavelengths corresponding to predominantly continuum radiation.

output measured at any wavelength where there is a prominent line will of course consist of both line and continuum radiation, and it is therefore helpful to compare the pulse shape for the wavelength at the centre of the line with that for the wavelength corresponding to the base of the peak. Figure 11.20 compares the pulse shapes for 485 nm (Xe(I) line) with that for 481 nm (base of peak). The difference approximates to the shape for pure line radiation.

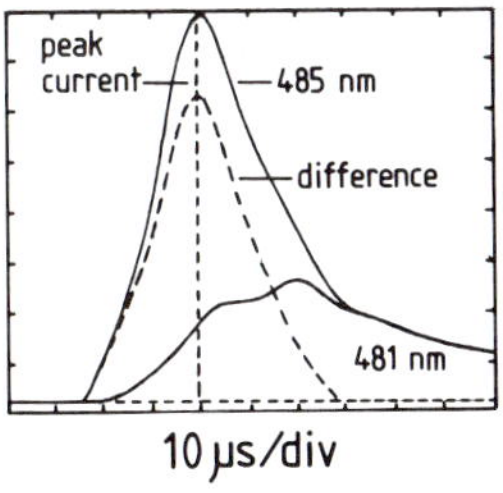

Fig. 11.20. Light pulses for 481 and 485 nm and difference between them.

It can be seen that under the conditions examined the duration of line radiation is rather less than that for continuum radiation, and also that while the peak of line radiation corresponds closely to the peak of current through the lamp, the peak for continuum radiation lags behind the peak current. There appears to be some variation in the duration of the line radiation pulses, Xe(II) lines not surprisingly having shorter duration than Xe(I) lines. The duration of light pulses has generally been found to increase with wavelength (Charnaya *et al.*, 1969; Basov *et al.*, 1975), because of the higher opacity of the discharge at longer wavelengths.

11.3.10 Efficiency

The energy consumed from the mains supply is dissipated as follows:

(1) Ohmic loss in the capacitor bank charging resistor
(2) Ohmic loss in the capacitor bank
(3) Ohmic loss in the leads
(4) Ohmic loss in the inductor (if used)
} circuit losses
(5) Electrode losses
(6) Wall losses
} tube losses
(7) Radiation.

The very high currents involved mean that unless the circuit is well designed, with heavy conductors and short leads, a high proportion of the energy from the mains supply will be lost in the circuit. Typical figures for circuit losses lie between 15 and 35% (Konyukhov *et al.*, 1964; Balagurov and Kubyshkin, 1973). Circuit losses will be greatest for lamps having a low peak resistance, i.e. short, large diameter arcs (Oliver and Barnes, 1969). Electrode losses will normally be between 2 and 5%.

Balagurov and Kubyshkin (1973) studied wall losses calorimetrically for tubes of 7 and 16 mm bore and found that the wall loss, as a fraction of input energy, decreased with increasing xenon pressure, and was a function of the surface energy density, passing through a minimum at about 20 J cm^{-2}. Wall loss was found to increase sharply as the limiting energy of the lamp was approached. This was due to increased UV output from the arc accompanied by reversible opacity of the tube. Ivanov *et al.* (1974) made a similar study with a 7 mm bore lamp, and found that wall loss was mainly determined by the power density (in megawatts per cubic centimetre), but was particularly large for short duration pulses.

Other authors have measured the radiative output of lamps in various spectral regions as a function of input energy. Radiative efficiency depends on pressure, pulse duration, tube bore and current density (Buck *et al.*, 1963; Trenholme and Emmett, 1970; Budnik *et al.*, 1970, 1971). Low efficiencies are found for short duration pulses and low pressures. The optimum current density for a particular pulse duration depends on the wavelength region (Budnik *et al.*, 1971). Shorter wavelengths are favoured by higher current densities. However, if the current density is too high, reversible opacity of the tube will occur. The conditions which provide maximum efficiency of UV output are not generally those that give the maximum *total* radiation.

In order to assess the UV output of a variety of lamps under a wide variety of operating conditions use was made of the International Light "Light Bug" described in Section 4.4. This integrating radiometer responds to wavelengths between 250 and 400 nm, with a peak response at 350 nm. Exposure to a flash pulse gives a reading in joules per square centimetre, although the calibration is only valid for monochromatic radiation at 350 nm. The lamps examined were all of 150 mm arc length and varied in bore from 3 to 12 mm, and were filled with 200 or 400 T of xenon. They are shown in Fig. 11.4. Input energies varied from 10 to 400 J. Figure 11.21 shows outputs measured for various combinations of capacitance,

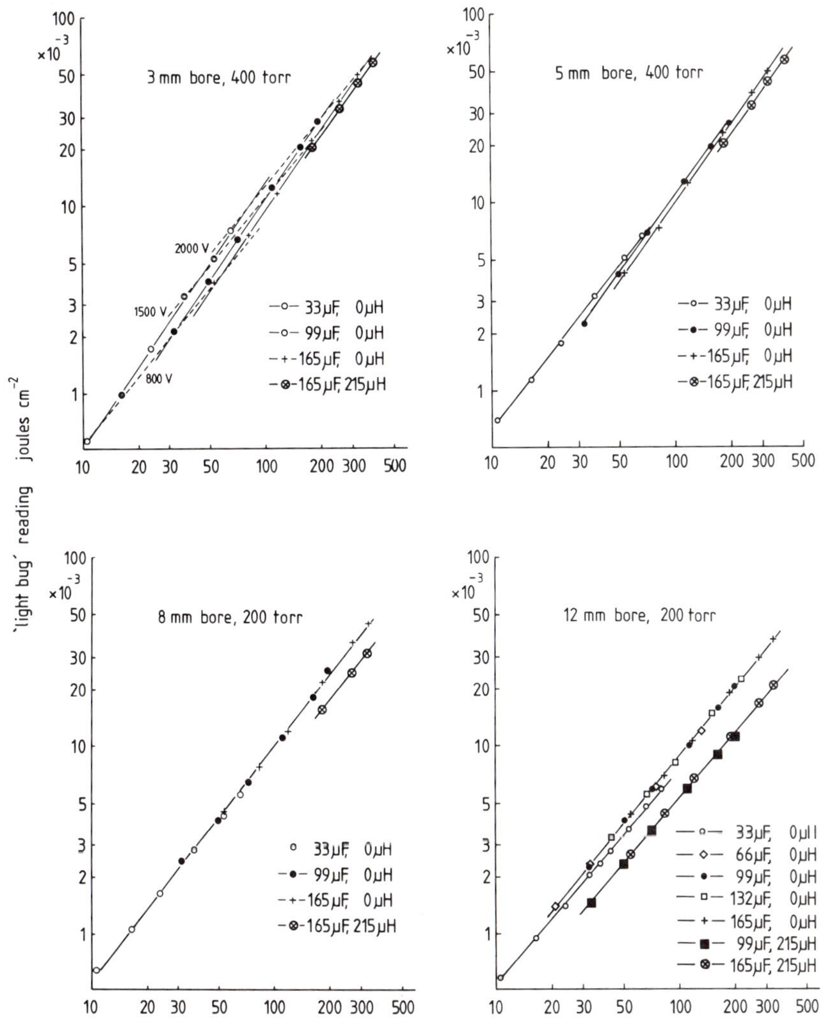

Fig. 11.21. Light Bug measurements for xenon flash lamps (all 150 mm bore). Horizontal axis gives capacitor bank energy, $\frac{1}{2}CV^2$ joules.

voltage and added inductance. Each set of points represents a constant capacitance, with energy being increased by increasing the voltage. Certain conclusions can be drawn from the results.

(1) The points lie on straight lines, as has been found by other authors for the visible region (Carlson and Pritchard, 1947). The gradients of the lines are always greater than unity, indicating that as the pulse energy is increased there is an increase in efficiency. As there is little change in pulse duration when the voltage is increased, the increase in energy implies also an increase in current density and an increase of the fraction of the total radiation which appears in the UV. Note that on the graph for the 3 mm bore tube, the broken lines show the effect of maintaining the voltage constant and increasing the capacitance (and the pulse width). It is interesting that the gradient for constant voltage is less than the gradient for constant capacitance.
(2) For the narrower tubes (3 and 5 mm bore), the efficiency for a particular pulse energy increases as the capacitance and therefore the pulse duration is reduced. The shorter the pulse width the higher the current density and the higher the UV output. The effect is different with the 12 mm bore lamp, and it may be that when the 33 μF capacitor is used the pulse duration is too short for the discharge to completely fill the tube.
(3) The addition of series inductance reduces the UV output by decreasing the current density. The effect of series inductance on the efficiency increases with the bore of the tube.

11.3.11 Double pulse and simmer mode

When flashlamps are used for laser pumping there is a need for an optical pulse with a fast rise time, a high peak and a short duration. A fast rise time is difficult to achieve if the capacitor bank is discharged into a "cold" tube. The inductance of the lamp is high during the period of arc expansion (Ornstein and Derr, 1974), and the arc expansion takes several microseconds. Also, the shock wave associated with rapid current rise may either destroy the lamp or at least reduce the number of flashes that can be obtained from it. In order to meet the requirements, and still maintain a reasonable lamp life, the *double pulse technique* was devised (Emmett and Hellwarth, 1962; Emmett and Schawlow, 1963; Avizonis and Legato, 1965; Goncz and Park, 1965; Ornstein and Derr, 1974). A low energy, long duration pulse is used to establish a discharge which fills the tube. After an optimum delay the main capacitor bank is discharged into the precon-

Colour temp

139
165
166
302
303
318
321

ditioned, low resistance tube, giving a pulse with faster rise, higher peak current density, and shorter duration than would have been obtained without the prepulse. The higher current density gives an enhanced UV output.

In the *simmer mode* a discharge of a few tens of milliamps is maintained continuously in the tube, and the capacitor bank is discharged through the lamp via a triggered spark gap or other high current switch. I have experimented with simmer mode using the 3, 5, 8 and 12 mm bore lamps shown in Fig. 11.4. A minimum 20 mA was required at a voltage of 2000 V to establish a simmer discharge in the 8 and 12 mm bore tubes, with 200 T xenon. The minimum simmer current for the 3 and 5 mm bore lamps with 400 T xenon was 40 mA. The simmer discharge formed a narrow streamer between the electrodes. When the lamps were horizontal the discharge was affected by convection currents and was displaced to near the top of the tube. Flashing the lamp destabilized the simmer discharge, and it wandered around the tube for several seconds before resuming its original position close to the top surface of the tube. Quite often the destabilization effect was sufficient to extinguish the simmer discharge completely. Various combinations of voltage and capacitance were tried with simmer mode. No changes in pulse shape or duration were observed, and in no case was any improvement seen in the light output at any wavelength.

Simmer mode has been found to give less variation from pulse to pulse and longer lamp lifetime (Stephens and Hug, 1972; Jethwa and Schäfer, 1974). It can be used in conjunction with the double pulse technique (Marotta and Argüello, 1976).

11.3.12 Life

Xenon flashlamps can be operated under a wide variety of conditions, and therefore the lifetime of a lamp, in terms of the number of satisfactory flashes which can be obtained from it, may be expected to vary greatly depending on the way in which it is used. Indeed, useful life may range between a single flash and more than 10^9.

If a sufficiently large pulse of energy is fed into a flashlamp, it will be destroyed. This is the explosion limit. The lamp will be destroyed either because the tube cannot withstand the maximum power per unit wall area, or because the rate of current rise is sufficient to cause a shock wave strong enough to fracture the tube (Perlman, 1967). Lamps are normally operated at well under the explosion limit, and it has been found that the percentage of the explosion limit is a reliable guide to the expected life of the lamp. Operation at pulse energies in excess of 50% of the explosion limit normally leads to catastrophic failure, while operation at lower energies leads to a gradual deterioration of the tube until it can no longer perform its function (Goncz, 1966).

As with other types of lamp, loss of material from the electrodes will cause a build up of deposits on the inside wall of the tube, which reduces the transmission, and may lead to crack formation (Edgerton, 1946). The shock wave may cause fine surface cracks to appear in the tube (Lang and Barnes, 1964). Direct plasma contact with the inside wall will cause local melting and evaporation (Rosolowski and Charles, 1965) and the formation of a milky white deposit of condensed silica. Outgassing of the tube and leakage through the seals may introduce impurities into the gas fill which will raise the ignition voltage.

Lamp suppliers provide a method of estimating useful lamp life (defined as the number of flashes which result in a 50% loss of light output). The method is strictly applicable only to linear flash tubes operating under critically damped conditions from a single mesh circuit. The principle is to measure the current pulse duration $t_{1/3}$ (between one-third peak amplitude points) for the tube concerned under the appropriate conditions of capacitance, voltage and inductance. It is then possible to read from a graph the value of $\frac{1}{2}CV^2$ which would be expected to cause the lamp to explode, i.e. to have a lifetime of a single flash. The graph is shown in Fig. 11.22, and is based on the work of Edgerton *et al.* (1963).

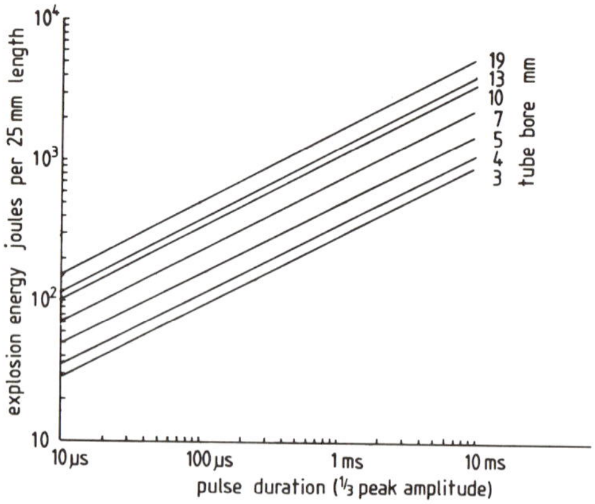

Fig. 11.22. Loadings (in joules per 25 mm length) at which linear silica flash tubes explode. (Adapted from data supplied by EG & G Electro-Optics Division, with permission.)

Figure 11.22 is based on the empirical expression

$$U/l = \kappa d\, t_{1/3}.$$

U is the explosion limit (in joules). κ is a constant which depends on the lamp construction and fill pressure. For linear tubes of normal construction and fill pressure, not operating in an enclosed space, κ has a numerical value of 3·5 if the interelectrode distance l and the bore of the tube d are expressed in millimetres, and the pulse duration is expressed in milliseconds (Goncz, 1966). The value for κ is lower for a helical configuration, or for operation in an enclosed space. The lifetime for a particular pulse energy is determined by its value as a percentage of the explosion limit, as shown in Fig. 11.23.

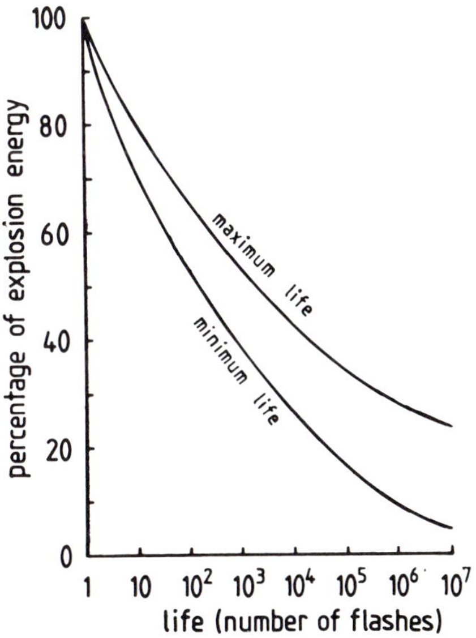

Fig. 11.23. Life to 50% of initial output as function of explosion energy. (Adapted from data supplied by EG & G Electro-Optics Division, with permission.)

If a flash tube is operated repetitively it is necessary that the time-averaged power consumption should not exceed a certain value otherwise the tube will overheat and its life will be shortened. The allowable wall loading depends on the method of cooling. Free air convection cooling allows a time averaged wall loading of about 5 W cm^{-2}. Forced air cooling increases this figure to 40 W cm^{-2}, while water cooling allows 300 W cm^{-2} to be dissipated (Goncz, 1966).

11.3.13 Uses

The main uses of xenon flash tubes have been for various forms of illumination, particularly in photography, and for laser pumping. As sources of actinic radiation they are generally less efficient than mercury or metal halide lamps. However, pulsed xenon lamps, with discharges taking place at mains frequency or a multiple of mains frequency, have found a small but significant place in UV curing (Bordzol, 1975; Michalski, 1975, 1976; Ryan, 1977; Blank, 1979, 1980). It has been claimed that the high intensity pulses provide superior cure in thick sections (Miller, 1975; Panico, 1975, 1976a,b). Pulsed xenon lamps have the attraction of instant start, and, if forced air or water cooling is used, a cold tube which does not radiate infra-red. Xenon flashlamps have been used for many years in the laboratory in flash photolysis, for the generation and analysis of shortlived species.

References

Aldington, J. N. and Meadowcroft, A. J. (1948). *J. IEE* **95,** 671–681.
Anan'ev, Yu. A., Bordachev, E. G., Irtuganov, V. M., Kalinin, V. P. and Sergeev, V. V. (1974). *Sov. J. Quant. Electron.* **4,** 656–658.
Anderson, W. T., Jr (1951). *J. Opt. Soc. Amer.* **41,** 385–388.
Andreev, S. I. and Gavrilov, V. E. (1974a). *J. Appl. Spectrosc.* **20,** 439–441.
Andreev, S. I. and Gavrilov, V. E. (1974b). *J. Appl. Spectrosc.* **20,** 588–593.
Andreev, Yu. P., Bogorodskii, M. M, Semiokhin, I. A. and Rimashevskaya, L. R. (1976). *Zh. Fiz. Khim.* **50,** 1532–1535.
Aspden, R. L. (1959). "Electronic Flash Photography". Temple Press, London.
Avizonis, P. V. and Legato, T. (1965). *J. Appl. Phys.* **36,** 3302–3307.
Baker, H. J. and King, T. A. (1975). *J. Phys. E* **8,** 219–223.
Balagurov, A. Ya., Ermakov, N. I. and Chivilev, V. A. (1972). *J. Appl. Spectrosc.* **17,** 1392–1395.
Balagurov, A. Ya. and Kubyshkin, V. V. (1973). *High Temp.* **11,** 451–455.
Barnes, F. S. (1964). *J. SMPTE* **73,** 569–573.
Basov, Yu. G., Morozova, S. G. and Tokareva, A. N. (1975). *J. Appl. Spectrosc.* **23,** 1309–1313.
Baum, W. A. and Dunkelman, L. (1950). *J. Opt. Soc. Amer.* **40,** 782–786.
Beck, G. (1974). *Rev. Sci. Instrum.* **45,** 318–319.
Belov, S. N., Demidov, M. I., Ogurtsova, N. N., Podmoshenskii, I. V., Rogovtsev, P. N. and Shelemina, V. M. (1969). *J. Appl. Spectrosc.* **10,** 275–277.
Blank, E. (1979). SME Tech. Pap. (Ser.) FC79-673.
Blank, E. (1980). *J. Radiat. Curing* **7** (4), 15–22.
Bokova, N. A., Gulyaev, Yu. N., Mel'chenko, V. S. and Pozdeev, V. V. (1974). *Sov. Phys. J.* **15,** 672–676.
Bordzol, L. (1975). SME Tech. Pap. (Ser.) FC75-338.

Brown, D. C. and Nee, T.-S. N. (1977). *IEEE Trans. Electron. Devices* **ED-24,** 1285–1287.
Buck, A., Erickson, R. and Barnes, F. (1963). *J. Appl. Phys.* **34,** 2115–2116.
Budnik, V. N., Kozlov, N. A. and Malashenkov, V. A. (1970). *J. Appl. Spectrosc.* **13,** 1573–1575.
Budnik, V. N., Kozlov, N. A. and Malashenkov, V. A. (1971). *J. Appl. Spectrosc.* **15,** 1299–1302.
Carlson, F. E. and Pritchard, D. A. (1947). *Illum. Eng.* **42,** 235–248.
Charnaya, F. A., Bykhovskaya, L. N. and Gurevich, I. M. (1969). *J. Appl. Spectrosc.* **11,** 1289–1293.
Demenik, I. V., Mnuskin, V. E., Skvortsov, B. V. and Fedorov, V. B. (1968). *Sov. Phys. Tech. Phys.* **13,** 829–832.
Dishington, R. H., Hook, W. R. and Hilberg, R. P. (1974). *Appl. Opt.* **13,** 2300–2312.
Edgerton, H. E. (1946). *J. Opt. Soc. Amer.* **36,** 390–399.
Edgerton, H. E. (1970). "Electronic Flash/Strobe". McGraw-Hill, New York.
Edgerton, H. E. and Cahlander, D. A. (1961). *J. SMPTE* **70,** 7–9.
Edgerton, H. E., Bonazoli, R. and Lamb, J. T. (1954). *J. SMPTE* **63,** 15–18.
Edgerton, H. E., Goncz, J. H. and Jameson, P. W. (1963). *In* "Proceedings of the 6th International Congress on High Speed Photography" (Ed. J. G. A. Degraaf and P. Tegelaar), p. 143. H. O. Tjeek Willink and Zoon NV, Haarlem.
Elenbaas, W. (1949). *Philips Res. Rep.* **4,** 221–231.
Elenbaas, W. (1972). "Light Sources", Philips Technical Library. Macmillan, London.
Emmett, J. L. and Hellwarth, R. W. (1962). *Bull. Amer. Phys. Soc.* **7,** Series 2, 615.
Emmett, J. L. and Schawlow, A. L. (1963). *Appl. Phys. Lett.* **2,** 204–206.
Emmett, J. L., Schawlow, A. L. and Weinberg, E. H. (1964) *J. Appl. Phys.* **35,** 2601–2604.
Finkelnburg, W. (1934). *Z. Phys.* **88,** 297–310.
Genkin, V. N. and Mel'nikov, V. V. (1966). *J. Appl. Spectrosc.* **5,** 328–330.
Goncz, J. H. (1965). *J. Appl. Phys.* **36,** 742–743.
Goncz, J. H. and Park, S. W. (1965). *Microwaves* **4** (4), 34, 36, 37.
Goncz, J. H. (1966). *ISA Trans.* **5,** 28–36.
Goncz, J. H. and Newell, P. B. (1966). *J. Opt. Soc. Amer.* **56,** 87–92.
Guenther, K. (1970). *Beitr. Plasmaphys.* **10,** 469–485.
Gusinow, M. A. (1973). *J. Appl. Phys.* **44,** 4566–4568.
Gusinow, M. A. (1975). *J. Appl. Phys.* **46,** 4847–4851.
Heard, H. G. (1963a). *Appl. Opt.* **2,** 1329.
Heard, H. G. (1963b). *Proc. IEEE* **51,** 1234–1245.
Hodgson, B. W. and Keene, J. P. (1972). *Rev. Sci. Instr.* **43,** 493 496.
Holzrichter, J. F. and Emmett, J. L. (1969). *Appl. Opt.* **8,** 1459–1465.
Hook, W. R., Dishington, R. H. and Hilberg, R. P. (1972). *IEEE Trans. Electron. Devices* **ED-19,** 308–314.
Hoyt, G. D. and McCormick, W. W. (1950). *J. Opt. Soc. Amer.* **40,** 658–663.
Ignat'ev, V. G. and Podgaetskii, V. M. (1971). *Sov. J. Quantum Electron.* **1,** 415–418.
Ivanov, V. V., Nikiforov, V. G. and Rozanov, A. G. (1974). *Sov. J. Quantum Electron.* **4,** 717–718.
Jethwa, J. and Schäfer, F. P. (1974). *Appl. Phys.* **4,** 299–302.

Klein, L. (1968). *Appl. Opt.* **7,** 677–685.
Konyukhov, V. K., Kulevskii, L. A. and Prokhorov, A. M. (1964). *Zh. Prikl. Spektrosk.* **1,** 51–56.
Lang, K. R. and Barnes, F. S. (1964). *J. Appl. Phys.* **35,** 107–110.
LaPorte, M. (1937). *Compt. Rend.* **204,** 1240–1242.
LeCompte, G. W. and Edgerton, H. E. (1956). *J. Appl. Phys.* **27,** 1427–1430.
Leonard, S. L. (1965). *In* "Plasma Diagnostic Techniques" (Ed. R. H. Huddlestone and S. L. Leonard), pp. 7–67. Academic Press, New York.
Markiewicz, J. P. and Emmett, J. L. (1966a). *IEEE J. Quantum Electron.* **QE2,** 707–711.
Markiewicz, J. P. and Emmett, J. L. (1966b). *Appl. Opt.* **5,** 1687–1688.
Marotta, A. and Argüello, C. A. (1976). *J. Phys. E* **9,** 478–481.
Marshak, I. S. (1963). *Appl. Opt.* **2,** 793–799.
Mavroyannakis, E. G. (1971). *IEEE Trans. Electron. Devices* **ED-18,** 115–118.
Michalski, M. (1975). SME Tech. Pap. (Ser.) FC75-337.
Michalski, M. (1976). *TAGA Proc.* **1976,** 302–314.
Miller, B. (1975). *Plastics World* **33** (2), 38–40.
Oliver, J. R. and Barnes, F. S. (1969). *IEEE J. Quantum Electron.* **QE5,** 232–237.
Ornstein, M. H. and Derr, V. E. (1974). *Appl. Opt.* **13,** 2100–2104.
Panico, L. R. (1975). SME Tech. Pap. (Ser.) FC75-326.
Panico, L. R. (1976a). SME Tech. Pap. (Ser.) FC76-499.
Panico, L. R. (1976b). *Paperboard Packaging* **61** (6), 50, 52.
Perlman, D. E. (1966). *Rev. Sci. Instr.* **37,** 340–343.
Perlman, D. E. (1967). *Rev. Sci. Instr.* **38,** 68–72.
Podmoshenskii, I. V., Polozova, L. P., Chirkov, V. N. and Yakovleva, A. V. (1968). *J. Appl. Spectrosc.* **9,** 707–708.
Popp, H. P. and Weninger, G. (1973). *Lichttechnik* **25,** 527–530.
Rehmet, M. (1980). *IEE Proc. A* **127,** 190–195.
Rosolowski, J. H. and Charles, R. J. (1965). *J. Appl. Phys.* **36,** 1792–1793.
Ryan, R. M. (1977). *Electro-Opt. Syst. Des.* **9** (11), 32–39.
Schirmer, H. (1959). *Z. Angew. Phys.* **11,** 357–361.
Schulz, P. (1944). *Reichsberichte Phys.* **1,** 147–153.
Schulz, P. (1947). *Ann. Phys.* **6,** 95–106; *Z. Naturforsch.* **2a,** 583a–584a.
Searle, N. Z., Giesecke, P., Kinmoth, R. and Hirt, R. C. (1964). *Appl. Opt.* **3,** 923–927.
Stearn, J. W. and Colliver, D. J. (1966). *J. Sci. Instr.* **43,** 52–54.
Stephens, R. L. and Hug, W. F. (1972). *Laser Focus* **8** (7), 38–40.
Striganov, A. R. and Sventitskii, N. S. (1968). "Tables of Spectral Lines, Neutral and Ionized Atoms". Plenum, New York.
Trenholme, J. B. and Emmett, J. L. (1970). *In* "Proceedings of the 9th International Congress on High Speed Photography" (Ed. W. G. Hyzer), pp. 299–302. SMPTE, New York.
White, H. E. (1934). "Introduction to Atomic Spectra". McGraw-Hill, New York.

12 Lasers

12.1 Population Inversion

The three possible radiative transitions which connect two energy states E_1 and E_2 were described briefly in Section 2.11. They can be shown diagramatically, as in Fig. 12.1.

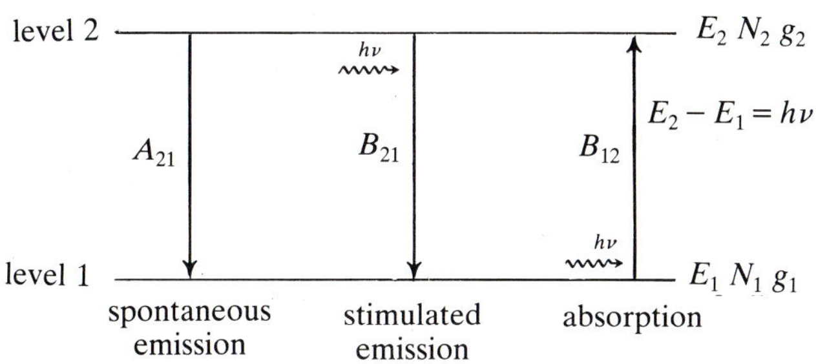

Fig. 12.1. Radiative transitions between energy levels. A_{21} is the Einstein coefficient for spontaneous emission; B_{21} is the Einstein coefficient for stimulated emission; B_{12} is the Einstein coefficient for (stimulated) absorption; N = population density; g = statistical weight.

If there is an incident beam of photons of frequency ν and radiation density u_ν, then the number of photons generated by *stimulated emission* in unit volume and in unit time will be $N_2 u_\nu B_{21}$. The number of photons lost by *absorption* will be $N_1 u_\nu B_{12}$. $N_2 A_{21}$ photons will be generated by *spontaneous emission*.

The three Einstein coefficients are related by the *Einstein relations*:

$$A_{21} = (8\pi h \nu^3 n^3 / c^3) B_{21},$$

where n is the index of refraction, and

$$g_2B_{21} = g_1B_{12}.$$

If the two energy levels are non-degenerate or at least have the same statistical weight, then $B_{21} = B_{12}$. The incident beam will suffer a net loss of $(N_1-N_2)u_\nu B_{21}$ photons per unit volume per unit time.

If a collection of particles (atoms, ions or molecules) is in thermal equilibrium, the population of each excited state can be calculated by application of the Boltzmann distribution law:

$$N_2 = N_1 \exp[-(E_2 - E_1)/kT]$$

or

$$N_2 = N_1 \exp[-h\nu/kT].$$

In the UV and visible parts of the spectrum and at moderate temperatures $h\nu \gg kT$ and therefore $N_1 \gg N_2$. This means that the photons of an incident beam are much more likely to be absorbed than to stimulate emission of further photons. Absorption exceeds emission by a factor $\exp(h\nu/kT)$, so that even as ν approaches zero, absorption is still greater than emission. The photon emission by the particles, at appropriate values of kT, is largely by spontaneous emission. The phase of a wave emitted by any one particle has no relation to the phase of a wave emitted by any other particle, and so the light emitted by a collection of particles fluctuates randomly from moment to moment and from place to place, and is therefore *incoherent.* All the radiation sources discussed in previous sections of this book generate incoherent radiation. Incoherent radiation sources have certain limitations (Lengyel, 1971, p. 6):

(1) The radiating surface emits into a hemisphere, often according to Lambert's law, so that light from a small area diverges, making collimation difficult for an extended source, and limiting the brightness of images.
(2) The energy radiated by an intense source is spread over a broad spectral region.

If a non-equilibrium situation can be created in which $N_2 > N_1$, then the loss of photons from an incident beam will be negative; and in other words there will be an amplification of the incident beam. The non-equilibrium situation in which $N_2 > N_1$ is known as a *population inversion.* It allows stimulated emission to exceed absorption. Not only is there a net increase in photons of frequency ν, but the amplified beam is coherent with the incident beam, and has the same directional properties.

12.2 Laser Principles

Light Amplification by Stimulated Emission of Radiation is the principle of the *laser*. Population inversion is easier to obtain if $E_1 - E_2$ is small. Also, the ratio of A_{21} to B_{21} is proportional to ν^3, and energy loss from spontaneous emission is severe at short wavelengths. For these reasons lasers producing long wavelength radiation are easier to design than lasers producing short wavelength radiation. It is not surprising that the first device which generated radiation by amplification by stimulated emission was the *maser*, which operated in the microwave region (Gordon *et al.*, 1955).

12.2.1 Laser components

The laser has three essential components, as shown in Fig. 12.2. There is an *active medium* which contains a pair of energy levels radiatively

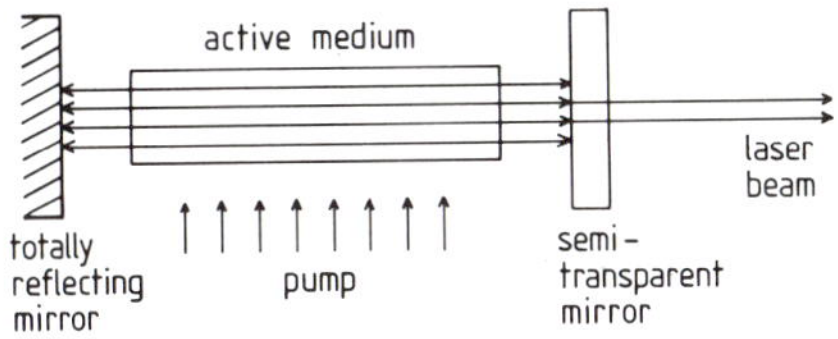

Fig. 12.2. The essential components of a laser.

connected by a photon energy of the required wavelength. This active medium is *pumped* by an energy source which is capable of generating a population inversion between the two energy levels. The active medium is placed in a *resonant cavity*. At one end of the cavity is a highly reflective mirror and at the other end a partly reflective and partly transmissive (but non-absorbent) mirror. Photons bounce back and forth between the mirrors, and as long as a population inversion is maintained by the pump the photon flux is amplified by stimulated emission. Some of the photons pass through the partly reflecting mirror to form the laser beam. Lasers may be operated in a CW or a pulsed mode.

12.2.2 Energy levels

The active medium may be a solid, a liquid, or a gas. The energy levels between which the lasing transition occurs may be as shown in the following table:

Energy levels	Solid	Liquid	Gas
Atomic : electronic			He/Ne, Cu vapour
Ionic : electronic	Cr^{3+}/Al_2O_3 (ruby)		He/Cd, He/Se
	Nd^{3+}/YAG		Ar ion, Kr ion
	Nd^{3+}/glass		
Molecular : electronic		Dye	N_2
Lattice : electronic	GaAs (semiconductor)		
Molecular : vibrational/ rotational			CO_2, CO, N_2O, NH_3
Excimer : electronic			Xe_2, XeF, ArF

12.2.3 Optical pumping

The common method of creating population inversion in the active media of solid and liquid lasers is optical pumping. An intense source of energetic photons, which may be incoherent (e.g. a xenon or krypton flash tube) or coherent (another laser) is used. It is not possible to create a population inversion between a pair of energy levels by optical pumping directly from the lower level to the higher level. It is necessary to employ a pumping band situated above the higher laser level and use this band to populate the higher laser level. For this reason the pumping wavelength must always be shorter than the wavelength of the laser transition. The simplest system is the three-level system as shown in Fig. 12.3(a). An example is provided by the ruby laser, in which Cr^{3+} ions in a crystalline Al_2O_3 matrix are the active medium.

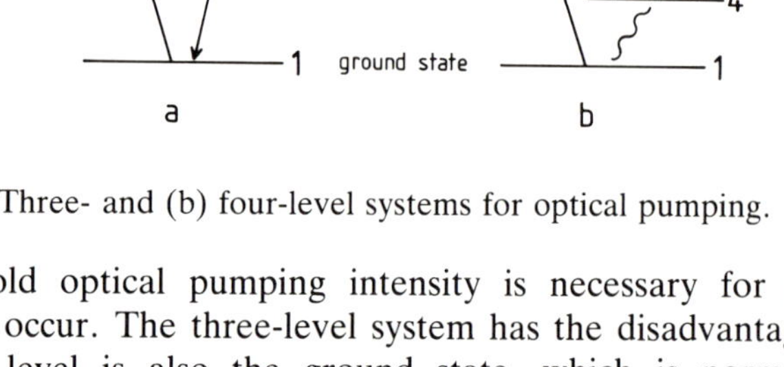

Fig. 12.3. (a) Three- and (b) four-level systems for optical pumping.

A threshold optical pumping intensity is necessary for population inversion to occur. The three-level system has the disadvantage that the lower laser level is also the ground state, which is normally highly populated, and therefore the threshold intensity is high, and the ruby laser is normally operated in the pulsed mode, using flashlamp pumping. This disadvantage is overcome in the more common four-level system as exemplified by the neodymium laser, where the active medium is Nd^{3+} ions in a matrix of glass, or a crystal such as yttrium aluminium garnet (YAG). The lower laser level is above the ground state (Fig. 12.3(b)), and is kept empty by rapid non-radiative relaxation to the ground state, so population inversion is easier to achieve.

12.2.4 *Pump cavities*

When incoherent light sources (which usually behave as Lambertian radiators) are used to pump solid and liquid lasers careful design of the pump cavity is required if a high transfer efficiency is to be achieved (Schuldt and Aagard, 1963; Bowness, 1965; Röss, 1969, pp. 383–455; Lengyel, 1971, pp. 120–124; Koechner, 1976, pp. 300–343). An elliptical geometry is normally used, with the light source at one focus and the active medium at the other (Ciftan *et al.*, 1961). There are analogies with UV curing systems and elliptical photoreactors (see Chapter 14).

12.2.5 *Discharge pumping*

In some gas lasers the population inversion is created by chemical reaction. For instance, in the hydrogen chloride laser, hydrogen chloride molecules in an excited state are formed by reaction between hydrogen and chlorine (Kasper and Pimentel, 1965). However, the general pumping method in gas-phase lasers involves energetic electrons, either produced in an electric discharge through the active medium, or in the form of an electron beam. The active medium may contain more than one component, and the excitation of one component may be used to create population inversion in the other (Fig. 12.4).

12.2.6 *Longitudinal modes*

Although the initial spontaneously generated photons are radiated in all directions, it is only those which travel parallel to the axis of the cavity which have the opportunity to be amplified by multiple passes through the active medium. Furthermore, the ability to make multiple passes without serious energy loss is only possible for wavelengths which satisfy the

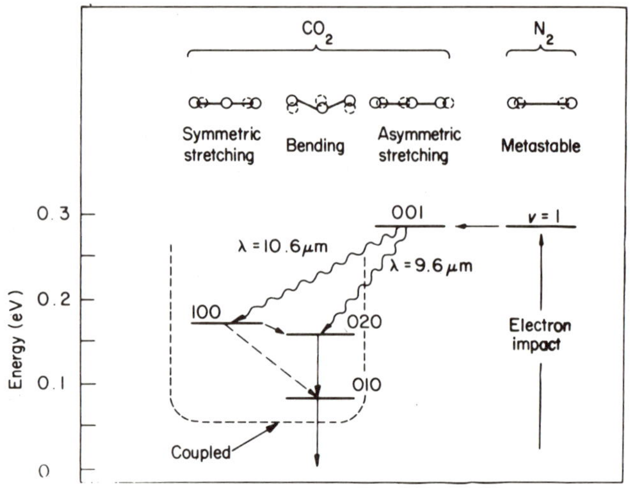

Fig. 12.4. Vibrational energy levels involved in the pumping of the carbon dioxide laser. (From Spalding, 1980, with permission.)

condition of constructive interference. The optical length of the cavity (D) must be an integral number of half wavelengths, i.e. $N\lambda = 2D$. If λ is in the UV or visible, $D \gg \lambda$ and therefore N is very large. There are therefore many different wavelengths which satisfy the resonance condition. These are known as *longitudinal modes.*

12.2.7 Transverse modes

Variations in the pattern of radial intensity distribution are possible (Kogelnik and Rigrod, 1962; Polanyi and Watson, 1963). These are known as transverse modes and are identified by the letters TEM (transverse electromagnetic) followed by a subscript code of two integers, which indicate the numbers of nodes in the pattern in two directions perpendicular to each other.

12.2.8 Q switching

If a laser is pumped by a d.c. electric field or a d.c. light source, the laser output is constant with time. If an a.c. field or light source is used the laser output has a periodicity of twice mains frequency, as found with a.c. operated discharge lamps. When a laser is pumped by a flashlamp the total duration of the flashlamp pulse is a few milliseconds. The laser emission begins when the active medium has absorbed enough energy from the light pulse to create the threshold population inversion. Laser action continues as long as the output of the pump lamp is high enough to maintain the required inversion. However, the laser output appears not as a single smooth pulse, but as a train of sub-pulses which represent a series of inversion–emission cycles. Stimulated emission is a very rapid process, and therefore as soon as the first population inversion is created it is rapidly destroyed by stimulated emission and laser action ceases until further excitation creates a further inversion. If one of the reflectors is prevented from feeding back photons into the cavity during the major part of the flash tube's pulse, it is possible to build up a population inversion far greater than the threshold value. If the reflector is then suddenly reinstated the energy built up in the laser medium is released in a "giant" pulse of very short duration (McClung and Hellwarth, 1962). This technique is known as "Q switching" or "Q spoiling". Various devices (Q switches) may be used, including mechanical, acousto-optic and bleachable dye types (Koechner, 1976, pp. 397–455).

12.2.9 Oscillators and amplifiers

The laser as described is an oscillator. Without the mirrors, but with the active medium pumped, it can act as an amplifier of incoming radiation. Some very high power installations use an initial oscillator to produce a collimated beam which is boosted by passage through a number of amplifiers.

12.2.10 Harmonic generation

It is possible to generate exceptionally intense beams of light using lasers. These beams have very high electric fields associated with them which create non-linear optical effects in certain crystalline dielectric materials causing the generation of second harmonics having twice the frequency of the incident beam (Franken *et al.*, 1961). It is also possible by using two laser beams of frequencies ν_1 and ν_2 to obtain two new frequencies having values $\nu_1 + \nu_2$ and $\nu_1 - \nu_2$.

12.3 The Unique Properties of Laser Radiation

Lasers are important light sources because it is possible to generate light beams having quite exceptional properties. These properties are (a) high peak beam power (in watts), (b) directionality (low divergence), (c) high peak beam power density (in watts per square centimetre), (d) spatial coherence, (e) monochromaticity (temporal coherence), (f) pulses of very short duration.

Because stimulated emission is a much more rapid process than spontaneous emission it is possible to dissipate the energy stored in a laser medium very quickly and so generate a beam of *exceptionally high peak power*, in some cases exceeding 10^{12} W.

Whereas the radiating surface of an incoherent source emits in all directions into a hemisphere, the radiation from a laser is *highly directional.* The beam is almost perfectly parallel and collimated. This is because a photon generated by stimulated emission has the same directional properties as the stimulating photon, and because the amplification in the laser cavity is greatest parallel to the axis. Diffraction effects do cause some divergence, depending on the wavelength and the width of the beam as it emerges from the laser. As the beam is generally small in diameter the *beam power density can be very high.* Because the beam divergence is usually very small (except in the case of semiconductor lasers) the beam power density can be maintained over large distances. A parallel beam is very easy to manipulate by optical components. It can be expanded in diameter, or can be focused to a spot having a diameter approximately equal to the wavelength of the light, allowing *exceptionally high values of irradiance.*

The wavelength emitted by a laser is determined by two main factors, the energy levels in the active medium and the spacing between the mirrors. The bandwidth for each longitudinal mode of the cavity is usually so narrow that it is this which determines the *monochromaticity* of the laser. The discrete wavelengths emitted by lasers have much narrower bandwidths than those emitted by incoherent line sources. "The product of pulse duration and frequency spread is physically limited by the uncertainty principle to be greater than π^{-1}. This limit is closely approached in a number of laser systems. The highest resolutions, of the order of 1 part in 10^{-12}, are usually achieved with continuous lasers" (Leone and Moore, 1974).

It is possible to obtain laser pulses of *very short durations.* Q switching produces pulses lasting only a few nanoseconds. However, even these pulses may have a substructure of even narrower "mode-locked" pulses (Deutsch, 1965) caused by interference between longitudinal modes of the cavity. It is not difficult to isolate by an electro-optic switch a single pulse having a duration of only a few picoseconds (Didomenico *et al.*, 1966). Even more sophisticated techniques have recently been developed (Shank *et al.*, 1982) capable of producing pulses having durations of only a few femtoseconds. A femtosecond is 10^{-15} s.

12.4 Laser Types

The commercially available laser types are extremely well documented in numerous excellent textbooks and specialist journals, and it is inappropriate to give more than a brief description of the more important types here. In the previous chapters on incoherent sources the emphasis has been on the generation of wavelengths short enough to bring about chemical change by a photochemical process. The situation is more complicated with lasers for two reasons. First, it is possible to use frequency doubling devices based on non-linear optics to generate short wavelengths from a long wavelength visible or infra-red laser beam. Secondly, the extremely high photon flux density of highly monochromatic radiation which is available from laser sources allows multiphoton absorption processes which produce chemical changes which are not possible with even the most intense incoherent sources at the same wavelength.

12.4.1 Optically pumped solid lasers

Optically pumped solid lasers employ ions of transition or rare earth elements as impurity dopants in crystalline (Kiss and Pressley, 1966) or glass (Snitzer, 1966) matrices. The two examples of commercial importance, the ruby and neodymium lasers, have already been mentioned. The ruby laser was the first laser to be demonstrated (Maiman, 1960). In the ruby laser a small percentage of the Al^{3+} ions in crystalline Al_2O_3 is replaced by Cr^{3+} ions. The overall efficiency of the ruby laser is very low, and its high threshold means that it is almost always operated in the pulsed mode. It has emission at 694·3 nm. The neodymium laser emits at 1·06 μm. It is significantly more efficient and is easier to operate CW. An active medium in the solid state allows good energy storage capability. These lasers are also particularly suitable for Q switching and mode locking, so that they are able to generate pulses of very high peak power and very short duration. The active medium is normally in the form of a cylindrical rod. If the host material is crystalline it must be of high optical quality, and there are limitations in the sizes of crystals that can be grown. There are no size limitations with glass, and when extremely high peak power is required, as in laser fusion work, the neodymium glass laser is used. However, unlike YAG, glass has poor thermal conductivity, and is therefore unsuitable for rapid repetitive pulsing. The limits to the operation of solid lasers are normally determined by thresholds of optical or thermal material damage. Sometimes, a glass laser is in the form of a series of discs in a zig-zag configuration to increase the surface area available both for optical pumping and for heat dissipation. Because the pump radiation must necessarily be at shorter wavelengths than the laser emission optical pumping is not generally suitable for UV lasers. However Ehrlich *et al.* (1980a) have demonstrated an optically pumped Ce : LaF_3 laser emitting at 286 nm.

12.4.2 Semiconductor diode injection lasers

At the other extreme of solid state lasers is the semiconductor diode injection laser. When a flat p–n junction is formed between p- and n-doped semiconductor crystal a forward voltage (positive on the p side) brings holes and electrons into the same region and their recombination is accompanied by emission of photons. If the current density is high enough, stimulated emission may become the dominant process, despite the fact that the transition is not intra-atomic. There is a direct conversion of electrical energy into coherent radiation, and efficiency may be as high as 50%. Semiconductor lasers are very small, usually with no dimension being more than 1 mm. The active medium is very narrow, and therefore diffraction effects cause the laser beam to fan out, and collimation is poor. By suitable choice of materials laser action can be obtained from the near UV to the far infra-red. The most common semiconductor laser uses gallium arsenide (Gooch, 1969). The wavelength emitted depends on temperature and varies between 845 nm at liquid helium temperature to 905 nm at room temperature. It is therefore a tunable laser. It is normally operated in the pulsed mode.

12.4.3 Dye lasers

Liquids have considerable attractions as laser materials. They provide, like solids, a high density of active material. Because they are fluid they can be circulated to allow dissipation of heat. They are easier to prepare than solid state laser materials and not prone to the same types of damage. Their composition can be readily changed. The only common type of liquid laser is the organic dye laser which uses an organic fluorescent material dissolved in a common solvent (Sorokin and Lankard, 1966; Sorokin et al., 1967; Sorokin, 1969). The solution may be used in a cell (for pulsed operation), or in the form of a jet flowing from a nozzle for CW. The laser is pumped by a flashlamp or by another laser. The emission from the active medium corresponds to the fluorescent emission of the dye, and therefore is in the form of a wide band rather than a narrow line. Bands with centres stretching from the near UV to the near infra-red can be obtained by suitable choice of dye. The most significant feature of the dye laser is its tunability. Any wavelength required can be selected by for instance the use of a prism, grating or filter within the resonant cavity.

12.4.4 Gas lasers

Most commercial lasers, and in particular those emitting in the UV and blue regions of the spectrum, use gases as active media. Gases are optically homogenous and can be circulated for cooling. Optical pumping is seldom suitable because gases absorb only weakly. A wide variety of alternatives is available. Some involve a chemical reaction (Pimentel, 1966; Chester, 1973; Gross and Bott, 1976). "In the chemical laser, elementary chemical reactions lead to an inverted population by selectively partioning exothermic reaction energy among non-equilibrium vibrationally excited nascent products" (Kimel and Speiser, 1977). Reactions are normally of the form $A + BC \longrightarrow AB^* + C$. In the photodissociation laser (Kasper and Pimentel, 1964), intense UV light is used to split a molecule so that one of the products is in an excited state. The gas dynamic laser (Gerry, 1970) uses the sudden expansion of a hot molecular gas through a nozzle to create a population inversion by briefly producing a non-equilibrium situation in which a lower energy level is depopulated more quickly than a higher level. However, the majority of gas lasers use some form of electrical discharge through the gas to create the population inversion. As in a discharge lamp, energy is absorbed from the electric field by the acceleration of electrons (and ions) and these transfer energy to the components of the gas mix by collisions. It is easy, by this method, to transfer enough energy to an atom, ion or molecule to allow population inversion between energy levels which are far enough apart to provide lasing action in the UV. The lasing particle may be excited directly by collision with an electron (direct excitation) or by resonant energy transfer from another component (indirect excitation). The gas is normally in a tube with transparent end windows tilted to the Brewster angle to minimize reflection losses, and the mirrors are external.

The discharge may be axial, or transverse. Sometimes an electrodeless, radio-frequency discharge is used. The system may be sealed or there may be a constant flow of gas to aid cooling and avoid the build-up of contaminants.

The first gas laser to be developed, and still the most widely used of all lasers, is the helium–neon laser (Javan *et al.*, 1961). This laser uses the electronic energy levels of the neutral neon atom to produce laser emission at a variety at wavelengths in the visible and infra-red, the most commonly used wavelength being 632·8 nm. Population inversion is created by resonant energy transfer from excited helium atoms. Operation is CW, at very low power, usually only a few milliwatts. A glow discharge is used with a current of a few milliamps.

The energy levels of the inert gases in their ionic forms allow laser transitions which generate shorter wavelengths. A high current discharge (10–100 A) at low pressure gives a high degree of ionization and population inversion. Argon ion and krypton ion lasers (Bridges, 1964; Bridges and Chester, 1965) are common commercial types and both can lase at various wavelengths in the UV and visible regions. They can be operated either pulsed or CW. CW power is quite high compared with that of the helium–neon laser, typically between one and ten watts but efficiency is low.

The carbon dioxide laser is the most widely used molecular gas laser. It contains a mixture of carbon dioxide, nitrogen and helium. There is resonant energy transfer from nitrogen molecules in their first vibrationally excited state to carbon dioxide molecules (Patel, 1964). The efficiency of the process is high because being homonuclear the nitrogen molecule cannot lose its energy by radiation, and the energy levels involved are well matched for resonant energy transfer. Laser transitions are possible between vibrational–rotational levels of the carbon dioxide molecule in two bands centred at 9·6 and 10·6 μm (Fig. 12.4). The 10·6 μm band is normally favoured as it provides the higher gain. However, a wavelength selective device within the laser cavity can provide tunability between 9·3 and 11·0 μm, and this range can be slightly extended if carbon dioxide with isotopes of carbon and oxygen are used. The presence of helium is found to increase power output by a large factor (Patel *et al.*, 1965) by enhancing the excitation of the nitrogen by collisional energy transfer, and by increasing the rate of depopulation, i.e. quenching, of the terminal laser level. Low powered carbon dioxide lasers use an axial gas flow and an axial discharge. Higher power can be obtained by arranging both gas flow and discharge perpendicular to the optical path. The carbon dioxide laser is capable of producing more CW power than any other laser, with outputs in the multikilowatt range being possible. One version of the carbon dioxide laser is used in the pulsed mode. This is the TEA (transverse-excited, atmospheric pressure) laser (Beaulieu, 1970). Molecular gas lasers are generally more efficient than atomic or ionic lasers, and the carbon dioxide laser operates at up to 20% efficiency.

Gas-phase lasers using metallic elements in their vapour phase (metal vapour lasers) are becoming increasingly important. The helium–cadmium laser is the most common, using cadmium ions to emit at 325 and 442 nm (Goldsborough, 1969; Silfvast, 1969). It is a CW laser with output in the microwatt region and with excellent beam quality. The copper vapour laser, giving atomic lines at 510 and 578 nm (Grove, 1982) has recently become available commercially.

The nitrogen laser, emitting at 337 nm, is an example of a molecular laser using transitions between electronic energy levels (Heard, 1963; Leonard, 1965; Singh and Thakur, 1980). The lifetime of the upper laser level is only about 5 ns, whereas that of the lower laser level is several microseconds. The lasing action is therefore "self-terminating" and the laser can only operate in the pulsed mode, giving pulses in the nanosecond region. The normal configuration uses a capacitor discharge between long parallel electrodes. The laser beam is parallel to the electrodes and perpendicular to the discharge path. The nitrogen laser is peculiar in that the gain is very high, and there is no need for the normal cavity feedback. Usually one mirror is used to provide a double pass through the medium. The beam that emerges has a rectangular cross section and has a high divergence because there is no opportunity for the normal mode structure to develop. The nitrogen laser is often used for pumping other lasers.

12.4.5 Excimer lasers

In recent years a major goal of laser research has been to develop high powered efficient lasers with emission in the UV. The impetus has not only been possible applications in photochemical processes but also the desire to generate laser beams which are less divergent and can be focused to smaller spot sizes. Gas ion lasers are inherently inefficient because the upper laser levels are populated from the neutral atom by multiple electronic collisions. Maximum power is limited because if the current is too high the upper laser level is depopulated by electronic collisions. Molecular lasers are more promising if the population inversion can be maintained either by collisional depopulation of the lower laser level or by dissociation of the molecule. Various types of molecular lasers with transitions between electronic energy levels have been developed (Green, 1978). There are two main types. In the first both the upper and lower levels are bound states of the molecule (bound–bound), although the population of the lower state may be kept low by facile dissociation of the molecule. Examples are the halogen dimer lasers. F_2 lases at 157 nm (Rice *et al.*, 1977), Br_2 at 292 nm (Murray *et al.*, 1976) and I_2 at 342 nm (Bradford *et al.*, 1975; Ewing and Brau, 1975a).

The second type is the "excimer" laser. The term "excimer" (Stevens and Hutton, 1960) properly denotes a *homonuclear* molecule which is bound in an electronically excited state but dissociative in the ground state, whereas an excited *heteronuclear* molecule dissociative in the ground state is properly known as an "exciplex". However, when excimers and exciplexes are used as active media in lasers, the distinction if often overlooked, and both homonuclear and heteronuclear systems are referred to as excimers (Hutchinson, 1980).

An excimer may be formed between a ground state atom or molecule and an electronically excited atom or molecule:

$$A + B \longrightarrow AB^*$$

The excimer decays radiatively to the ground state and then dissociates:

$$\begin{array}{c} AB^* \longrightarrow (AB) + h\nu \\ \downarrow \\ A + B \end{array}$$

This is clearly an attractive system for laser operation as a population inversion can easily be established between the excimer level and the dissociative ground state. The ground state may be repulsive, so that the lifetime is no longer than about 10^{-13} s, the period of vibration, or may be very weakly bound. The transition is therefore termed "bound–free".

The first excimer laser used the Xe_2 excimer, either in liquid xenon (Basov *et al.*, 1970) or in high pressure xenon (Koehler *et al.*, 1972). Excimer lasers based on Ar_2 and Kr_2, on the oxides of the rare gases and on mercury monohalides have also been demonstrated, but the excimer lasers which are of most technological interest are those using rare gas halides (Brau, 1978, 1979; Ewing, 1978, 1979; Shaw, 1979; Mace, 1981). These excimers, e.g. XeCl, ArF, have ionic structures, e.g. Xe^+Cl^-, Ar^+F^- (Brau and Ewing, 1975a). Laser action has not been observed with all the rare gas halides, but only with those formed from the higher atomic weight rare gases and the lower atomic weight halogens, as shown in the following table of wavelengths.

The rare gas halide excimer laser has the potential for providing UV laser emission at high powers and with relatively high efficiency. They are tunable in so far as the high power and short wavelength allow the stimulated Raman scattering effect to be used to shift the wavelength. The

Halogens	Rare gases				
	He	Ne	Ar	Kr	Xe
F	—	—	193	248	352
Cl	—	—	175	222	308
Br	—	—	—	—	283
I	—	—	—	—	—

All wavelengths are given in nanometres.
References: ArF, Hoffman *et al.* (1976); ArCl, Waynant (1977); KrF, Ewing and Brau (1975b); KrCl, Murray and Powell (1976); XeF, Ault *et al.* (1975), Brau and Ewing (1975b); XeCl, Ewing and Brau (1975b); XeBr, Searles and Hart (1975).

versatility of the laser has allowed a wide range of applications (Rhodes and Hoff, 1979; McKee and Nilson, 1982).

12.4.6 *Free electron lasers*

The free electron laser is a novel laser type now in the early stages of development (Deacon *et al.*, 1977; Brau, 1981; Green, 1981). The electrons are not confined to an atom or molecule or crystal lattice, but are in the form of an electron beam which is made to "wiggle" by passage through a spatially varying magnetic field. The radiation emitted is stimulated *Bremsstrahlung*. There is potential for a high power high efficiency laser tunable, like a synchrotron, over a very wide wavelength range, from 1 mm to 100 nm.

12.5 Laser Applications

Laser photons are very expensive. The ratio of the capital cost of lasers to their mean output power is very much higher than it is for conventional incoherent sources. With few exceptions mean power outputs and overall efficiencies are very low, and operational and maintenance costs high. However, a large number of techniques and processes have been developed to exploit the unique properties of laser radiation. These can be divided into the broad areas of metrology, information processing, chemical analysis, thermal beam techniques and laser photochemistry. The established techniques have been well documented in numerous texts, for example those of Ross (1971), Charschan (1972), Goldman (1973), Harry (1974), Ready (1978) and Burroughs (1982), and therefore will be mentioned only briefly here.

12.5.1 Metrology, information processing and chemical analysis

Applications in metrology include the precise measurement of dimensions and distances, of flow rates by Doppler anemometry, of small dimensional changes by holographic interferometry, of movement by laser gyroscopes, and alignment techniques. Lasers are used to record, read and transmit information (e.g. laser marking, bar code scanning, fibre optics communications). The exceptional monochromaticity and high beam intensities available from tunable lasers have allowed significant improvements in the capabilities of established spectroscopic techniques, and have allowed novel techniques to be developed. The ability to generate subpicosecond pulses from tunable lasers has facilitated the study of the kinetics and mechanisms of very fast reactions. Robinson (1980) has commented that ". . . it is now possible to examine the course of chemical reactions to any desired level of comprehension . . .". The directionality of laser beams has allowed analysis to be carried out over large distances, and lasers are now being used to investigate the atmosphere and the hydrosphere (Hinkley, 1976; Svanberg, 1980).

12.5.2 Thermal beam techniques

Thermal beam techniques, using the high power of a focused beam, ease of beam manipulation and spatial selectivity of the laser, are widely used for machining processes such as cutting, welding and drilling (Cohen *et al.*, 1982; Ready, 1982). Spatially selective rapid heating (and cooling) of surfaces allows techniques such as case hardening of metals, surface alloying and surface annealing of semiconductor material after ion implantation (Narayan, 1980). Metal films can be deposited on surfaces by decomposition of vapours, for instance triethylaluminium, in contact with the surface (Deutsch *et al.*, 1979). The laser has become an important tool for the surgeon, for whom it can act both as a knife and a coagulator (Koebner, 1980; Dixon, 1982). Internal surgery on highly vascular organs, the delicate procedures of ophthalmology, and the treatment of pigmentary disorders of the skin have been particular areas to benefit. The most spectacular application of very high powered laser beams is in the attempt to raise the temperature of hydrogen and tritium contained in a tiny glass sphere to a high enough value ($\sim 10^6$ K) for fusion to helium to take place, and obtain a net gain in energy (Stickley, 1978; Witkowski and Kompa, 1979; Spalding, 1980). The pressure exerted by the laser beams on the target is an essential part of the technique. Carbon dioxide, iodine and neodymium lasers are being used in experimental programmes, and there is growing interest in excimer lasers. Powers in the region of 10^{14}–10^{15} W will be required for the process to be viable.

12.5.3 Laser photochemistry

An essential feature of conventional photochemistry is that ultraviolet or visible light is used to excite molecules selectively into electronically excited states, so creating a situation in which the molecules of one type in a mixture have much more energy than the others, and will retain that energy long enough to undergo chemical reaction without necessarily affecting other molecules. The conventional broad band, low intensity, grey body infra-red source is often capable of bringing about chemical reaction, but the photons have only enough energy to excite the vibrational modes of molecules. The energy contained in vibrational modes is very quickly equilibrated between the molecules in a mixture, and any reaction takes place in a system which is in thermal equilibrium. The reaction is essential thermal rather than photochemical, as it lacks the feature of selectivity. The introduction of lasers has required some rethinking of the definition of photochemistry, because infra-red lasers have beams which are monochromatic enough to excite the vibrational modes of molecules very selectively, and powerful enough to allow multiphoton absorption, so that many infra-red photons can be absorbed successively by a molecule to raise a vibrational mode to a very high energy level (Ambartzumian and Letokhov, 1977; Bloembergen and Yablonovitch, 1978; Golden *et al.*, 1981). As infra-red lasers, unlike conventional infra-red sources, can be used to bring about very selective reactions in systems not in thermal equilibrium, it is proper that the concept of laser photochemistry should include the use of ultraviolet, visible and infra-red lasers to bring about both electronic and vibrational excitation, and reactions from both excited and ground electronic states.

Because laser photons are so expensive, it is unlikely that they will be used for large scale chemical synthesis of low value materials (Hall, 1982). Photochemical syntheses involving electronic excitation of molecules with broad absorption spectra are generally well served by conventional discharge lamps with their high CW outputs and high efficiencies, although in some cases filtering of the radiation to eliminate unwanted side reactions may be desirable. However, Malatesta *et al.* (1981) were able to obtain very high conversions of 7-dehydrocholesterol into previtamin D_3 by using 248 nm radiation from a KrF laser followed by 337 nm radiation from a nitrogen laser.

The lower energy levels of the vibrational modes of molecules are equi-spaced like rungs of a ladder, and the spacing is very precise. It is therefore possible, by carefully tuning an infra-red laser, to energize selectively a particular vibrational mode in a particular type of molecule. Perhaps the most significant debate in current laser photochemistry concerns the degree of selectivity that can be obtained in chemical reactions following laser irradiation, and whether "mode-selective" or "bond-selective" chemistry is possible (Letokhov, 1980; Zewail, 1980; Brookfield, 1981). If a molecule can react in a variety of different ways, it may be possible to select the reaction pathway by choosing a particular vibrational mode to energize. However, when energy is fed into a vibrational mode, there is a very rapid sharing of that energy with the other vibrational modes of the molecule, and so mode-selective chemistry may be possible only under very special circumstances. Less rapid than intramolecular energy sharing is the transfer of vibrational energy to other molecules, particularly if the molecule is in the gas phase at low pressure, or in a solid matrix at low temperature.

Species-selective chemistry is now well established and it is likely that this will be one of the most promising areas of industrial laser photochemistry (Kaldor and Woodin, 1982). Species-selective techniques may involve vibrational or electronic excitation, or both, and atomic or molecular species. It is desirable that species should have narrow absorption lines to provide the necessary discrimination, and for this reason processing normally takes place in the gas phase at low pressure. The area of greatest interest has been the separation of isotopes (Letokhov and Moore, 1976, 1977; Zare, 1977; Denning, 1978; Cantrell *et al.*, 1979; Hartford 1979; Letokhov, 1979; Jensen *et al.*, 1980). The selective excitation of one particular atomic isotope, or molecule containing this isotope, is the first step in the separation process. Numerous methods have been devised to separate the excited from the unexcited species including, for atoms, selective photoionization, and for molecules, two-step (infra-red + UV) and multiphoton infra-red photodissociation. These methods are well documented in the review articles which have been referenced above. The isotope separation of most commercial significance is that of ^{235}U from ^{238}U, and the laser photochemistry of atomic uranium vapour and the gaseous compounds of uranium has been actively investigated (Davis and Davis, 1977; Jensen and Robinson, 1977; Cox *et al.*, 1979; Cox and Maas, 1979; Robinson and Jensen, 1979; Horsley *et al.*, 1980; Davis *et al.*, 1982).

Lasers have been used to decompose selectively small amounts of contaminants in gas mixtures and allow their removal (Hartford and Clark, 1980; Kaldor and Woodin, 1982). Merritt and Robertson (1977) used a CO_2 laser to remove phosgene ($COCl_2$) from boron trichloride. Clark and Anderson (1978) removed phosphine, arsine and diborane from silane by selective photolysis using 193 nm radiation from a ArF laser. Chen and Borzileri (1980) used the same laser type to remove H_2S from synthesis gas (a mixture of H_2 and CO). Lasers have potential applications in separating the components of mixtures of high value materials, for instance the rare earths (Donohue, 1979) and nuclear waste (Eden *et al.*, 1979).

The spatial selectivity provided by the focused laser beam allows spatially selective chemistry, particularly at surfaces. Exposure of photoresists (Dunn, 1979) and the coating and etching of surfaces, particularly in the electronics industry (Ehrlich *et al.*, 1980b), are examples.

References

Ambartzumian, R. V. and Letokhov, V. S. (1977). *Acc. Chem. Res.* **10,** 61–67. *In* "Chemical and Biochemical Applications of Lasers", Vol. III (Ed. C. B. Moore), pp. 167–316. Academic Press, New York.

Ault, E. R., Bradford, R. S., Jr and Bhaumik, M. L. (1975). *Appl. Phys. Lett.* **27,** 413–415.

Basov, N. G., Danilychev, V. A., Popov, Yu. M. and Khodkevich, D. D. (1970). *JETP Lett.* **12,** 329–331.

Beaulieu, A. J. (1970). *Appl. Phys. Lett.* **16,** 504–505.

Bloembergen, N. and Yablonovitch, E. (1978). *Phys. Today* **31** (5), 23–30.

Bowness, C. (1965). *Appl. Opt.* **4,** 103–107.

Bradford R. S., Jr, Ault, E. R. and Bhaumik, M. L. (1975). *Appl. Phys. Lett.* **27,** 546–548.

Brau, C. A. (1978). *In* "High Power Laser Applications", Springer Series in Optical Science, Vol. 9, pp. 3–8. Springer Verlag, Berlin.

Brau, C. A. (1979). *In* "Excimer Lasers" (Ed. C. K. Rhodes), Topics in Applied Physics, Vol. 30, pp. 87–133. Springer Verlag, Berlin.

Brau, C. A. (1981). *Laser Focus* **17** (5), 48, 50, 52–54, 56.

Brau, C. A. and Ewing, J. J. (1975a). *J. Chem. Phys.* **63,** 4640–4647.

Brau, C. A. and Ewing, J. J. (1975b). *Appl. Phys. Lett.* **27,** 435–437.

Bridges, W. B. (1964). *Appl. Phys. Lett.* **4,** 128–130.

Bridges, W. B. and Chester, A. N. (1965). *Appl. Opt.* **4,** 573–580.

Brookfield, R. (1981). *New Sci.* **91,** 540–543.

Burroughs, W. J. (1982). "Lasers". Longmans, Harlow.

Cantrell, C. D., Freund, S. M. and Lyman, J. L. (1979) *In* "Laser Handbook", Vol. 3 (Ed. M. L. Stitch), pp. 485–576. North Holland, Amsterdam.

Charschan, S. S. (1972). "Lasers in Industry", Van Nostrand Rheinhold, New York.

Chen, H.-L. and Borzileri, C. (1980). *IEEE J. Quantum Electron.* **QE16,** 1229–1232.

Chester, A. N. (1973). *Proc. IEEE* **61,** 414–422.

Ciftan, M., Luck, C. F., Shafer, C. G. and Statz, H. (1961). *Proc. IRE* **49,** 960–961.

Clark, J. H. and Anderson, R. G. (1978). *Appl. Phys. Lett.* **32,** 46–49.

Cohen, M. G., Kaplan, R. A. and Arthurs, E. G. (1982). *Proc. IEEE* **70,** 545–555.
Cox, D. M. and Maas, E. T., Jr (1980). *Chem. Phys. Lett.* **71,** 330–334.
Cox, D. M., Hall, R. B., Horsley, J. A., Kramer, G. M., Rabonowitz, P. and Kaldor, A. (1979). *Science* **205,** 390–394.
Davis, J. I. and Davis, R. W. (1977). *AIChE Symp. Ser.* **73** (169), 69–75.
Davis, J. I. Holtz, J. Z. and Spaeth, M. L. (1982). *Laser Focus* **18** (9), 49–54.
Deacon, D. A. G., Elias, L. R., Madey, J. M. J. Ramian, G. J., Schwettman, H. A. and Smith, T. I. (1977). *Phys. Rev. Lett.* **38,** 892–894.
Denning, R. G. (1978). *Phys. Technol.* **9,** 242–247.
Deutsch, T. F. (1965). *Appl. Phys. Lett.* **7,** 80–82.
Deutsch, T. F., Ehrlich, D. J. and Osgood, R. M., Jr (1979). *Appl. Phys. Lett.* **35,** 175–177.
DiDomenico, M., Jr, Geusic, J. E., Marcos, H. M. and Smith, R. G. (1966). *Appl. Phys. Lett.* **8,** 180–183.
Dixon, J. A. (1982). *Proc. IEEE* **70,** 579–588.
Donohue, T. (1979). *Opt. Eng.* **18,** 181–186.
Dunn, S. T. (ed.) (1979). *Proc. Soc. Photo-Opt. Instrum. Eng.* **169.**
Eden, G., Burnham, R., Champagne, L. F., Donohue, T. and Djeu, N. (1979). *IEEE Spectrum* **16** (4) 50–59.
Ehrlich, D. J., Moulton, P. F. and Osgood, R. M. Jr (1980a). *Opt. Lett.* **5,** 339–341.
Ehrlich, D. J., Osgood, R. M. Jr and Deutsch, T. F. (1980b). *IEEE J. Quantum Electron.* **QE16,** 1233–1243.
Ewing, J. J. (1978). *Phys. Today* **31** (5), 32–39.
Ewing, J. J. (1979). *In* "Laser Handbook", Vol. 3 (Ed. M. L. Stitch), pp. 135–197. North Holland, Amsterdam.
Ewing, J. J. and Brau, C. A. (1975a). *Appl. Phys. Lett.* **27,** 557–559.
Ewing, J. J. and Brau, C. A. (1975b). *Appl. Phys. Lett.* **27,** 350–352.
Franken, P. A., Hill, A. E., Peters, C. W. and Weinreich, G. (1961). *Phys. Rev. Lett.* **7,** 118–119.
Gerry, E. T. (1970). *IEEE Spectrum* **7** (11), 51–58.
Golden, D. M., Rossi, M. J., Baldwin, A. C. and Barker, J. R. (1981). *Acc. Chem. Res.* **14,** 56–62.
Goldman, L. (1973). "Applications of the Laser". CRC Press, Cleveland.
Goldsborough, J. P. (1969). *Appl. Phys. Lett.* **15,** 159–161.
Gooch, C. H. (1969). "Gallium Arsenide Lasers". Wiley–Interscience, London.
Gordon, J. P., Zeiger, H. J. and Townes, C. H. (1955). *Phys. Rev.* **99,** 1264–1274.
Green, J. M. (1978). *Opt. Laser Technol.* **10,** 289–300.
Green, J. M. (1981). *Opt. Laser Technol.* **13,** 245–251.
Gross, R. W. F. and Bott, J. F. (1976). "Handbook of Chemical Lasers". Wiley, New York.
Grove, R. E. (1982). *Laser Focus* **18** (7), 45–50.
Hall, R. B. (1982). *Laser Focus* **18** (9), 57–62.
Harry, J. E. (1974). "Industrial Lasers and Their Applications". McGraw-Hill, London.
Hartford, A., Jr (1979). *AIChE Symp. Ser.* **75** (186), 44–49.
Hartford, A., Jr. and Clark, J. H. (1980). *In* "Chemical and Biochemical Applications of Lasers", Vol. V (Ed. C. B. Moore), pp. 217–237. Academic Press, New York.
Heard, H. G. (1963). *Nature* **200,** 667.

Hinkley, E. D. (1976). "Laser Monitoring of the Atmosphere", Topics in Applied Physics, Vol. 14. Springer Verlag, Berlin.
Hoffman, J. M., Hays, A. K. and Tisone, G. C. (1976). *Appl. Phys. Lett.* **28,** 538–539.
Horsley, J. A., Rabinowitz, P., Stein, A., Cox, D. M., Brickman, R. O. and Kaldor, A. (1980). *IEEE J. Quantum Electron.* **QE 16,** 412–419.
Hutchinson, M. H. R. (1980). *Appl. Phys.* **21,** 95–114.
Javan, A., Bennett, W. R., Jr and Herriott, D. R. (1961). *Phys. Rev. Lett.* **6,** 106–110.
Jensen, R. J. and Robinson, C. P. (1977). *AIChE Symp. Ser.* **73,** (169), 76–82.
Jensen, R. J., Sullivan, J. A. and Finch, F. T. (1980). *Separation Sci. Technol.* **15,** 509–532.
Kaldor, A. and Woodin, R. L. (1982). *Proc. IEEE* **70,** 565–578.
Kasper, J. V. V. and Pimentel, G. C. (1964). *Appl. Phys. Lett.* **5,** 231–233.
Kasper, J. V. V. and Pimentel, G. C. (1965). *Phys. Rev. Lett.* **14,** 352–354.
Kimel, S. and Speiser, S. (1977). *Chem. Rev.* **77,** 437–472.
Kiss, Z. J. and Pressley, R. J. (1966). *Appl. Opt.* **5,** 1474–1486.
Koebner, H. K. (1980). "Lasers in Medicine", Vol. 1. Wiley–Interscience, Chichester.
Koechner, W. (1976). "Solid State Laser Engineering". Springer Verlag, Berlin.
Koehler, H. A., Ferderber, L. J., Redhead, D. L. and Ebert, P. J. (1972). *Appl. Phys. Lett.* **21,** 198–200.
Kogelnik, H. and Rigrod, W. W. (1962). *Proc. IRE* **50,** 220.
Lengyel, B. A. (1971). "Lasers", 2nd edn. Wiley–Interscience, New York.
Leonard, D. (1965). *Appl. Phys. Lett.* **7,** 4–6.
Leone, S. R. and Moore, C. B. (1974). *In* "Chemical and Biochemical Applications of Lasers", Vol. I (Ed. C. B. Moore), pp. 1–27. Academic Press, New York.
Letokhov, V. S. (1979). *Nature* **277,** 605–610.
Letokhov, V. S. (1980). *Phys. Today* **33** (11), 34–41.
Letokhov, V. S. and Moore, C. B. (1976). *Sov. J. Quantum Electron.* **6,** 129–150; **6,** 259–276.
Letokhov, V. S. and Moore, C. B. (1977). *In* "Chemical and Biochemical Applications of Lasers", Vol. III (Ed. C. B. Moore), pp. 1–165. Academic Press, New York.
Mace, P. N. (1981). *Opt. Eng.* **20,** 935–940.
McClung, F. J. and Hellwarth, R. W. (1962). *J. Appl. Phys.* **33,** 828–829.
McKee, T. and Nilson, J. A. (1982). *Laser Focus* **18** (6), 51–55.
Maiman, T. H. (1960). *Nature* **187,** 493–494.
Malatesta, V., Willis, C. and Hackett, P. A. (1981). *J. Amer. Chem. Soc.* **103,** 6781–6783.
Merritt, J. A. and Robertson, L. C. (1977). *J. Chem. Phys.* **67,** 3445–3548.
Murray, J. R. and Powell, H. T. (1976). *Appl. Phys. Lett.* **29,** 252–253.
Murray, J. R., Swingle, J. C. and Turner, C. E. (1976). *Appl. Phys. Lett.* **28,** 530–531.
Narayan, J. (1980). *J. Met.* **32** (6), 15–21.
Patel, C. K. N. (1964). *Phys. Rev. Lett.* **13,** 617–619.
Patel, C. K. N., Tien, P. K. and McFee, J. H. (1965). *Appl. Phys. Lett.* **7,** 290–292.
Pimentel, G. C. (1966). *Sci. Amer.* **214** (4), 32–39.
Polanyi, T. G. and Watson, W. R. (1963). *J. Appl. Phys.* **34,** 553–560.

Ready, J. F. (1978). "Industrial Applications of Lasers". Academic Press, New York.
Ready, J. F. (1982). *Proc. IEEE* **70,** 533–544.
Rhodes, C. K. and Hoff, P. W. (1979). *In* "Excimer Lasers" (Ed. C. K. Rhodes), Topics in Applied Physics, Vol. 30, pp. 175–188. Springer Verlag, Berlin.
Rice, J. K., Hays, A. K. and Woodworth, J. R. (1977). *Appl. Phys. Lett.* **31,** 31–33.
Robinson, C. P. (1980). *IEEE J. Quantum Electron.* **QE16,** 1150–1151.
Robinson, C. P. and Jensen, R. J. (1979). *In* "Uranium Enrichment" (Ed. S. Villani), Topics in Applied Physics, Vol. 35, pp. 269–290. Springer Verlag, Berlin.
Röss, D. (1969). "Lasers, Light Amplifiers and Oscillators". Academic Press, New York.
Ross, M. (1971). "Laser Applications", Vol. 1. Academic Press, New York.
Schuldt, S. B. and Aagard, R. L. (1963). *Appl. Opt.* **2,** 509–513.
Searles, S. K. and Hart, G. A. (1975). *Appl. Phys. Lett.* **27,** 243–245.
Shank, C. V., Fork, R. L., Yen, R. and Stolen, R. H. (1982). *Appl. Phys. Lett.* **40,** 761–763.
Shaw, M. J. (1979). *Prog. Quantum Electron.* **6,** 3–54.
Silfvast, W. T. (1969). *Appl. Phys. Lett.* **15,** 23–25.
Singh, J. P. and Thakur, S. N. (1980). *J. Sci. Ind. Res.* **39,** 613–624.
Snitzer, E. (1966). *Appl. Opt.* **5,** 1487–1499.
Sorokin, P. P. (1969). *Sci. Amer.* **220** (2), 30–40.
Sorokin, P. P. and Lankard, J. R. (1966). *IBM J. Res. Dev.* **10,** 162–163.
Sorokin, P. P., Lankard, J. R., Hammond, E. C. and Moruzzi, V. L. (1967). *IBM J. Res. Dev.* **11,** 130–148.
Spalding, I. J. (1980). *Opt. Laser Technol.* **12,** 187–197.
Stevens, B. and Hutton, E. (1960). *Nature* **186,** 1045–1046.
Stickley, C. M. (1978). *Phys. Today* **31** (5), 50–56, 58.
Svanberg, S. (1980). *Contemp. Phys.* **21,** 541–576.
Waynant, R. W. (1977). *Appl. Phys. Lett.* **30,** 234–235.
Witkowski, S. and Kompa, K.-L. (1979). *Recherche* **10,** 956–966.
Zare, R. N. (1977). *Sci. Amer.* **236** (2), 86, 87, 90–98.
Zewail, A. H. (1980). *Phys. Today* **33** (11), 27–33.

13 Miscellaneous Sources

13.1 Solar Radiation

During each year, solar radiation supplies to the surface of the Earth quantities of energy which are several orders of magnitude greater than all the energy produced artificially from fossil fuels, nuclear reactors and other sources. This is true even for highly developed countries well away from the equator. Success in exploiting the actinic component of solar radiation for non-agricultural purposes has been limited because the irradiance is low, and is subject to predictable hourly, daily and seasonal variations, and unpredictable variations due to atmospheric disturbances.

13.1.1 The source of solar radiation

The Sun is a sphere of gas $1{\cdot}39 \times 10^6$ km in diameter and of mean density $1{\cdot}41\ \mathrm{g\,cm^{-3}}$. The density varies from $90\ \mathrm{g\,cm^{-3}}$ at the core to $10^{-7}\ \mathrm{g\,cm^{-3}}$ at the surface. At the core the temperature is estimated to be 15×10^6 K, and the pressure over 100 atm. Under these extreme conditions thermonuclear chain reactions take place by which hydrogen nuclei are converted into helium nuclei. There is a mass loss of 0·7% which is released as energy, mainly in the form of gamma radiation. This energy is absorbed, re-emitted and scattered, and degraded to longer wavelengths, in a radiative zone which surrounds the core, and this is then transferred by convection in the outer layers to the photosphere, the visible surface of the Sun. The photosphere is not in fact a discrete surface, but rather a layer several hundred kilometres thick in which a transition takes place between the opacity of the interior and the transparency of the chromosphere which surrounds the visible surface. The total energy output of the Sun is $3{\cdot}86 \times 10^{33}$ erg s^{-1}. If it is assumed that the photosphere radiates as a black body, it can be calculated that the effective temperature of the surface of the Sun is 5780 K.

13.1.2 The solar constant

The normal irradiance of a surface outside the Earth's atmosphere, and at the Earth's mean distance from the Sun, is known as the solar constant, and has the value $0{\cdot}1374\,W\,cm^{-2}$. The spectral distribution of solar radiation outside the Earth's atmosphere approximates to that of a black body at about 5800–6000 K, although there are deviations due to absorption and emission processes in the solar atmosphere, particularly at shorter wavelengths. Some 98% of the radiation from the Sun lies between 200 nm and $3{\cdot}0\,\mu m$.

13.1.3 The effects of the Earth's atmosphere

Solar radiation is attenuated and its spectral distribution modified by its passage through the Earth's atmosphere. Figure 13.1 shows the spectral

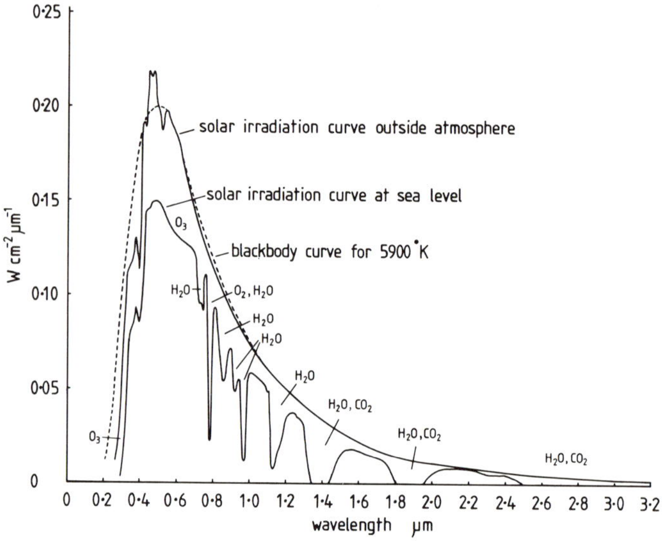

Fig. 13.1. Spectral irradiance of solar radiation (From Kondratyev, 1969, p. 247.)

distribution of solar radiation outside the Earth's atmosphere compared with that at sea level. Attenuation is by absorption and scattering. Solar irradiance at sea level is a combination of direct radiation from the Sun, and radiation reflected from the sky. Wavelengths below 290 nm are absent. This is because of the presence of small concentrations of ozone in the atmosphere at altitudes between 10 and 50 km. The peak concentration is about 10 p.p.m. at about 25 km. The total amount of ozone is very small. If it were compressed into a pure layer at standard temperature and pressure, its thickness would be only 2–3 mm. However, it is very effective at absorbing wavelengths which, if allowed to penetrate to the Earth's surface, would prejudice our continued existence.

13.1.4 Photodegradation by solar UV

Constructive use of the ultraviolet region of solar radiation has been limited, but its study is important for a number of industries because of its destructive nature. Its effects may be enhanced by the presence of atmospheric pollutants although environmental problems as severe as the photochemical smog of Los Angeles are rare. The long-term effects of solar UV on surface coatings, fabrics and plastics, such as chalking, fading and embrittlement, cause problems which are of widespread and major significance. Some degree of protection can be provided by careful selection of components in coatings and polymers and by the use of additives which absorb the UV and dissipate its energy harmlessly. Artificial environments involving the simulation of solar radiation have allowed accelerated test procedures. The xenon lamp is probably the most useful artificial solar simulator, although it does need correction by filters at both ends of the spectrum.

13.2 The Carbon Arc

Carbon arcs played a very important part in the development of artificial sources of light, and until fairly recently were very widely used for applications where a high power, very high intensity source is required. Typical applications include motion picture projection, studio lighting, searchlights, diazo printing, graphic arts, and solar simulators for the accelerated testing of resistance to photochemical degradation. However, for most of these applications doped and undoped mercury arcs, and xenon lamps, with their simpler and safer operation, have superceded the carbon arc, and their use has declined rapidly during the last few years.

The carbon arc is an example of an open arc. The gas involved is therefore air at atmospheric pressure, with the possibility of components derived from evaporation or chemical reaction of the electrode material. The electrodes are carbon rods 0·5–2 cm in diameter and the arc is started by bringing the electrodes into contact, and then separating them by a small distance. This distance must be maintained constant by continuously moving the rods towards each other as they are slowly eroded by the arc. The rods will typically last only a few hours before requiring replacement.

There are three main types of carbon arc:

13.2.1 The low intensity arc

The low intensity arc is normally operated on direct current with a current density of less than 50 A cm^{-2}, and a voltage of about 50 V. Most of the radiation from the low intensity arc is produced not by the discharge itself but by the incandescence of a crater formed in the end of the anode rod. The heating of the anode results from the large anode drop (Macpherson, 1942). The incandescent crater radiates as a grey body at about 3800 K, which is close to the sublimation temperature of carbon. The discharge between the rods contributes a band between 380 and 400 nm which is due to cyanogen formed by reaction with nitrogen in the air. The erosion of the rods is mainly by combustion. The anode rod is consumed at about 10 cm h^{-1}, the consumption of the cathode being slower. The rods are not necessarily made of solid carbon. There may be a hollow cylinder of hard carbon with a core of softer carbon to ensure that the crater is formed in the centre of the rod end, and to improve stability.

13.2.2 The high intensity arc

(See Finkelnburg, 1949.)

The high intensity arc operates in the current density range 100–400 A cm^{-2}. There is rapid evaporation of the anode, which is consumed at the rate of 25–100 cm h^{-1}. The radiation from the high intensity arc is emitted mainly by the hot gas in the anode crater, which is at the base of a brilliant anodic vapour flame. Whereas in the low intensity arc the electrode rods are on the same axis, the rods of the high intensity arc are at an angle to each other. The orientation of the anodic flame is determined by the effects of the magnetic fields produced by the arc, and therefore the high intensity arc may be considered to be "field stabilized". The spectral distribution approximates to that of a black body at 4500–6000 K. The radiative efficiency is much higher than that of the low intensity arc. The high intensity arc has a positive voltage current characteristic because an increase in current increases the rate of vaporization of the anode and this causes an increase in the anode drop, which contributes the major part of the total voltage between the electrodes. The high intensity arc can therefore be operated without a ballast, although some form of stabilizing circuitry may be desirable.

13.2.3 The flame arc

It is possible to modify the spectral distribution of the light from a carbon arc by doping the electrodes with metallic elements or their compounds. These elements enter the discharge and their line spectra dominate the total radiative output. By careful choice of elements enhancement can be obtained in any region of the spectrum as desired. Flame arcs normally have coaxial rods and operate on alternating current.

13.3 The Deuterium Lamp

13.3.1 Spectral output

The arc discharge in a low pressure hydrogen or deuterium lamp gives a line-free continuum between 170 and 350 nm, with a peak output at about 200 nm. A spectral distribution of a deuterium lamp is shown in Fig. 13.2. Published spectra do vary considerably, particularly with regard to the peak output wavelength. Deuterium is generally prefered to hydrogen for practical sources of continuum UV because it provides about 50% more output in the 200–250 nm region (Levikov and Shishatskaya, 1961; Smolkin and Berdnikov, 1963). The short wavelength cut-off depends on the envelope or window material. The continuum radiation is derived from transitions between electronic states of the molecule (Samson, 1967). Atomic lines appear at very short wavelengths and in the visible. The overall efficiency of the deuterium lamp is not high. According to Mimura *et al.* (1978) the Original Hanau 200 W watercooled deuterium lamp gives only 40 mW in the region 200–250 nm.

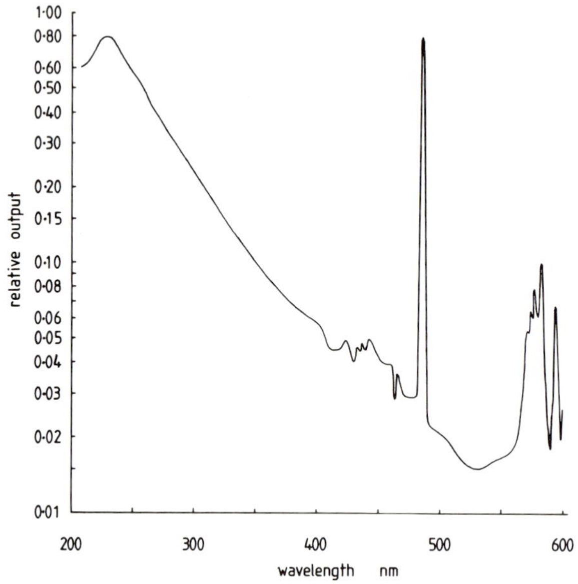

Fig. 13.2. Spectral output of a 50 W Cathodeon deuterium lamp. (Data taken, with permission, from Oriel Corporation Optical Systems and Components Catalogue.)

13.3.2 Construction and operation

The deuterium lamp in its most common form (Kern, 1954) is a low pressure (5–10 T), low power (20–200 W), very compact source of quite high intensity. The anode and cathode structures are supported in a vitreous silica tube about 9 cm in length and about 3 cm in diameter. Three leads, one for the anode and two for the cathode, emerge from one end of

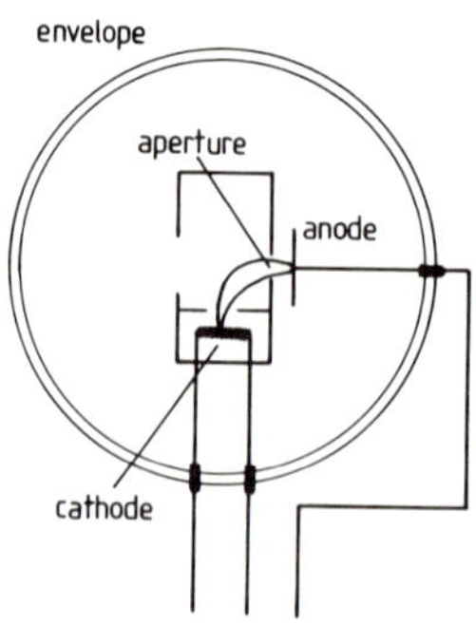

Fig. 13.3. Construction of deuterium lamp: schematic.

the tube. It is a d.c. arc. The cathode is a filament which is heated for a few seconds before the anode voltage is applied, to facilitate ignition. Operating voltage is 60–120 V, and the discharge current is 0·2–2·0 A. In many lamps a current of 3–5 A is maintained through the cathode filament during operation. A curious feature of the deuterium lamp is an aperture of area 1–4 mm^2 which is placed between the cathode and the anode. The arrangement is shown schematically in Fig. 13.3. This constricts and stabilizes the arc. The intensity is highest at the aperture, and this forms a "point" source which is easy to incorporate in an optical system.

13.3.3 Uses of deuterium lamps

The deuterium lamp is too weak a source to be of much interest for the processing of materials, although it has found some use in deep UV lithography (Lin, 1975). Its main application is as a source for spectrophotometers, spectrofluorimeters and spectropolarimeters. Its attractions are its point source characteristic, its line free continuum over a wide range of UV wavelengths, and its high UV : visible ratio. Other sources of appreciable continuum such as the compact xenon arc and the carbon arc have strong visible outputs which give problems with stray light when dispersing elements are used. These attractions are also the rationale for a minor, but significant application of deuterium lamps. They are used as secondary, or transfer, standards for spectral radiance and irradiance (Einfeld and Stuck, 1976; Bridges *et al.*, 1977; Saunders *et al.*, 1978).

13.4 Synchrotron Radiation

Synchrotron radiation is electromagnetic radiation emitted by charged particles (electrons or protons) moving in circular orbits in high vacuum in an annular chamber with velocities approaching that of light. The power radiated by particles of different mass varies inversely with the fourth power of the mass, and therefore if the production of radiation is the most important function of the synchrotron, an electron synchrotron is the natural choice.

Synchrotron radiation has interesting properties (Kunz, 1979):

(1) It is a pure continuum which stretches from the X-ray region through the UV and visible into the infra-red. The spectral distribution can be altered by changing the particle energy. It is an "ultraclean" source, and there are no lines in emission or absorption.

(2) If the electron velocity is near to the velocity of light, the emitted radiation is collimated in the instantaneous direction of flight of the emitting electron. If the orbit is horizontal the vertical angular spread is very small. The intensity available is high.
(3) All the properties of the source can be calculated absolutely (Schwinger, 1949). This allows synchrotron radiation to be used as a primary standard (Tomboulian and Hartman, 1956; Tomboulian and Bedo, 1958; Codling and Madden, 1965; Lemke and Labs, 1967; Pitz, 1969; Stuck and Wende, 1972).
(4) The radiation has a time structure with pulse lengths down to 100 ps.

The synchrotron is primarily a research tool. The intense pure continuum, particularly in the X-ray and vacuum UV parts of the spectrum, has allowed refinements of existing techniques, and the development of new analytical methods. The applications of synchrotron radiation have recently been reviewed by Winick and Bienenstock (1982).

13.5 High Pressure Mercury Lamps

Previous chapters have discussed the many technological applications of low and medium pressure mercury lamps. The high pressure lamps find fewer applications, but are particularly useful where very high intensity compact sources of spectrally continuous radiation is a requirement.

There is a possible source of confusion in the way in which mercury discharges are categorized. There are basically three pressure regions, each with its own characteristic operation and spectral distribution, as explained in Chapter 6. In the literature of illumination the three types of discharge are termed "low pressure", "high pressure" and "very high pressure", whereas in the literature of actinic radiation, the terms are "low pressure", "medium pressure" and "high pressure".

Lamps operating at pressures considerably in excess of 10 atm are normally small in size and are of two distinct and contrasting types, the electrode-confined compact arc and the wall-stabilized capillary arc. The high pressure mercury compact arc is similar to the xenon compact arc and operates at low voltage and high current. The capillary arc has a thick walled envelope with an internal bore of 1–2 mm, and an arc length up to 6 cm. The current is low, 1–2 A, but the voltage gradient is high, up to 500 V cm^{-1}. The capillary arc may be water cooled.

The high pressure mercury arc has a spectral output which shows strong line broadening, a strong continuum and pronounced self-absorption, particularly of 254 nm radiation. The capillary arc may be operated in a pulsed mode, using a capacitor discharge, and becomes a suitable source for dye laser pumping (Dal Pozzo *et al.*, 1975). Pulsing enhances the continuum, and some of the mercury lines are seen in absorption rather than emission. At very high current densities, or very high pressures (Noel, 1941) the spectral output of the high pressure mercury lamp approximates to that of a black body.

High pressure mercury capillary lamps, and compact arc lamps containing mercury, or a mixture of xenon and mercury (Xe–Hg lamps), are widely used as "point" sources in photolithography, and for monochromators. The Xe–Hg lamp gives a spectrum typical of mercury, but has faster warm up, and an arc which is narrower and more stable than that of an equivalent high pressure mercury lamp, which enhances its performance as a point source.

References

Bridges, J. M., Ott, W. R., Pitz, E., Schulz, A., Einfeld, D. and Stuck, D. (1977). *Appl. Opt.* **16,** 1788–1790.

Codling, K. and Madden, R. P. (1965). *J. Appl. Phys.* **36,** 380–387.

Dal Pozzo, P., Polloni, R. and Svelto, O. (1975). *Appl. Phys.* **6,** 341–344; 381–382.

Einfeld, D. and Stuck, D. (1976). *Opt. Commun.* **19,** 297–301.

Finkelnburg, W. (1949). *J. Appl. Phys.* **20,** 468–474.

Kern, J. (1954). *Z. Angew. Phys.* **6,** 536–538.

Kondratyev, K. Ya. (1969). "Radiation in the Atmosphere", Academic Press, New York and London.

Kunz, C. (1979). *In* "Synchrotron Radiation—Techniques and Applications" (Ed. C. Kunz), Topics in Current Physics, Vol. 10, pp. 1–23. Springer Verlag, Berlin.

Lemke, D. and Labs, D. (1967). *Appl. Opt.* **6,** 1043–1048.

Levikov, S. I. and Shishatskaya, L. P. (1961). *Opt. Spectrosc.* **11,** 371–372.

Lin, B. J. (1975). *J. Vac. Sci. Technol.* **12,** 1317–1320.

Macpherson, H. G. (1942). *J. Appl. Phys.* **13,** 97–102.

Mimura, Y., Ohkubo, T., Takeuchi, T. and Sekikawa, K. (1978). *Jap. J. Appl. Phys.* **17,** 541–550.

Noel, E. B. (1941). *Illum. Eng.* **36,** 243–256.

Pitz, E. (1969). *Appl. Opt.* **8,** 255–259.

Samson, J. A. R. (1967). "Techniques of Vacuum Ultraviolet Spectroscopy", John Wiley and Sons, New York.

Saunders, R. D., Ott, W. R. and Bridges, J. M. (1978). *Appl. Opt.* **17,** 593–600.

Schwinger, J. (1949). *Phys. Rev.* **75,** 1912–1925.

Smolkin, M. N. and Berdnikov, N. B. (1963). *Opt. Spectrosc.* **14,** 220–221.

Stuck, D. and Wende, B. (1972). *J. Opt. Soc. Amer.* **62,** 96–100.
Tomboulian, D. H. and Bedo, D. E. (1958). *J. Appl. Phys.* **29,** 804–809.
Tomboulian, D. H. and Hartman, P. L. (1956). *Phys. Rev.* **102,** 1423–1447.
Winick, H. and Bienenstock, A. (1982). *In* "McGraw-Hill Encyclopedia of Science and Technology", 5th edn, Vol. 13, pp. 450–457. McGraw-Hill, New York.

14 Ultraviolet Irradiators

14.1 Irradiation of Materials

The practical application of UV radiation sources requires that the radiation from the source is efficiently transmitted to the material to be irradiated, in a way which satisfies the requirements of the chemical processes taking place.

In most cases the *physical* attributes of materials to be irradiated will fit one of the categories shown in Table 14.1. All the situations listed require an appropriate irradiator, and it is helpful to think of each situation involving a *photochemical reactor.* It is an entertaining, but quite useful thought, that when one lies on a sunny beach one is setting up a photochemical reactor, and the consequences of not applying the correct photochemical and engineering concepts can be both painful and damaging. This may appear a frivolous comment, but it serves to emphasize the point that the general concepts which are involved in the design of, say, an efficient UV drier for printing inks are much the same as those involved in the design of a sterilization unit for water, or a unit for the phototherapy of psoriasis.

The possible benefits of appreciating this point are illustrated by the following situation. There have been few, if any, serious attempts to set up mathematical models for UV driers, although in the literature of printing and coating technology there are many descriptions of production-scale installations. In the case of photochemical reactors for chemical synthesis, the reverse is true. There is a large literature, emanating almost exclusively from academic institutions, of sophisticated and detailed mathematical models of the various types of photoreactor (a list of references is given at the end of the chapter), while descriptions of the use of these models in the

Table 14.1 Physical attributes of materials to be irradiated

Situation	Examples
(1) A thin fluid film on a substrate. The film may be a liquid or a liquid–solid mix. The film is *stationary* relative to the substrate. The substrate is usually thicker than the film, and is usually opaque, although it could be transparent. The substrate is flat, and is moving as part of a continuous production line.	Curing of inks and varnishes on sheets and webs; woodfinishing; curing of adhesives.
(2) As (1) but the substrate is convex.	Curing of inks, varnishes, etc., on preformed cans, pots, bottles, wire, glass fibres, etc.
(3) As (1) but the substrate is stationary.	Processing of printing plates.
(4) A thick viscous section on a supporting substrate. The section is stationary relative to the substrate. The substrate may or may not be removed after processing.	Curing of glass fibre resin composites; curing of dental restorations.
(5) A thin fluid film on a substrate. The film is *moving* relative to the substrate. The substrate may be transparent or opaque.	Thin film photoreactor (see Fig. 14.20(c)).
(6) A thin fluid film confined between two plates, one or both of which are transparent. The fluid is moving relative to the plates.	Thin film photoreactor (see Fig. 14.20(d)).
(7) A bulk moving fluid confined in a vessel.	Batch photoreactor with stirring; continuous flow photoreactor (see Figs 14.19(b), 14.20(a),(b)).
(8) A heterogeneous irregular three-dimensional photosensitive surface.	Phototherapy.
(9) An impregnated paper or film.	Diazo copying.
(10) An unconfined (?) gas.	Air sterilization.

design of full scale plant are conspicuous by their absence in the chemical engineering literature. Indeed there are few descriptions of production-scale installations, perhaps on account of proprietary reticence, and perhaps, one suspects, because there are few large installations in existence. The common type of UV drier, with its semi-elliptical reflector, has much in common with the elliptical photoreactor, and the mathematical analysis of the latter has much that is relevant to the design of UV driers.

14.2 Source Shapes

Practical sources of UV radiation almost all fall into one of the following categories: (a) compact source, high intensity arcs, such as high pressure mercury lamps, deuterium and hydrogen lamps, xenon lamps or carbon arcs; (b) lamps of linear tube design, and in particular low and medium pressure mercury lamps, metal halide lamps and pulsed xenon lamps; (c) helical or U-shaped lamps, particularly low pressure mercury (Braendli, 1975), and pulsed xenon lamps—a helical design allows a long tube to be formed into a compact source. High intensity compact sources of UV radiation are used in the laboratory, where they are valuable as sources for scientific instruments and for experimentation in photochemistry on a laboratory scale. They are also used for the exposure of photoresists when a "point" source is required. Carbon arcs were once a commonly used source for platemaking, but for reasons explained elsewhere they have largely disappeared.

14.3 Light Distribution Characteristics of Tubular Lamps

The majority of UV sources for applications in which chemical change takes place are of linear tube design and therefore have an axis of symmetry formed by a line between the electrodes. In principle therefore, linear tubes radiate uniformly in all directions perpendicular to the tube axis. For a long lamp, ideally, except near the ends, areas of uniform irradiance form concentric cylinders around the tube axis. In practice there is a variety of reasons why deviations from this simple model occur. In metal halide lamps there is often incomplete vaporization of the metal halide and the deposits on the inside of the tube cause local reductions in emittânce (Levin and Lemons, 1971). Areas of devitrification and obscuration by sputtered electrode materials will have a similar effect. Non-uniformity of plasma composition, particularly in vertically operated metal halide lamps, and bowing of the arc, particularly in horizontally operated lamps, also cause deviations from the simple model.

The mathematical modelling of irradiation systems requires an assumption or an understanding of the way in which a small element of the surface of the lamp radiates energy. There are three basic models which have been proposed and used (Akehata and Shirai, 1972). These are shown in Fig. 14.1. *The radial model* makes the (obviously erroneous) assumption that

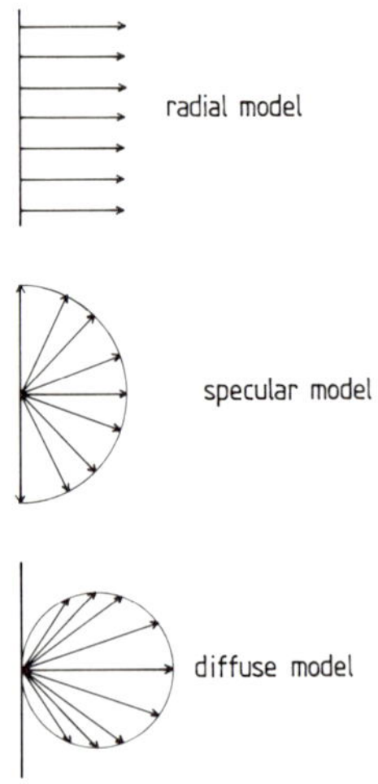

Fig. 14.1. Radial, specular and diffuse models for tubular lamps.

all the photons are emitted in a direction perpendicular to the length of the lamp. This model has the attraction of allowing a three-dimensional reality to be simplified to a two-dimensional model. Despite its dubious basis, this model is often successful in predicting the irradiance of a surface in the vicinity of a tubular lamp. The radial model is quite inadequate in predicting the irradiance of such a surface due to a *small element* of the surface of the lamp, but the errors tend to cancel when the irradiance is integrated for the entire lamp, particularly if the surface of interest is close to the lamp. The radial model does not predict accurately the *angles* at which photons strike surfaces. The radial model has been widely used in the mathematical modelling of photochemical reactors, and was used in almost all work before 1970.

The *specular* and *diffuse* models provide closer approximations to the way in which real lamps radiate. The *specular* model makes the assumption that a very small element of the *length* of the lamp acts as a point source and radiates uniformly into a solid angle of 4π. The *diffuse* model assumes that each small element of the surface of the lamp acts as a Lambertian radiator.

Fluorescent tubes approximate closely to Lambertian radiators. Other discharge lamps have a more complex characteristic, with the radiation pattern depending on the optical density of the arc at the relevant wavelength. The higher the optical density the closer is the approximation to a Lambertian radiator (see Section 4.2). However, even for the non-resonance lines of a low pressure mercury discharge, the diffuse model is superior to the specular model (Funayama *et al.*, 1977).

Mathematical modelling can be much simplified if the lamp can be treated as a line source (i.e. as a point in cross section). This is reasonable if the diameter of the lamp is small compared with the distance from the lamp to the nearest reflective or absorbing surface.

Sometimes the radiation pattern from a linear lamp is ideal. This is the case when the material to be irradiated presents a concave surface which can completely surround the lamp. This situation is found in the various types of immersion well photochemical reactor. Examples are shown in Fig. 14.19. When the material to be irradiated presents a flat or convex surface the radiation pattern around a linear lamp is not ideal, and therefore it is necessary, for efficient use of the available radiation, to modify the radiation pattern from the lamp. This is done by using a reflector.

14.4 Reflector Requirements

When *surface coatings* are to be irradiated, as in UV curing, and the processing of printing plates, two situations may be distinguished. In the first situation, (3) in Table 14.1, both the lamp and the surface are stationary, and the exposure time is controlled either by a shutter mechanism, or by switching the lamp on and off. This is the normal arrangement for platemaking. The requirement is that as much as possible of the total radiation from the lamp should be directed on to the surface, and that the irradiance of the surface should be as uniform as possible. If the surface is large, the necessity for uniform irradiance of the entire surface inevitably reduces the magnitude of the irradiance available.

In the second situation, (1) and (2) in Table 14.1, there is relative motion between the lamp and the surface, which usually means in practice that the lamp is stationary and the surface moves on some type of conveyor system. This situation is clearly necessary if the continuous processing of a large quantity of material is involved, as on a printing press. In this situation it is possible to achieve uniformity of irradiance without necessarily paying the penalty of a low magnitude of irradiance. The relevance of this point was explained in Section 3.2.4.

In the case of sheet- or web-fed applications it is normal to situate a linear lamp across the direction of travel of the sheet or web. All that is required for uniform irradiation of the surface is that the cross section of the reflector should be constant for the entire arc length, and that the arc length is a little longer than the width of the sheet or web. However, if the coating is not on a flat substrate, but, for example, on a preformed can, a plastic bottle or a wire, the orientation of the lamp, and the reflector design, present greater problems if uniform irradiation of the coating is to be achieved.

The efficiency of a reflector depends on two main factors. The first requirement is that it should have a reflecting surface of a material which has a high reflectivity in the UV region, and will maintain this reflectivity for long periods (several thousands of hours if possible) in what may be a moderately hostile environment. The second requirement is that the size and shape of the reflector, and the positioning of the lamp relative to it, should be chosen carefully to provide the ideal irradiance pattern on the surface, and to direct as much as possible of the radiation from the lamp to the surface.

In most cases the material to be irradiated is a solid or liquid surface and it is essential that the radiation should be absorbed by the surface and not reflected by it. For this reason glancing angles of irradiation should be avoided. This causes problems with irregularly shaped objects (Wicks, 1974).

14.5 Reflector Materials

14.5.1 Specular and diffuse reflectance

The term "reflectance" requires careful definition in this context.† It is convenient to first consider two extreme types of reflectance. A perfectly smooth surface behaves as a "specular" reflector, and the angle of reflectance equals the angle of incidence. At the other extreme a "diffuse" reflector has a surface which is microscopically rough, i.e. the

† Note that the term "reflectance" refers to a property of an actual surface. The term "reflectivity" is a property of a *material* in a section thick enough to be opaque (Grum and Becherer, 1979, p. 18).

surface irregularities are large compared with the wavelength of light. The reflected rays show little or no dependence on the angle of incidence. Perfectly diffuse reflectors obey the Lambert cosine law. Whatever the angle of incidence, the intensity of the reflected radiation is proportional to the cosine of the angle of reflectance, as shown in Fig. 14.2.

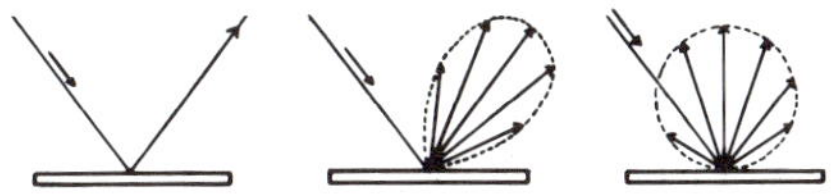

Fig. 14.2. Specular and diffuse reflectance.

In practice, reflector materials lie somewhere between the two extremes, and even with the smoothest available surface the reflected rays are spread over a finite range of angles. Furthermore, the specularity of a material lessens as the wavelength is reduced and becomes more comparable with the scale of the surface imperfections.

14.5.2 UV reflectors

There are not many materials which exhibit a high reflectivity in the UV. Magnesium oxide and barium sulphate, for instance, have a high diffuse reflectivity, and are used to coat the interiors of integrating spheres (Grum and Luckey, 1968). For many photochemical applications, a fairly high degree of specularity is required. Aluminium is the only corrosion-resistant material available which is capable of providing a high level of specular reflectance throughout the region between 200 and 400 nm (Sabine, 1939; Mondolfo, 1976, p. 108).

14.5.3 Metallic reflectivity

The high absorption and high reflectivity of metals are related to their high electrical conductivity. Light is absorbed by the free electrons, and then re-emitted as reflected radiation at the same frequency. The basic physics of the reflectivity of metal surfaces is quite complex, and is derived from Maxwell's equations of electromagnetism with the application of suitable boundary conditions (see, for instance, Born and Wolf, 1980, Chapter XIII). The reflectivity depends on the two principal optical constants of the material, n, the refractive index, and k, the absorption coefficient, both of which are functions of the wavelength. The reflectivity also depends on the angle of incidence and, except in the special case of normal incidence, on the nature of the polarization of the incident ray. Exact solutions are cumbersome, but can be considerably simplified if $n^2 + k^2 \gg 1$. If the light is polarized so that the electric vector is perpendicular to the plane of incidence, then the reflectivity R_s is given (Heavens, 1955, p. 54) by

$$R_s = \frac{(n^2 + k^2) - 2n\cos\phi_0 + \cos^2\phi_0}{(n^2 + k^2) + 2n\cos\phi_0 + \cos^2\phi_0},$$

where ϕ_0 is the angle of incidence. If the light is polarized so that the electric vector is parallel to the plane of incidence, then the reflectivity R_p is given by

$$R_p = \frac{(n^2 + k^2)\cos^2\phi_0 - 2n\cos\phi_0 + 1}{(n^2 + k^2)\cos^2\phi_0 + 2n\cos\phi_0 + 1}.$$

If $\phi_0 = 0$, then

$$R_s = R_p = \frac{(n - 1)^2 + k^2}{(n + 1)^2 + k^2}.$$

For natural unpolarized light,

$$R(\phi_0) = \tfrac{1}{2}[R_S(\phi_0) + R_p(\phi_0)]$$

(Abelès, 1967).

14.5.4 Evaporated metal films

The most perfect metal surfaces for reflectance are prepared by the vacuum deposition of thin films (Hass, 1965). Aluminium gives the highest reflectance in the UV (Fig. 14.3). The best results are obtained with the purest aluminium, a rapid rate of deposition, high vacuum conditions and a low substrate temperature (Holland and Williams, 1955; Hass *et al.*, 1957; Hass, 1965; Hutcheson *et al.*, 1971). Values of n and k have been measured for opaque aluminium films by Hass and Waylonis (1961). Selected values, and the calculated reflectivity for each wavelength, are shown in Table 14.2.

The values of n and k for a wavelength of 340 nm were used to calculate values of R_s and R_p for angles of incidence between 0° and 90°. The results

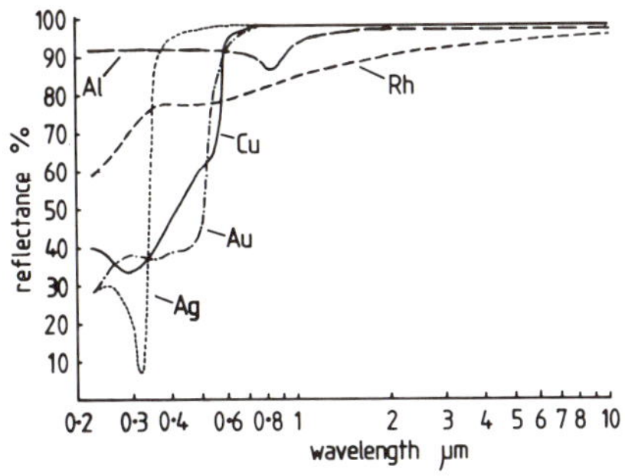

Fig. 14.3. Reflectance of freshly deposited films of Ag, Al, Au, Cu and Rh as a function of wavelength from 220 to 1000 nm. (From Hass, 1955).

are shown in Fig. 14.4. Note that R_p passes through a shallow minimum at about 75°. This is analogous to Brewster's angle for dielectrics. The minimum value of R_p/R_s is a function of k/n and decreases as k/n decreases (Hass, 1965). For unpolarized light the reflectivity varies little up to about 80° and then increases up to 100% at 90°.

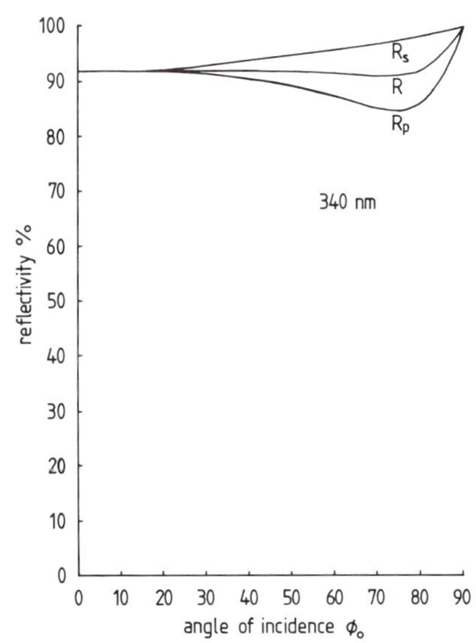

Fig. 14.4. R_s and R_p as a function of angle of incidence, for evaporated aluminium at 340 nm. R, the reflectivity for unpolarized light, is also shown.

Table 14.2 Optical constants of evaporated aluminium, and calculated reflectivity at normal incidence. (From Hass and Waylonis, 1961)

λ(nm)	n	k	$R_{\phi_0=0}$ (%)
220	0·14	2·35	91·8
240	0·16	2·60	92·1
260	0·19	2·85	92·0
280	0·22	3·13	92·2
300	0·25	3·35	92·1
320	0·28	3·56	92·2
340	0·31	3·80	92·3
360	0·34	4·01	92·4
380	0·37	4·25	92·6
400	0·40	4·45	92·6

14.5.5 Atmospheric oxidation of aluminium

A pure smooth aluminium surface does not tarnish visibly under normal ambient conditions because of the protective qualities of the natural oxide film which grows in dry air to an ultimate thickness of about 4·5 nm, depending on the quality of the surface (Hass, 1946). The oxidation rate is independent of the temperature up to about 400 °C (Hass, 1946, 1949). The natural oxide film on pure aluminium, if it is only a few nanometres thick, has little effect on the reflectivity at wavelengths above 200 nm (Berning *et al.*, 1960). Hass and Waylonis (1961) recorded a decrease of only 1·5% in the reflectance at 220 nm of an aluminium film exposed to air for a year.

Film growth is faster and continues for longer in humid conditions (Hart, 1956; Godard, 1967). Film growth is faster in the presence of UV radiation (Coblentz, 1924; Coblentz and Stair, 1930; Cabrera *et al.*, 1947; Hass *et al.*, 1957; Berning *et al.*, 1960), and in the presence of ozone (Holland, 1956, p. 345). Excess film growth causes a significant reduction in reflectance, particularly at short wavelengths.

14.5.6 Aluminium alloys for reflectors

Although the very best specular reflectors are prepared by vacuum deposition on optically smooth substrates, it is often necessary to construct aluminium reflectors from massive aluminium, either in the form of sheet or as an extrusion. The best results are obtained with superpurity aluminium (99·99%), although a cheaper and more readily available alternative are the bright trim alloys such as BTR.S1 and BTR.S2 (BS 4300/2, 1973).

These are alloys of aluminium of commercial purity with small amounts of magnesium (up to 1%).

14.5.7 Brightening processes

In order to produce a surface from massive aluminium with a high specular reflectance in the UV it is necessary to "brighten" the surface by either chemical or electrochemical means. After any necessary mechanical polishing, the aluminium surface is cleaned and smoothed by either chemical polishing or electrobrightening. A typical method of chemical polishing is the use of mixtures of phosphoric acid, sulphuric acid and nitric acid, with a small addition of Cu ions (the Phosbrite 159 process). Careful control of composition, temperature and immersion time allows the preparation of surfaces of high specularity (Culpan and Arrowsmith, 1973). With superpurity aluminium, electrobrightening gives somewhat better results. Two important processes are those using an aqueous solution of sodium carbonate and trisodium phosphate (the Brytal process), or an aqueous solution of hydrofluoroboric acid (the Alzak process). The aluminium is made the anode in an electrochemical system. Oxidation of the surface takes place. Most of the oxide dissolves in the electrolyte, and what remains on the surface (the smudge) is easily removed.

14.5.8 Surface protection

Although a pure aluminium surface is very resistant to tarnishing in clean, dry conditions, pitting corrosion may take place in the long term in damp polluted atmospheres. UV and ozone produced by high intensity UV lamps cause an increase in the oxide layer and a decrease in reflectivity. Also, a pure aluminium surface is quite soft, and therefore easily damaged, and difficult to clean. For these reasons it is desirable to protect the aluminium surface by coating it with an impervious transparent abrasion-resistant coating. Some very sophisticated processes have been developed for the protection of vacuum-deposited aluminium mirrors. Silicon monoxide and magnesium fluoride are the two most common materials to be evaporated on the aluminium surface before it is exposed to the atmosphere. Magnesium fluoride is transparent down to very short wavelengths. Silicon monoxide as deposited has rather poor transparency in the UV and must be converted to SiO_2 by exposure to UV irradiation in the presence of oxygen (Bradford and Hass, 1963). SiO_2 films can also be deposited by electron beam evaporation of water-free fused silica, as can films of Al_2O_3 (Hass *et al.*, 1969). By carefully controlling the thickness of the dielectric layer it is sometimes possible to obtain a reflectance at a chosen wavelength which is actually higher than that obtainable with an uncoated aluminium surface (Berning *et al.*, 1960). An even more sophisticated approach is to use a "multilayer quarter wave stack" (Turner, 1946). This involves employing two dielectric materials, one of low refractive index (e.g. MgF_2, SiO_2, Al_2O_3) and one of high refractive index (e.g. ZnS, TiO_2, CeO_2). The low n material is first deposited until a minimum reflectance is obtained at the chosen wavelength. The high n material is then deposited until the reflectance is maximized. The process may be repeated until in principle the reflectance approaches a theoretical maximum of 100% at the chosen wavelength. The process is rather similar to that for producing multilayer interference filters (see Chapter 4). It depends of course on the availability of dielectric materials which are transparent in the required wavelength region.

When reflectors are constructed from massive aluminium, the most commonly used protective coating is produced by anodizing. After brightening, the aluminium is anodized under controlled conditions to give an oxide film of several micrometres thickness. When aluminium is made the anode in an electrolytic cell, the discharged OH^- and O^{2-} ions oxidize the metal at the surface, and hydrogen is evolved at the cathode. At first a thin non-conducting layer of amorphous or microcrystalline Al_2O_3 is formed. The further development of the film depends on the electrolyte used. If alumina has a limited solubility in the electrolyte (e.g. sulphuric acid), further growth gives a hard, porous film. For bright trim and reflector applications it is normally desirable to seal the pores by immersing the film in hot water or steam. This converts some of the alumina into hydrated forms. $Al_2O_3.3H_2O$ (bayerite) is formed at temperatures below 80 °C, while $Al_2O_3.H_2O$ (boehmite) is formed above 80 °C (Giles, 1974). The hydrates fill the pores.

14.6 Measurement of Reflectance

A complete analysis of the reflectance characteristics of a surface would involve the measurement of the angular distribution of the reflected light for every possible angle of incidence and for every possible wavelength. Such a complete analysis is seldom either feasible or necessary, and the choice of a method for measuring reflectance depends on what part of the total pattern is of interest. Methods are characterized by the angle (or range of angles) of incidence, and the range of angles over which reflected light is measured.

I have used a simple technique for measuring spectral reflectance in the UV and blue regions of the spectrum for a plane reflector material. This technique was designed to give a value of reflectance which is relevant to the use of the material when fair, but not total specularity is required. The equipment used is shown in Fig. 14.5. The light source used is a Cathodeon

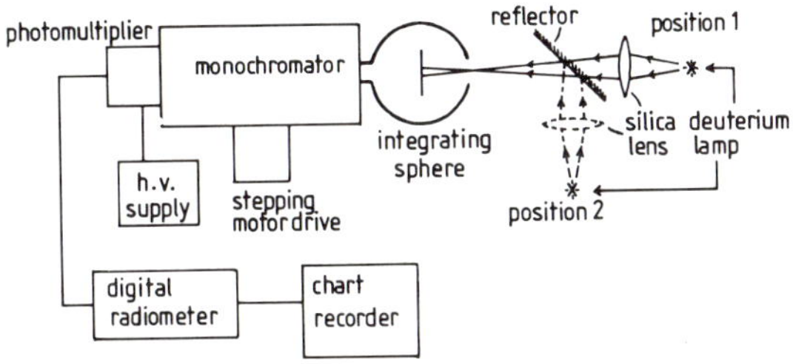

Fig. 14.5. Measurement of spectral reflectance (Phillips, 1982).

deuterium lamp. This provides a very small intense source of continuum radiation between 200 and 500 nm with a few superimposed lines at the higher wavelengths (see Fig. 13.2). Light from this almost point source is focused on the entrance port of a 10 cm diameter integrating sphere internally coated with Eastman Kodak barium sulphate. The entrance port is 2 cm in diameter. Substantially all the light focused by the lens passes through the entrance port and falls on the internal baffle. Light from the exit port of the integrating sphere is split into its constituent wavelengths by the Applied Photophysics grating monochromator and is measured by an RCA 1P28 photomultiplier. A chart recorder records the variation of photomultiplier response with wavelength. The first measurements are taken with the lamp in position 1, and then the lamp is moved to position 2. The reflector material is interposed at 45° to redirect the focused beam into the entrance port of the integrating sphere and a second recorder trace is made. The spectral reflectance curve is then calculated from the ratio of the recorder traces.

An integrating sphere is employed in the method because in principle it should ensure that any photon passing through the entrance port has an equal chance of passing through the entrance slit of the monochromator. In practice it was found that the attenuation of a ray passing into the integrating sphere depended somewhat on the position at which it struck the baffle. Samples with an abnormally high diffuse component for which the reflected rays would spread over the entire baffle surface gave spuriously high values of reflectance. It was also found that if the alignment of the components was less than perfect and the reflected rays struck the baffle off centre, then again spuriously high values were obtained. An optical bench was therefore used to ensure correct alignment.

I have applied this method of measuring spectral reflectance to a variety of bright anodized sheet materials which may be suitable for the fabrication of reflectors:

(1) Two commercially available materials, Alzak Type 1 Lighting Sheet, specular grade, and Coilzak Lighting Sheet, specular grade, available from Alcoa. Alzak is a premium quality product. An aluminium alloy is clad with high purity aluminium and then brightened by the Alzak process and anodized in sheet form. Coilzak is a less expensive material based on a bright trim alloy and is brightened and anodized in a continuous coil. The results for these two materials are shown in Fig. 14.6.

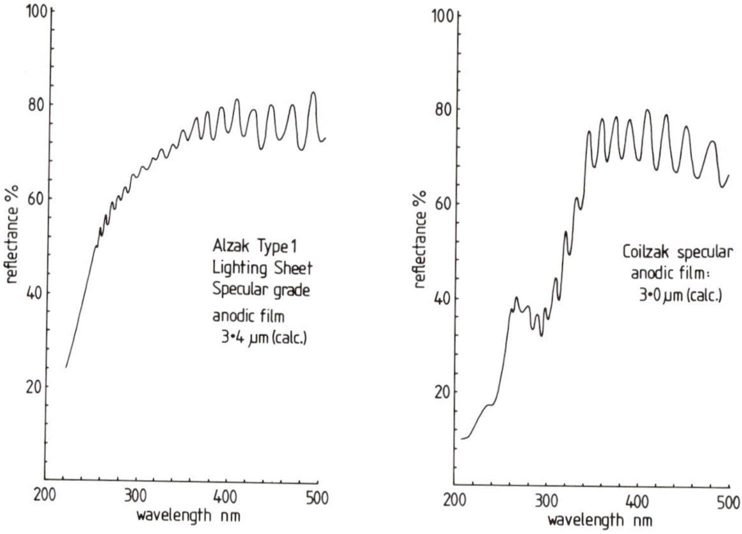

Fig. 14.6. Spectral reflectance measurements: Alzak and Coilzak (Phillips, 1982).

(2) A bright trim alloy BAC212 (an alloy of 99·85% pure aluminium with 0·4% magnesium), of half hard (H4) temper was chemically polished by the Phosbrite 159 process and bright anodized in sulphuric acid and sealed. Figure 14.7 shows the influence of anodic film thickness on spectral reflectance.

(3) Samples of superpurity aluminium (99·99%) were electrobrightened by the Brytal process and bright anodized to various nominal anodic film thicknesses and sealed. Electrobrightening is less capable of

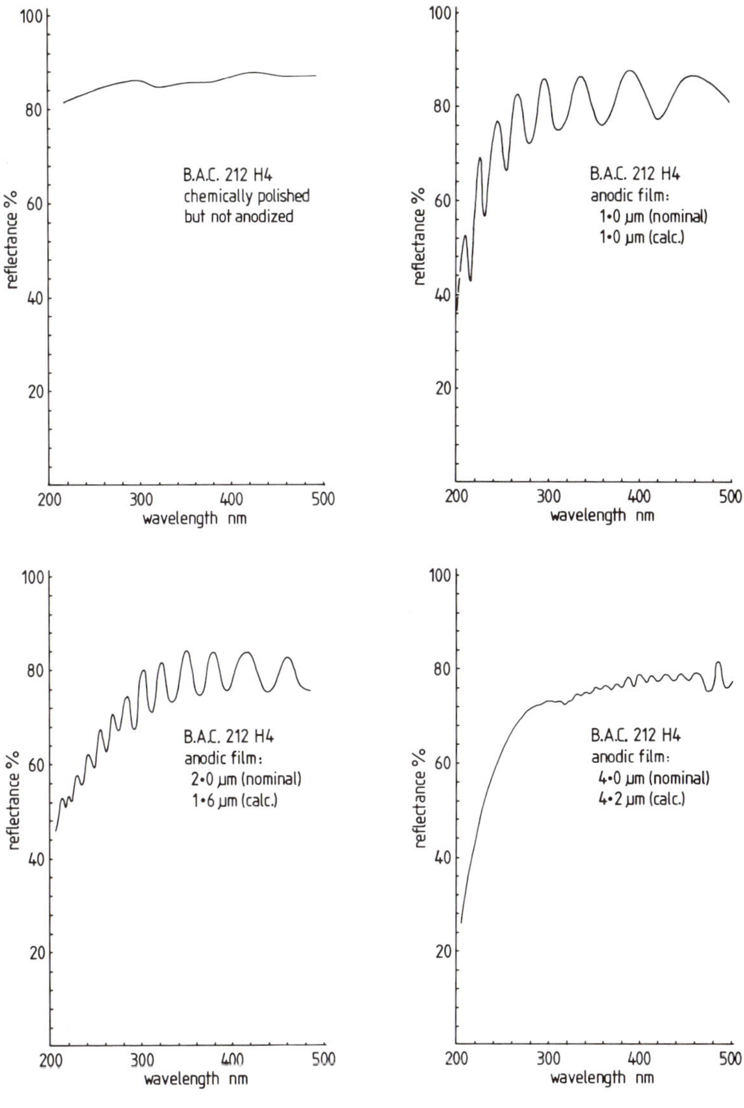

Fig. 14.7. Spectral reflectance measurements: bright trim alloy BAC212 (Phillips, 1982).

removing surface scratches and other defects than is chemical polishing (Chalkley, 1973), and therefore it is important that the aluminium is rolled and/or mechanically polished to a high standard prior to electrobrightening. Three types of superpurity sheet were examined. The first (coded SP (I)) had been rolled to "anodizing quality". The second (SP (II)) had significant rolling defects and was not of anodizing quality. The third (SP (III)) had been rolled to an exceptionally high standard, and required no mechanical polishing prior to electrobrightening. SP (I) and SP (II) both required mechanical polishing. It was necessary to mechanically polish SP (II) for a prolonged period to remove the rolling defects. Figure 14.8

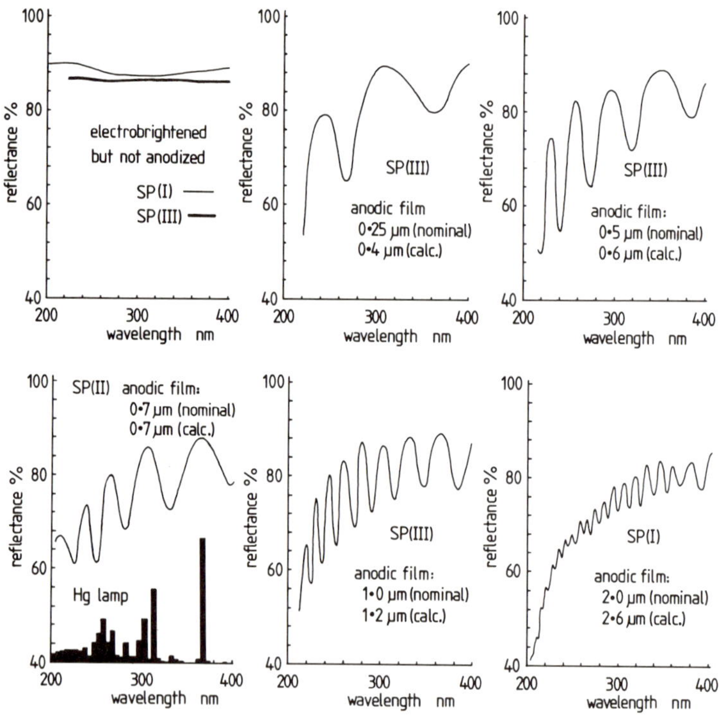

Fig. 14.8. Spectral reflectance measurements: superpurity aluminium (Phillips, 1982).

shows a selection of spectral reflectance measurements for the superpurity samples, chosen from a larger collection of data. It was found that if the mechanical polishing of sheet with rolling defects was adequate, there was no significant difference in the final result.

The data shown in Figs 14.6, 14.7 and 14.8 tend to confirm what has been found before, that the higher the purity of the aluminium the higher the specular reflectivity (Wernick and Pinner, 1972; Chalkley, 1973), and that anodizing causes a significant drop in reflectance (Jackson and Thomas, 1979). The data for Alzak is in close agreement with that recently published by Blank (1980a).

One of the striking features of the spectral reflectance data is the very distinct modulations which can be seen in the curves. These modulations have been observed before when aluminium has been protected by a transparent dielectric film (Hass, 1949; Bradford and Hass, 1963; Palmer *et al.*, 1977). Note that the peaks and troughs are closer together for the thicker films. If the intervals between the peaks (or the troughs) are examined, it can be seen that the peaks (or troughs) are equi-spaced in terms of reciprocal wavelength (i.e. frequency).

The modulation phenomenon is a result of interference between rays which are reflected from the surface of the anodic film, and those which pass into the film and are reflected from the metallic aluminium. The peaks are due to constructive interference and the troughs to destructive interference.

The source of the interference phenomenon is shown in Fig. 14.9.

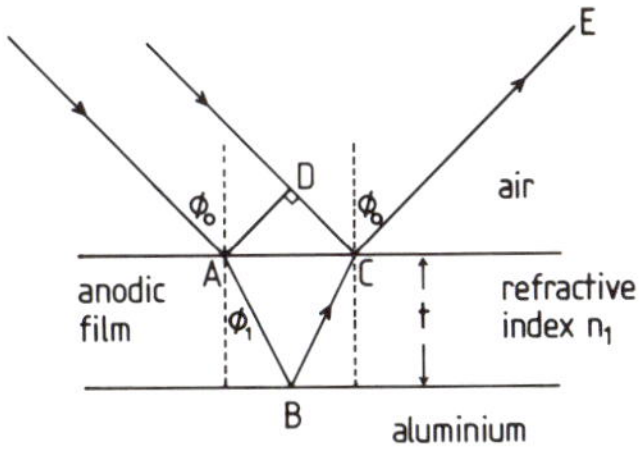

Fig. 14.9. Interference phenomena in anodic films.

A wavefront approaches the anodic film surface at an angle of incidence ϕ_0. Part of the light enters the surface of the film with an angle of refraction ϕ_1, and is then reflected from the aluminium surface. The reflected ray CE is composed of components reflected from both surfaces. If the optical path difference between DC and ABC is an integral number of wavelengths, constructive interference will take place. We must take into account the absolute phase changes which take place at the two boundaries (δ_1 for air–Al_2O_3; δ_2 for Al_2O_3–Al). Constructive interference will take place for wavelength λ_1 if

$$2n_1 t \cos\phi_1 + \delta_1 - \delta_2 = m\lambda_1,$$

where m is an integer. If the wavelength is changed to λ_2 such that the path difference becomes $m + 1$ times the wavelength, constructive interference will again take place, i.e.

$$2n_1 t \cos\phi_1 + \delta_1 - \delta_2 = (m + 1)\lambda_2.$$

Elimination of m between these two equations gives an expression relating the wavelengths of the adjacent peaks:

$$t = \frac{1}{2n_1 \cos\phi}\left[\frac{\lambda_1\lambda_2}{\lambda_1 - \lambda_2} - (\delta_1 - \delta_2)\right] \qquad (14.1)$$

Hass (1949) has shown that $(\delta_1 - \delta_2)$ expressed in nanometres is almost independent of wavelength and has a value of about 50 nm. This is very small compared with $\lambda_1\lambda_2/(\lambda_1 - \lambda_2)$, and can be neglected. Equation (14.1) can then be converted to

$$t \sim (2n_1 \cos\phi_1 \Delta\lambda^{-1})^{-1},$$

where $\Delta\lambda^{-1}$ is the reciprocal wavelength interval between peaks (or troughs) in the reflectance curve.

The measurements were made with an angle of incidence of 45°. The correct value of n_1 to use is debatable. Alumina has a refractive index of 1·76, but this is reduced by hydration. A value of 1·60 was used, based on figures for $Al_2O_3.H_2O$ of 1·62 and for $Al_2O_3.3H_2O$ of 1·58 (*Handbook of Chemistry and Physics*, Chemical Rubber Co.). This gives a value of ϕ_1 of 26·2° and $\cos\phi_1 = 0{\cdot}897$. Note that the value of $\cos\phi_1$ is not particularly sensitive to the value of ϕ_0 when ϕ_0 is about 45° or less. Therefore

$$t \sim 0{\cdot}348/\Delta\lambda^{-1}.$$

Anodic films after sealing are by no means homogenous (Baker *et al.*, 1976; John and Shenoi, 1976), and the effective refractive index may vary with film thickness and also with wavelength. Nevertheless, the calculated values of film thickness were in reasonable agreement with the nominal values.

The anodic film enhances the reflectance at certain wavelengths and reduces it at others. It may be possible, for a particular angle of incidence, to choose a film thickness which will maximize the reflectance for the lines emitted by a line source. A 5 nm histogram of the output of a medium pressure mercury lamp has been drawn under one of the curves in Fig. 14.8, to demonstrate this possibility, which is of academic interest rather than practical significance.

Two techniques were devised for examining reflectors fabricated for use in UV driers. The "Light Bug", described in Section 4.4, provides an excellent method of comparing the total reflectances of reflector surfaces over a wide range of wavelengths in the UV, between about 250 and 400 nm. A series of reflectors of semi-elliptical cross section were fabricated from SP (II)-type superpurity aluminium sheet and were mechanically polished, electrobrightened and anodized after bending to the correct shape, to various nominal film thicknesses. The reflectors were tested in a lamp module, with a medium pressure mercury lamp being placed at the focus, and with the lamp module oriented across a conveyor belt running at a constant speed of 0·5 m s^{-1}. Light Bug measurements are shown in Fig. 14.10, as a function of lamp current. Note that the anodic

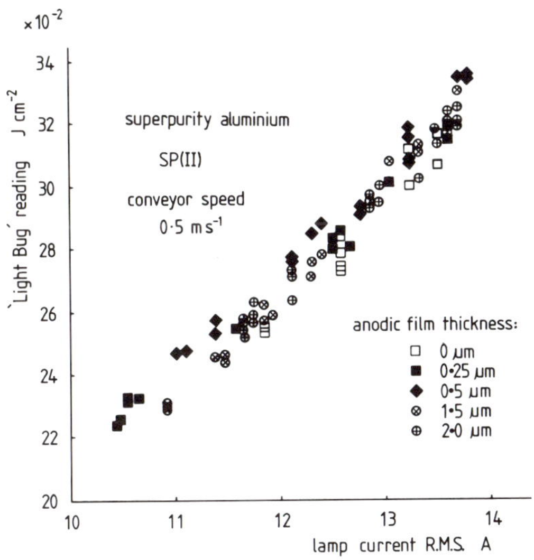

Fig. 14.10. Superpurity aluminium anodized to various nominal film thickness: "Light Bug" measurements (Phillips, 1982).

film thickness has very little effect on the results obtained. Similar measurements were made with a reflector fabricated to the same dimensions from a sheet of BAC212 alloy chemically polished and anodized to a nominal thickness of 4 µm prior to bending, and with reflectors made from Alzak and Coilzak. The results are shown in Fig. 14.11. The results given by this technique are broadly compatible with the spectral reflectance data, except that superpurity aluminium electrobrightened only is no better than the anodized materials. The reason for this is that the combination of UV and ozone causes within a few minutes an increase in the thickness of the natural oxide layer sufficient to reduce the UV reflectance.

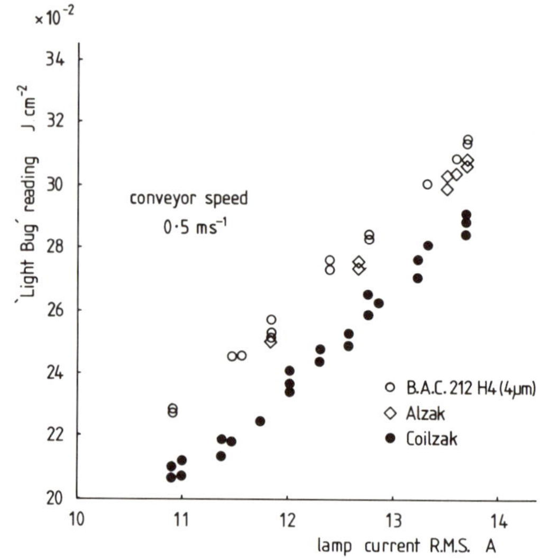

Fig. 14.11. BAC212 (4 μm), Alzak Type 1 Lighting Sheet, Coilzak: "Light Bug" measurements (Phillips, 1982).

The interior of the lamp module was painted matt black and when the lamp was operated without a reflector present the Light Bug reading at a lamp current of 13·4 A was found to be $11{\cdot}2 \times 10^{-2}$ J cm^{-2}, which indicates the importance of the reflector. The geometry of the system was such that 25% of the radiation from the lamp left the module directly, and 75% after reflection.

The second technique was devised to investigate the ability of a semi-elliptical reflector to focus short wavelength radiation to the second focus of the ellipse (see Section 14.8.1). A tubular germicidal lamp was placed at the first focus, i.e. in the normal lamp position, and a detector sensitive to 254 nm radiation was placed at the second focus, facing the lamp. The detector used was an International Light PT171D photodiode fitted with a cosine correction diffuser and an ACT270 wide band interference filter (see Section 15.8). Measurements with and without reflectors in position allowed the percentage increase in intensity at the focus position due to the presence of each reflector to be measured. Some of the data obtained by this technique are shown in Table 14.3. Note the high value of enhancement obtained with brightened-only superpurity aluminium before exposure to the UV and ozone from a high intensity lamp, and the much lower value afterwards. This technique provides a quick and easy method of assessing reflectors. However, it should be remembered that the discharge in a low pressure mercury lamp is wider in diameter than in a medium pressure lamp. A narrow flash tube would provide an alternative tubular source if an integrating radiometer is available.

Table 14.3. Semi-elliptical reflectors: enhancement of irradiance at the second focus (Phillips, 1982)

Material	Anodic film thickness (nominal) (μm)	Increase in intensity (%)
SP (I)†	0	390
SP (II)‡	0	300
SP (II)‡§	0·25	330
SP (II)‡§	0·5	320
SP (II)‡§	1·5	330
SP (II)‡§	2·0	330
SP (II)‡¶	2·0	320
BAC212‡¶	4·0	280
Alzak‡		280
Coilzak‡		230

† Measurement made *prior* to use of reflector in high intensity lamp module.
‡ Measurement made *after* use of reflector in high intensity lamp module.
§ Reflector anodized *after* fabrication.
¶ Reflector anodized *prior* to fabrication.

14.7 Irradiators for Photolithography

In the processing of printing plates and silk screens the image mask is normally a photographic emulsion in fairly close contact with the photoresist. The main requirement is to confine the radiation to the surface being processed, but to ensure uniform irradiation of that surface, which may be quite large in area. Platemaking lamp modules which use a single high powered lamp often have a reflector as shown in Fig. 14.12. The shape of the reflector confines the reflected radiation to the surface of the plate. Uniformity of irradiation is obtained mainly by having a reflector surface which acts as a diffuse reflector. Although in principle it would be possible to use a matt diffuse reflector such as magnesium oxide, or barium sulphate, the difficulty of protecting such surfaces against soiling, and the difficulty of cleaning them, has made the use of textured anodized aluminium more appropriate. An alternative configuration uses a number

Fig. 14.12. An irradiator for exposing printing plates. (Thiemer Violight unit. Photograph supplied by Sericol Group Ltd.)

of low powered fluorescent tubes, close to the plate surface. A simple flat reflector, situated behind the tubes, is used.

There are some similarities between the irradiation of printing plates and the irradiation of the human body in phototherapy. Levin and Parrish (1975) have discussed the design of a radiation unit for the treatment of vitiligo, and concluded that for a large irregular three-dimensional subject like the human body a large source is required. They advocated banks of linear fluorescent tubes, these tubes having built-in cylindrical reflectors formed by a reflective coating between the phosphor and the tube wall over part of the circumference (Balder and van de Weijer, 1956). Diffey *et al.* (1977) have used manikins equipped with polymer film dosimeters to assess the uniformity of irradiance provided by a phototherapy unit.

Irradiator systems for the photolithography used in the fabrication of microcircuits require extremely careful design and construction. They fall into four main categories (Bruning, 1980; Lin, 1980; Nagel, 1981).

(1) *Direct write.* It is possible to "write" a complex pattern on a photoresist surface using a computer controlled focused laser beam. This method obviously dispenses with the need for an image mask.
(2) *Contact.* There is, in principle, hard contact between the image mask

and the wafer. Good edge acuity and a high resolution of detail are possible if the mask and wafer are extremely clean and flat and intimate contact is possible. However, repeated contact causes damage to both mask and wafer and the formation of debris. This, coupled with deviations from perfect flatness, cause there to be a small non-uniform spacing between mask and wafer. A well collimated beam of the appropriate wavelength is required for exposure. The most widely used source for deep UV lithography is the Xe–Hg compact arc. This is spectrally inefficient and if long exposure times are required there is the possibility that mask and wafer will overheat, causing distortions. Long exposure times also introduce problems if there is the possibility of mechanical movement or vibration. For these reasons there has been considerable interest in the use of excimer lasers, which allow very short times, little heat, and a highly collimated beam (Jain *et al.*, 1982; Kawamura *et al.*, 1982).

(3) *Proximity.* This is very similar to the contact system, except that a larger, more uniform separation (e.g. 50 μm) is maintained by the use of shims. These prevent contact damage, but degrade resolution by producing a penumbral region, the width of which is inversely proportional to the square root of the product of wavelength and separation (Bruning, 1980).

(4) *Projection.* This is a system analogous to a slide projector or photographic enlarger, except that the projected image, instead of being enlarged, is either the same size as the mask, or is reduced. The mask is not subject to wear, and this saves replacement costs and improves the uniformity of the product. The mask can be larger in scale than the image on the photoresist, and therefore easier to prepare. Projection lenses of exceptional quality are required, with aberrations corrected to diffraction limits. There is still a limit of about 10^8 to the number of distinct features which can be resolved within an image field, and therefore if high resolution is required over a large wafer area, sophisticated multiple imaging techniques, involving scanning, or step and repeat, are required.

14.8 UV Driers

14.8.1 The semi-elliptical reflector

In UV curing, obtaining uniform irradiation of the surface is easier than in platemaking because a radiation pattern uniform in only one dimension is required, uniformity in the second dimension being provided by the uniform speed of all parts of the surface. A linear lamp with a reflector of uniform cross section provides uniformity in one dimension and there is no restriction on the irradiance distribution in the direction of movement of the surface. The kinetics of free radical polymerization, discussed in Section 3.2.4 and the need to minimize the effects of oxygen inhibition, have led to the use of reflector designs which give the highest possible intensity of irradiation of the surface, and in most instances this arrangement gives the best results. The reflector cross section used is normally a semi-ellipse. An ellipse has two foci, and the lamp, considered a point in cross section, is placed at the first focus of the ellipse, as shown in Fig. 14.13. The distance between the two foci is known as the focal length. Note that the ray diagram shown on Fig. 14.13 is based on a line source, radial model for the lamp. There are UV curing situations where the whole ellipse may be used, for example the curing of coatings on wire and optical fibres.

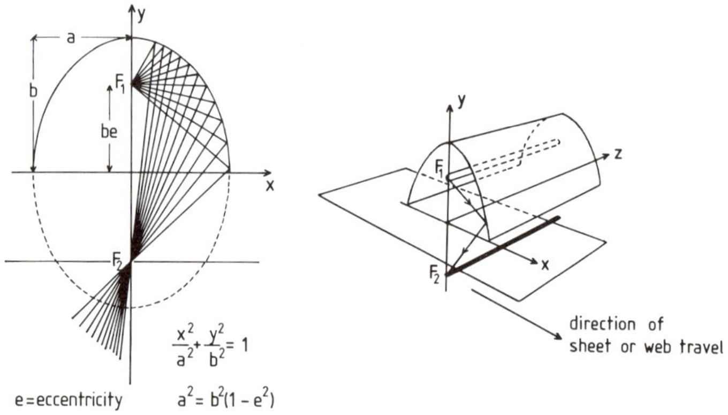

Fig. 14.13. The semi-elliptical reflector.

The geometry of the ellipse is such that ideally all reflected radiation is focused to the second focus of the ellipse. In practice, of course, the lamp arc has a finite diameter and is therefore not a point in cross section. Because of convection currents the arc may not be coaxial with the tube. These effects have some influence on the point of optimum focus. Schooley (1976) found that when a medium pressure mercury lamp was located so that its axis was at the first focus of an elliptical reflector of focal length 12·1 cm, the position of highest irradiance was at a point 10·2 cm from the axis of the lamp. Reflector materials, and particularly those made from massive aluminium, are not perfectly specular, so the best focused beam which can be achieved on the coating surface is about 1 cm in width. In practice it is found that the cure is not adversely affected if the lamp–coating distance is less than the focal length, but deteriorates if it is significantly greater than the focal length (Bassemir and Bean, 1974; Phillips, 1978).

If a pulsed xenon lamp is used as a radiation source, the pulsing supplies the necessary intensity of irradiation. If the pulses have a duration of only 10^{-5}–10^{-3} s, it is necessary to use a reflector giving a diffuse radiation pattern to prevent only bands of the coating being irradiated. In this case it might be appropriate to use a parabolic reflector with the lamp at the focus of the parabola.

14.8.2 Calculation and measurement of irradiance distributions

The irradiance pattern that a lamp–reflector combination produces in the plane of the coating is of considerable importance. It is possible to calculate this to a first and rough approximation from a knowledge of the geometry of the reflector, and its specularity. If reflector shapes of uniform cross section are considered, and a line source, radial model is assumed, it is possible to reduce the problem to two dimensions, as in Fig. 14.14. The

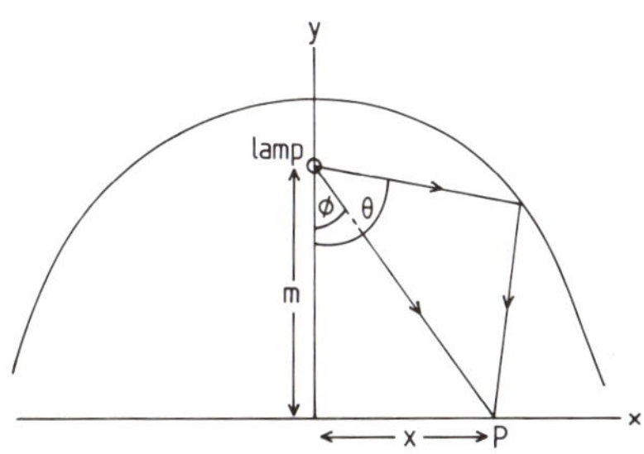

Fig. 14.14. Cross section of lamp–reflector combination (reflector considered perfectly specular).

coating at position P receives direct radiation from the lamp at angle ϕ and reflected radiation leaving the lamp at angle θ. The intensity at position P is then given by

$$I = \frac{W}{2\pi}\left[\frac{d\phi}{dx} + R\left(\frac{d\theta}{dx}\right)\right] \qquad \text{watts cm}^{-2},$$

where W is the output of the lamp in the spectral range of interest in watts per centimetre of arc length. R is the reflectance.

$$\frac{d\phi}{dx} = \frac{m}{m^2 + x^2}$$

The calculation of irradiance distributions requires the determination of $d\theta/dx$ as a function of x. This can be done either mathematically or by graphical construction. To check the validity of the method I constructed a polished aluminium reflector of semicircular cross section and radius 10 cm. A Sylvania black light fluorescent tube with a peak output at 350 nm and diameter 1·5 cm was placed 5 cm above the centre of the semicircle, as shown in Fig. 14.15. $d\theta/dx$ was calculated as a function of x, for a plane passing through the centre of the circle. The irradiance distribution was measured using a vacuum photodiode fitted with a 365 nm narrow band filter, with and without the reflector is position. The differences between the two sets of results gave the the irradiance due to the reflected radiation, and the comparison of this with the calculated values of $d\theta/dx$ is shown in Fig. 14.15.

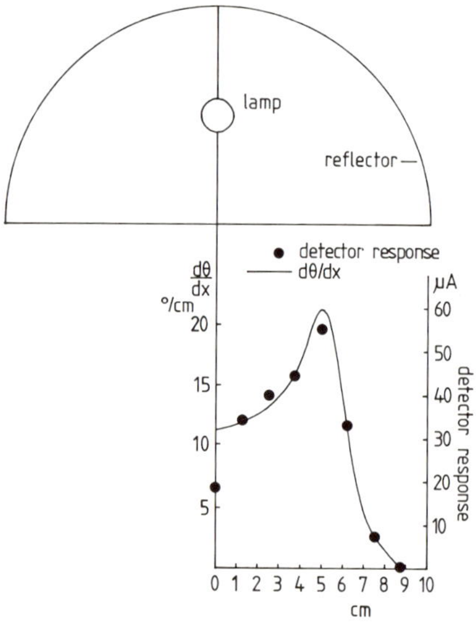

Fig. 14.15. Comparison of calculated and measured irradiance distributions.

When the sources of error are considered—in the geometry and specularity of the reflector, the positioning of the lamp, the finite size of the lamp cross section, and the limited angular response of the photodetector—the results show a pleasing correlation. Note, however, the large discrepancy at $x = 0$, where the measured irradiance is less than that calculated. This is because of "lamp re-entry depletion". Radiation reflected to a point directly below the lamp must pass back through the lamp, and part will be absorbed, reflected, or refracted in the process.

For a more detailed and wide ranging treatment of reflector design the reader is referred to the text of Elmer (1980) on the optical design of reflectors.

The mathematical prediction of irradiance distributions is no substitute for actually measuring them. This poses few problems if the level of irradiance is only moderate, as in platemaking, and a main concern is to ensure uniform irradiation. In the UV curing of surface coatings it is often desirable to produce the maximum possible irradiance in the plane of the coating by the use of a focusing elliptical reflector. The optimum position of the coating surface relative to the lamp–reflector module is of considerable importance. It is also worth knowing how uniform the irradiance pattern is along the length of the lamp, i.e. in the z direction in Fig. 14.13, and how near to the ends of the lamp the irradiance falls off. Irradiance levels are very high, and use of a static detector requires not only attenuating filters, but also thermal insulation (Bassemir and Bean, 1974). An alternative is to use a fused silica light pipe to conduct light from the position of interest to a remote detector (Rubin, 1974). These authors have examined the irradiance distribution around a linear lamp in an elliptical reflector and it is instructive to compare the results of their measurements with predictions from the ideal geometry. The problems of high irradiance levels and the risk of overheating the detector can be overcome by traversing the detector rapidly across the plane of interest. I have been able to examine irradiance distributions in the x direction by passing a filtered silicon diode detector on a conveyor belt underneath a lamp–reflector combination. The detector had a voltage output and was connected by trailing leads to a storage oscilloscope. As the conveyor speed was fixed, the voltage–time trace on the oscilloscope screen could be used to derive a plot of irradiance against x.

Photosensitive papers and films are of great value in the analysis of irradiance distributions, particularly if a semi-quantitative result is adequate. They have particular value in checking the uniformity of irradiation of containers, or other objects with non flat surfaces (see Section 4.5.1).

14.8.3 Lamp modules

Lamps for UV curing are housed in modules which provide the electrical connections, suitable cooling and extract facilities, a reflector, and a

(a)

(b)

shutter mechanism if this is necessary. The complete drier must be designed so that the radiation is adequately screened and cannot be deliberately or accidently viewed. This involves the use of interlocks on any moveable section, so that the equipment is made virtually foolproof (Lott, 1976).

(c)

Fig. 14.16 (facing page and above). Four lamp set stack UV drier for sheet metal decorating.

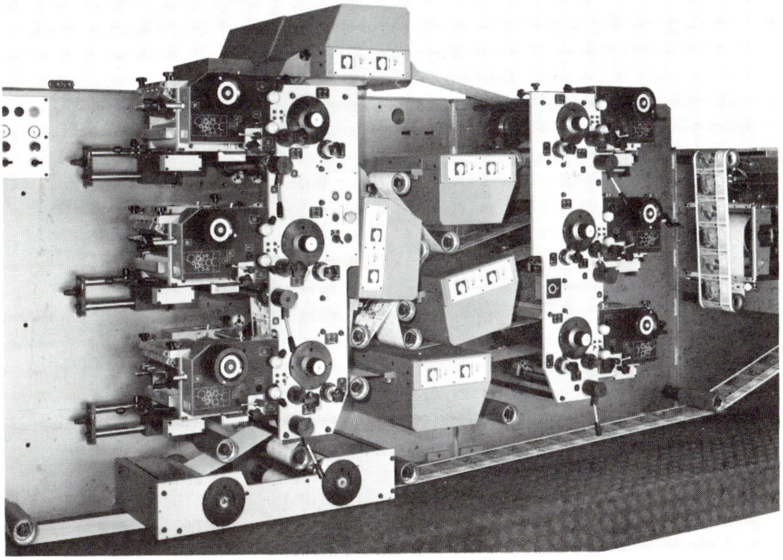

Fig. 14.17. Gallus R160B rotary letterpress web label printing machine fitted with UV curing units to allow drying between colours. (Photograph by courtesy of Edlon Machinery Ltd. Reproduced with permission.)

While most pulsed xenon lamps are water cooled, medium pressure mercury and metal halide lamps are normally air cooled. The foil seals in particular must be carefully cooled. The discharge tube itself needs to be at a fairly high temperature in order to maintain the lamp fill in the vapour state. At low loadings the loss of heat from the tube by radiation and convection in still air is adequate to prevent the tube overheating. However, at ratings of 80 $W\,cm^{-1}$ or higher, or if the tube has a narrow bore, some forced air cooling may be desirable to reduce the rate of devitrification of the silica. Cooling is via holes cut in the back of the reflector. There is a choice between positive cooling, i.e. blowing air over the lamp, or negative cooling, i.e. sucking air over the lamp. Positive cooling has the advantage that it is easy to filter the cooling air, which keeps the lamp clean. Positive cooling also allows hot air blown off the lamp to warm the coating, usually with beneficial results. The advantage of negative cooling is that ozone produced is extracted and prevented from reaching a toxic concentration. When negative cooling is used, care must be taken to prevent the suction created lifting lightweight substrates up towards the lamp.

The installation of lamp modules on printing and coating lines has been discussed by Hankins (1977), Kacmarcik (1975), Szpitalak (1975), Kastelic (1976) and Pray (1976). In some cases, in order to avoid oxygen inhibition, inerting of the coated surface has been used. Inerting systems have been described by Rothchild (1978).

Figure 14.16 shows three views of a UV drier designed and manufactured by Metal Box Engineering Ltd, Automation and Controls Division, Worcester, UK, for set stack metal decorating. The drier is shown after completion of construction prior to delivery to the customer. Four lamp modules are oriented across the sheet conveyor, each housing a single medium pressure mercury lamp (Fig. 14.16(a)). Outlets can be seen which during installation will be connected to extract ducting, allowing an air flow which by careful design enables each component in the system, i.e. reflector, lamp tube, lamp ends, to operate at a temperature which ensures efficiency and long life. The extract also removes ozone produced by the lamps. In the event of a sheet "crashing" under the lamp units, access to the conveyor is achieved by raising one end of the hinged lamp assembly. The lamp modules are fitted with shutters (see Section 14.8.4), and these can be seen in the closed position in Fig. 14.16(b). The lamp modules have lids which hinge open to allow easy removal of the cradles which hold the lamp assemblies (Fig. 14.16(c)).

Figure 14.17 shows a Gallus R160B rotary letterpress web label printer, fitted with UV curing units to allow between colour drying.

14.8.4 Substrate protection

For most UV curing coatings, the dose requirement of UV radiation for adequate cure is less than $0{\cdot}5\,J\,cm^{-2}$. The most commonly used radiation source, the medium pressure mercury lamp, radiates considerable quantities of non-actinic radiation in the visible and infra-red. While this non-actinic radiation is ineffective in initiating cure it may be absorbed by the coating. However, during normal curing procedures, the temperature rise of the coating and its substrate, due to the absorption of radiation and to the exotherm of the polymerization reaction, is only a few degrees (Clark, 1974), and this seldom causes any problems. There are occasionally situations when the substrate is very temperature sensitive. Certain plastic films are liable to distortion, and large sheets of paper or board may show dimensional changes as a result of a change in moisture content. In these situations it may be desirable to use cooling platens under the substrate and/or direct cooling air on to the substrate (Knight, 1981). Alternatively a filter may be used to prevent the infra-red from the lamp reaching the substrate. This is normally pure water in a silica jacket (Schooley, 1976; Bird, 1979). Use of a water filter may retard the cure (Fuhr, 1977). Dichroic mirrors which reflect the UV but transmit the infra-red have also been used (Blank, 1979, 1980a, b).

Conveyor systems are liable to mechanical malfunction and occasionally a coated substrate will come to an inadvertent halt under a lamp. If the system is not designed to allow for this possibility, there will be a very rapid temperature rise which may cause distortion of the substrate, or even conflagration. If the lamp is automatically extinguished when the conveyor stops, there may still be a problem because although radiation from the arc ceases immediately, the thermal inertia of the tube will cause it to radiate infra-red for a significant time. The problem of residual infra-red can be minimized either by using a narrow bore electrodeless lamp (see Chapter 10), or by using a positive cooling air flow which continues after the arc is extinguished. However, even if switching the lamp off immediately removes the risk of damage to the substrate, the on–off–on cycle of most medium pressure mercury lamps is several minutes, and this constitutes an unacceptable down time.

The most commonly employed solution to this problem is to use a shutter system which operates automatically when the conveyor stops, and forms an optical barrier between the lamp and the coated substrate (Szpitalak, 1975). The lamp remains on, but at a reduced "idling" power input. Several different mechanisms of shutter are described in the patent literature. The more significant are shown in Fig. 14.18.

14.8.5 Curing of coatings on containers

The curing of coatings on containers is more difficult than the curing of coatings on flat sheets or webs. Simply handling containers at high speed in such a way that they can be presented to the radiation from a lamp may involve considerable mechanical difficulties. While webs are continuous, and sheets can be handled so that the gaps between them are quite small compared with the sheet dimensions, the gaps between containers on a conveyor system tend to be a greater proportion of the container dimensions, and therefore radiation tends to be lost through the gaps. A third problem is that of obtaining uniform irradiation of the coated area. Many container curing units have a long lamp mounted alongside the conveyor system so that radiation is focused on the side of the container. The container moves continuously on the conveyor, and the lamp is angled so that the focus moves from one end of the coated area to the other. The container is rotated about its own axis, so that the radiation pattern forms a spiral around the surface of the container and uniform irradiation of the entire coated area is ensured (see, for example, US Patent 3,894,237 (Coors Container Co., 1975)).

A particularly elegant solution to the drying of print on containers is exemplified by the recent introduction of *curing on mandrel* (de Saussure, 1978). A lamp module is installed on the printing machine itself. Containers are held on mandrels attached to a turret which rotates with intermittent motion. In each stationary position an operation can be performed on the container, e.g. delivery, flame treatment, printing, take-off. If spare positions are available a short lamp module can be installed to irradiate the container after the printing operation. Container printing speeds are so high that the demands both on the inks used and the lamp are severe. The time for irradiation of the whole container while it rotates in front of the lamp is limited to the time the printing takes, and as the container will be handled within a very short time cure to a scuff-resistant surface must be extremely rapid. The benefits of curing on mandrel are as follows: (a) there is no need for a separate conveyor system with its mechanical complexity, capital investment and space requirement; (b) the energy requirement is much lower because the radiation is used more efficiently; (c) if there is a misfeed and the mandrel is printed, the ink on the mandrel is cured before it can be transferred to the inside of the next container. This is a very significant advantage when containers for food packaging are being printed.

Various methods of curing coatings on containers have been devised which eliminate the need to spin the container in front of a lamp. A system devised by Continental Can Corp. (US Patent 4,050,412) for curing

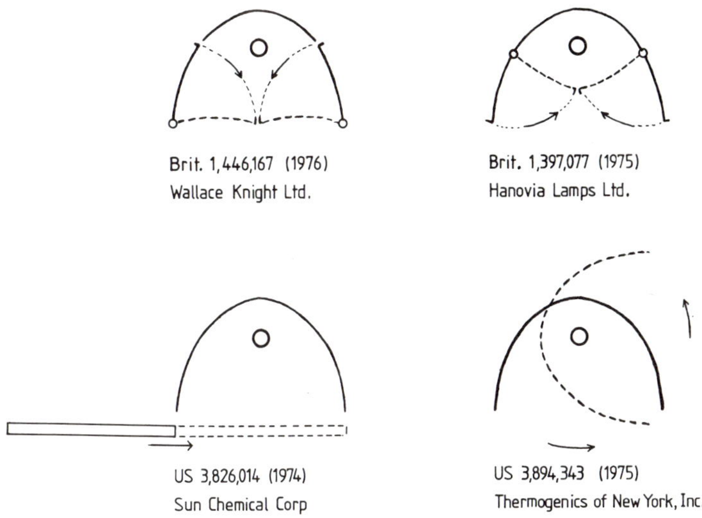

Fig. 14.18. Shutter systems for UV drier lamp modules.

coatings on can bodies passes cans without rotation through a lamp chamber containing two lamps. The can passes between the lamps, which are oriented parallel to the direction of motion of the cans. After leaving the lamp chamber the cans are turned through 90° and are then passed through a second lamp chamber similar to the first.

14.9 Photochemical Reactors for Chemical Synthesis

Photochemical reactors for chemical synthesis may be classified as either batch reactors or continuous flow reactors. Batch reactors are normally used for reactions of low quantum yield for which the necessary irradiation times are quite long, while continuous flow reactors are appropriate for reactions of high quantum yield. The various designs of reactor in common use have been described by Doede and Walker (1955), Mullin (1968), Harano and Matsuura (1972) and Fischer (1978). A batch reactor is shown schematically in Fig. 14.19(a). The lamp, surrounded by a cooling jacket, is immersed in the vessel containing the reactants. No reflector is needed, and substantially all the radiation from the lamp is available to be absorbed by the reactants. However, the light intensity falls off rapidly with increasing distance from the lamp, partly due to absorption by the

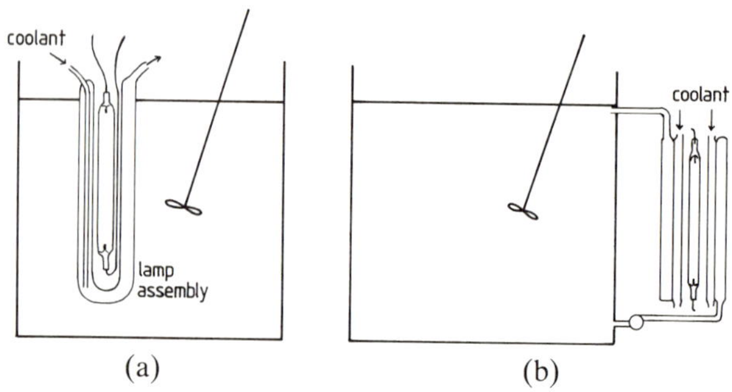

Fig. 14.19 (a) Batch photoreactor. (b) Batch photoreactor with circulation.

reactants, and partly due to the divergence of the light rays (Fischer, 1978). Therefore an efficient mixing system is necessary to allow uniform irradiation of the reactants, and to prevent local hot spots being formed when the reaction is exothermic. Another version of the batch reactor is shown in Fig. 14.19(b). The lamp is now separate from the vessel, and the reactants circulate from the vessel, through an irradiation jacket around the lamp, and then back into the vessel. Figure 14.20(a) shows a continuous flow reactor of the annular type. The reactants flow between concentric glass or silica cylinders, with the lamp placed at the centre of the inner cylinder. Figure 14.20(b) shows a continuous flow reactor of the elliptical type, which is analogous to the most common type of UV drier. A tube through which the reactants flow is placed at the second focus of a reflector which is an ellipse in section, while the lamp is placed at the first focus. Radiation from the lamp is therefore concentrated on the tube containing the reactants. An attraction of this system is that the light from the lamp tends to converge towards the centre of the tube containing the reactants. As this converging light is attenuated by the absorption by the reactants, the light intensity is reasonably uniform across the cross section of the tube (Fischer, 1978). If the reactants are relatively opaque to the radiation from the lamp, or the fluid is turbid due to the presence of suspended colloidal or solid matter, it is necessary to use a thin film photoreactor. The fluid may flow between two parallel plates a small distance apart (Fig. 14.20(c)) or over the outside of a transparent tube containing the lamp (Fig. 14.20(d)).

In principle, the photochemical reactors used for chemical synthesis can also be used for the sterilization of fluids, or for the treatment of waste water (Matsuura and Smith, 1970). The short wavelengths used for some of these processes create difficulties in achieving adequate penetration, and

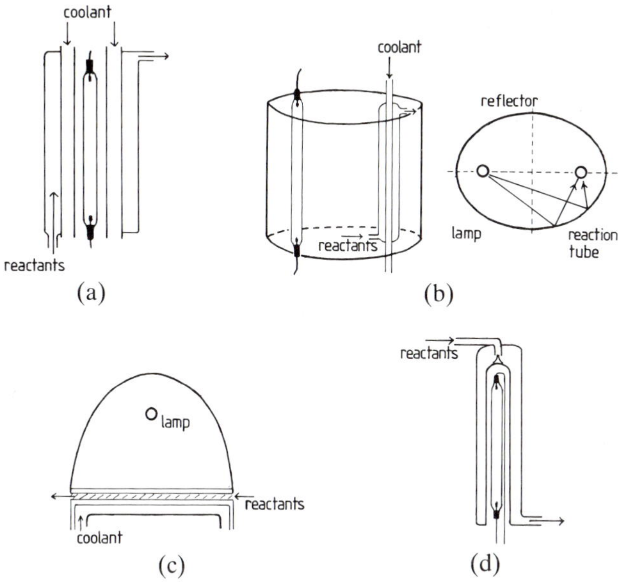

Fig. 14.20. Continuous flow photoreactors: (a) annular flow photoreactor; (b) elliptical photoreactor; (c) and (d) thin film photoreactors.

special care must be taken to ensure that all parts of the fluid are subjected to an adequate dose of radiation.

The photochemical reactors shown in Figs 14.19 and 14.20 differ from UV driers, and irradiators for platemaking, phototherapy and photocopying, in that there is in all cases a solid transparent wall which confines the fluid, and which lies between the lamp and the fluid. This wall creates a problem of considerable significance in the design of photochemical reactors, because there is often the possibility that opaque solid material will be deposited from the fluid phase on to the wall, so attenuating the light passing into the reactor. Various approaches have been made to overcome this problem. Examples are as follows: (1) mechanical scraping of the reactor wall (Isaacson and Ting, 1970); (2) elimination of short wavelengths (Hulme and Turner, 1968; Ito and Matsuda, 1969); (3) annular reaction chambers may be rotated so that a hollow zone is created between the wall and the fluid, so preventing the possibility of deposits being formed (Kobayashi *et al.*, 1977; Kaspar, 1978); (4) when the photosensitive reactant is a gas, and in particular chlorine, it may be possible to have separate irradiation and reaction zones (Lucas, 1974; Dworkin and Dranoff, 1978); (5) it is possible to design a thin film reactor so that the confining wall for the fluid does not lie between the fluid and the lamp (Cohen *et al.*, 1967; Williams and Grisdale, 1968).

References

Abelès, F. (1967). *In* "Advanced Optical Techniques" (Ed. A. C. S. van Heel), pp. 145–188. North Holland, Amsterdam.
Akehata, T. and Shirai, T. (1972). *J. Chem. Soc. Jap.* **5,** 385–391.
Baker, B. R., Balser, J. D., Pearson, R. M. and Strahl, E. O. (1976). *Aluminium* **52,** 241–245.
Balder, J. J. and van de Weijer, M. H. A. (1956). *Philips Tech. Rev.* **17,** 198–205.
Bassemir, R. W. and Bean, A. J. (1974). Paper presented at 26th Annual Meeting of TAGA, St Paul, Minn., 13–15 May; *TAGA Proc.*, **1974,** 133–146.
Berning, P. H., Hass, G. and Madden, R. P. (1960). *J. Opt. Soc. Amer.* **50,** 586–597.
Bird, J. (1979). *Folding Carton Industry* **1979** (Oct.), 22, 28, 30.
Blank, E. (1979). SME Tech. Pap. (Ser.) FC79-673.
Blank, E. (1980a). *Proc. Fifth Radiat. Curing Conf., Boston* 509–517.
Blank, E. (1980b). *J. Radiat. Curing* **7** (4), 15–22.
Born, M. and Wolf, E. (1980) "Principles of Optics", 6th edn. Pergamon, Oxford.
Bradford, A. P. and Hass, G. (1963). *J. Opt. Soc. Amer.* **53,** 1096–1100.
Braendli, G. (1975). *Brown Boveri Rev.* **62,** 202–205.
Bruning, J. H. (1980). *J. Vac. Sci. Technol.* **17,** 1147–1155.
Cabrera, N., Terrien, J. and Hamon, J. (1947). *Comptes Rendus* **224,** 1558–1560.
Chalkley, B. (1973). *Proc. 4th Symp. Electroplating* 193–205.
Clark, K. R. (1974). *Paint and Varnish Production* **64** (8), 44–46.
Coblentz, W. W. (1924). *Science* **60** (1542), 64.
Coblentz, W. W. and Stair, R. (1930). *US Bur. Stand. J. Res.* **4,** 189–193.
Cohen, S. D., Mijovic, M. V., Newman, G. A. and Pitts, E. (1967). *Chem. Ind.* **1967,** 1079–1081; see also Brit. Pat. 1,239,383 (Kodak Ltd) (1971).
Culpan, E. A. and Arrowsmith, D. J. (1973). *Trans. Inst. Met. Finish.* **51,** 17–22.
de Saussure, E. C. (1978). SME Tech. Pap. (Ser.) FC78-522.
Diffey, B. L., Harrington, T. R. and Davis, A. (1977). *Phys. Med. Biol.* **22,** 1014–1016.
Doede, C. M. and Walker, C. A. (1955). *Chem. Eng. (N.Y.)* **62,** 159–178.
Dworkin, D. and Dranoff, J. S. (1978). *AIChE J.* **24,** 1134–1137.
Elmer, W. B. (1980). "The Optical Design of Reflectors" 2nd edn. Wiley-Interscience, New York.
Fischer, M. (1978). *Angew. Chem. Int. Ed. Eng.* **17,** 16–26.
Fuhr, K. (1977). *Polym. Paint Colour J.* **167,** 672–675.
Funayama, H., Ogiwara, K., Sugawara, T. and Ohashi, H. (1977). *Kagaku Kogaku Rombunshu* **3,** 354–358; *Chem. Abstr.* **90,** 112920.
Giles, C. H. (1974). *Rev. Prog. Coloration* **5,** 49–64.
Godard, H. P. (1967). *J. Electrochem. Soc.* **114,** 354–356.
Grum, F. and Becherer, R. J. (1979). "Optical Radiation Measurements", Vol. 1, "Radiometry". Academic Press, New York.
Grum, F. and Luckey, G. W. (1968). *Appl. Opt.* **7,** 2289–2294.

Hankins, W. C. (1977). *J. Oil Colour Chem. Assoc.* **60,** 300–306.
Harano, Y. and Matsuura, T. (1972). *Int. Chem. Eng.* **12,** 131–143.
Hart, R. K. (1956). *Proc. Roy. Soc.* **A 236,** 68–88.
Hass, G. (1946). *Optik* **1,** 134–143.
Hass, G. (1949). *J. Opt. Soc. Amer.* **39,** 532–540.
Hass, G. (1955). *J. Opt. Soc. Amer.* **45,** 945–952.
Hass, G. (1965). *In* "Applied Optics and Optical Engineering", Vol. III (Ed. R. Kingslake), pp. 309–330. Academic Press, New York.
Hass, G. and Waylonis, J. E. (1961). *J. Opt. Soc. Amer.* **51,** 719–722.
Hass, G., Hunter, W. R. and Tousey, R. (1957). *J. Opt. Soc. Amer.* **47,** 1070–1073.
Hass, G., Ramsey, J. B., Heaney, J. B. and Triolo, J. J. (1969). *Appl. Opt.* **8,** 275–281.
Heavens, O. S. (1955). "Optical Properties of Thin Solid Films". Butterworth, London.
Holland, L. (1956). "Vacuum Deposition of Thin Films". Chapman and Hall, London.
Holland, L. and Williams, B. J. (1955). *J. Sci. Instr.* **32,** 287.
Hulme, P. and Turner, P. (1968). *Chem. Eng. (N.Y.)* **75** (7), 80–82.
Hutcheson, E. T., Hass, G. and Coulter, J. K. (1971). *Opt. Commun.* **3,** 213–216.
Isaacson, W. B. and Ting, S. T. (1970). *Ind. Eng. Chem. Fundam.* **9,** 171–173.
Ito, Y. and Matsuda, S. (1969). *Ann. N. Y. Acad. Sci.* **147,** 618–624.
Jackson, C. M. and Thomas, R. W. (1979). *Trans. Inst. Met. Finish.* **57,** 105–109.
Jain, K., Willson, C. G. and Lin, B. J. (1982). *IEEE Electron Device Lett.* **EDL-3,** 53–55.
John, S. and Shenoi, B. A. (1976). *Met. Finish.* **74** (7), 31–35.
Kacmarcik, T. J. (1975). SME Tech. Pap. (Ser.) FC75-350.
Kaspar, J. (1978). Swiss Pat. 595,133 (Sandoz AG); *Chem. Abstr.* **88,** 180230.
Kastelic, F. L. (1976). SME Tech, Pap. (Ser.) FC76-517.
Kawamura, Y., Toyoda, K. and Namba, S. (1982). *J. Appl. Phys.* **563,** 6489–6490; *Appl. Phys. Lett.* **40,** 374–375.
Knight, R. (1981). PIRA Seminar S1, 2 Sept. 1981.
Kobayashi, T., Okuda, T. and Seki, H. (1977). Japan Pat. 77 39, 386; *Chem. Abstr.* **88,** 180229.
Levin, R. E. and Lemons, T. M. (1971). *IEEE Trans. Ind. Gen. Appl.* **IGA-7,** 218–224.
Levin, R. E. and Parrish, J. A. (1975). *Light. Des. Appl.* **5** (3), 35–43.
Lin, B. J. (1980). *In* "Fine Line Lithography" (Ed. R. Newman), pp. 105–232. North Holland, Amsterdam.
Lott, A. D. (1976). *J. Oil Colour Chem. Assoc.* **59,** 141–145.
Lucas, G. (1974). US Pat. 3,800,159 (Société Nationale des Pétroles d'Aquitaine); see also US Pat. 3,853,729; *Chem. Abstr.* **81,** 8358; **83,** 18926.
Matsuura, T. and Smith, J. M. (1970). *AIChE J.* **16,** 1064–1071; *Ind. Eng. Chem. Fundam.* **9,** 252–260.
Mondolfo, L. F. (1976). "Aluminium Alloys, Structure and Properties". Butterworth, London.
Mullin, C. R. (1968). *In* "Kirk–Othmer Encyclopedia of Chemical Technology", 2nd edn, Vol. 15, pp. 331–354. Wiley–Interscience, New York.
Nagel, D. J. (1981). *Proc. Soc. Photo-Opt. Instr. Eng.* **279,** 98–110.
Palmer, R. E., Padrick, T. D. and Pettit, R. B. (1977). *J. Appl. Phys.* **48,** 3125–3126.

Phillips, R. (1978). *J. Oil Colour Chem. Assoc.* **61,** 233–240.
Phillips, R. (1982). *Trans. Inst. Met. Finish.* **60,** 57–62.
Pray, R. W. (1976). SME Tech. Pap. (Ser.) FC76-485.
Rothchild, R. D. (1978). *Radiat. Curing* **5** (1), 3–8.
Rubin, H. (1974). *J. Paint Technol.* **46** (588), 74–81.
Sabine, G. B. (1939). *Phys. Rev.* **55,** 1064–1069.
Schooley, W. A. (1976). SME Tech. Pap. (Ser.) FC76-528.
Szpitalak, W. J. (1975). SME Tech. Pap. (Ser.) FC75-313.
Turner, A. F. (1946). *J. Opt. Soc. Amer.* **36,** 711.
Wernick, S. and Pinner, R. (1972). "The Surface Treatment and Finishing of Aluminium and its Alloys", 4th edn. Robert Draper Ltd, Teddington.
Wicks, Z. W. (1974). SME Tech. Pap. (Ser.) FC74-501.
Williams, J. L. R. and Grisdale, P. J. (1968). *Chem. Ind.* **1968,** 1477–1479.

Selected Journal Articles on the Design and Analysis of Photochemical Reactors

Akehata, T. (1972). Cylindrical, diffusing surface source model for an elliptical photoreactor. *Bull. Tokyo Inst. Technol.* (110), 1–11.
Akehata, T. and Shirai, T. (1972). Effect of light source characteristics on the performance of circular annular photochemical reactor. *J. Chem. Eng. Jap.* **5,** 385–391.
Akehata, T., Shirai, T., Ishizaki, N., Osaka, S. and Ito, K. (1973). Average light intensity in an annular photochemical reactor. *Kagaku Kogaku* **37,** 1026–1031; *Chem. Abstr.* **80,** 102251.
Atlas, H., Hasson, D., Rubin, M. B. and Resnick, W. (1976). Annular centre-line light source photochemical reactor. Design equations and experimental results. *Ind. Eng. Chem., Process Des. Dev.* **25,** 68–75.
Bandini, E., Stramigioli, C. and Santarelli, F. (1977). A rigorous approach to photochemical reactors. *Chem. Eng. Sci.* **32,** 89–96.
Bhattacharya, A. and Deshpande, P. K. (1978). Analysis of an axially irradiated photochemical reactor. *J. Indian Inst. Sci.* **60,** 327–343.
Bhattacharya, A. and Deshpande, P. K. (1979). Analysis of an axially irradiated photoreactor. I. Plug flow model. *Chem. Eng. Sci.* **34,** 145–148.
Cassano, A. E. (1968). Use of actinometers in continuous tubular reactors. *Rev. Fac. Ing. Quim. Univ. Nac. Litoral (Argentina)* **37,** 469–502; *Chem. Abstr.* **72,** 73115.
Cassano, A. E., Silveston, P. L. and Smith, J. M. (1967). Photochemical reaction engineering. *Ind. Eng. Chem.* **59,** 18–38.
Cassano, A. E. and Smith, J. M. (1966). Photochlorination in a tubular reactor. *AIChE J.* **12,** 1124–1133.
Cerda, J., Irazoqui, H. and Cassano, A. E. (1973). Radiation fields inside an elliptical photoreflector with a source of finite spatial dimensions. *AIChE J.* **19,** 963–968.
Cerda, J., Marchetti, J. L. and Cassano, A. E. (1978). The use of simple radiation

models for the case of direct irradiation of photochemical reactors. *Rev. Latinoam. Ing. Quim. Quim. Appl.* **8,** 15–25; *Chem. Abstr.* **90,** 79075.

Esplugas, V. S. (1981). The elliptical photoreactor. *Afinidad* **38,** 315–318; *Chem. Abstr.* **95,** 178521.

Funayama, H., Ogiwara, K., Sugawara, T. and Ohashi, H. (1977). Light intensity profiles in photoreactors applied by low pressure mercury lamp. Effect of spectral distribution of light source. *Kagaku Kogaku Rombunshu* **3,** 354–358; *Chem. Abstr.* **90,** 112920.

Harano, Y. and Matsuura, T. (1972). Problems in designing photochemical reactors. *Int. Chem. Eng.* **12,** 131–143.

Harano, Y. and Smith, J. M. (1968). Tubular flow reactors for complex, nonchain kinetics. *AIChE J.* **14,** 584–591.

Harris, P. R. and Dranoff, J. S. (1965). A study of a perfectly mixed photochemical reactor. *AIChE J.* **11,** 497–502.

Hartig, H. (1970). Simple design of photochemical reactors. *Chem. Ing. Tech.* **42,** 1241–1245.

Huff, J. E. and Walker, C. A. (1962). The photochlorination of chloroform in continuous flow systems. *AIChE J.* **8,** 193–200.

Irazoqui, H. A., Cerda, J. and Cassano, A. E. (1970). Rational analysis of an *a priori* model of photochemical reactors. I. Lambert's law and radiation balances for energy sources of punctiform, lineal and finite geometry. *Rev. Fac. Ing. Quim., Univ. Nac. Litoral* (*Argentina*) **39,** 373–393; *Chem. Abstr.* **77,** 90450.

Irazoqui, H. A., Cerda, J. and Cassano, A. E. (1970). Rational analysis for an *a priori* model of photochemical reactors. II. Emission model for a coaxial, finite, cylindrical source with a nonabsorbing annular reactor. *Rev. Fac. Ing. Quim., Univ. Nac. Litoral* (*Argentina*) **39,** 395–409; *Chem. Abstr.* **77,** 90451.

Irazoqui, H. A., Cerda, J. and Cassano, A. E. (1973). Radiation profiles in an empty photoreactor with a source of finite spatial dimensions. *AIChE J.* **19,** 460–467.

Jacob, S. M. and Dranoff, J. S. (1966). Radial scale up of perfectly mixed photochemical reactors. *Chem. Eng. Prog. Symp. Ser.* **62** (68), 47–55.

Jacob, S. M. and Dranoff, J. S. (1968). Design and analysis of perfectly mixed photochemical reactors. *Chem. Eng. Prog. Symp. Ser.* **64** (89), 54–63.

Jacob, S. M. and Dranoff, J. S. (1969). Light intensity profiles in the elliptical photoreactor. *AIChE J.* **15,** 141–144.

Jacob, S. M. and Dranoff, J. S. (1970). Light intensity profiles in a perfectly mixed photoreactor. *AIChE J.* **16,** 359–363.

Magelli, F. and Santarelli, F. (1978). The modelling of batch photoreactors. *Chem. Eng. Sci.* **33,** 611–614.

Matsuura, T. and Smith, J. M. (1970). Light distribution in cylindrical photoreactors. *AIChE J.* **16,** 321–324.

Pasquali, G. and Santarelli, F. (1978). Radiant energy transfer in batch-photoreacting media. *Chem. Eng. Commun.* **2,** 271–274.

Ragonese, F. P. and Williams, J. A. (1971). Application of empirical rate expressions and conservation equations to photoreactor design. *AIChE J.* **17,** 1352–1359.

Roger, M. and Villermaux, J. (1975). Light distribution in cylindrical photoreactors. *AIChE J.* **21,** 1207–1209.

Schechter, R. S. and Wissler, E. H. (1960). Photochemical reactions in an isothermal laminar-flow reactor. *Appl. Sci. Res. A* **9,** 334–344.

Shendalman, L. H. and Hill, F. B. (1969). Some limiting radical concentration profiles in a batch photoreactor. *Chem. Eng. Sci.* **24,** 909–911.

Shirotsuka, T. and Nishiumi, H. (1971). Design factors of photochemical reactors. Rate of absorption of light energy. *Kagaku Kogaku* **35,** 1364–1372; *Chem. Abstr.* **76,** 134112.

Shirotsuka, T. and Nishiumi, H. (1972). Design factors for photochemical reactors. Power model of an absorption coefficient. *Kagaku Kogaku* **36,** 300–306; *Chem. Abstr.* **76,** 147211.

Shirotsuka, T. and Nishiumi, H. (1972). Design factors for photochemical reactors. Photochemical reactions induced by polychromatic radiation. *Kagaku Kogaku* **36,** 307–313; *Chem. Abstr.* **76,** 147208.

Shirotsuka, T. and Nishiumi, H. (1972) Design factors for photochemical reactors. Mixing and models of the absorption rate of radiation energy. *Kagaku Kogaku* **36,** 328–330; *Chem. Abstr.* **76,** 147212.

Skarbo, R. and Williams, J. A. (1973). Effect of internal light filtering on photoreactor performance. *Chem. Eng. Sci.* **28,** 83–89.

Smith, J. M. (1967). Reactor design for photochemical systems. *ALTECH* **16,** 7–56; *Chem. Abstr.* **71,** 8383.

Spadoni, G., Stramigioli, C. and Santarelli, F. (1980). Rigorous and simplified approach to the modelling of continuous photoreactors. *Chem. Eng. Sci.* **35,** 925–931.

Stramigioli, C., Santarelli, F. and Foraboschi, F. P. (1975). Photosensitized reactions in an annular photoreactor. Influence of the effective volume on the conversion. *Quad. Ing. Chim. Ital.* **11,** 143–147; *Chem. Abstr.* **88,** 91627.

Sugawara, T., Omori, K. and Ohashi, H. (1976). The effect of internal light filtering on the performance of plate-type photochemical reactors. *Kagaku Kogaku Rombunshu* **2,** 304–309; *Chem. Abstr.* **90,** 14600.

Sugawara, T., Omori, K. and Ohashi, H. (1979). Effects of radiation profile and internal light filtering by a product on the reaction characteristics in plate-type photoreactors. *J. Chem. Eng. Jap.* **12,** 143–147; *Chem. Abstr.* **91,** 12023.

Sugawara, T., Tadaki, T. and Maeda, S. (1971). Analysis of the plate-type photochemical reactor. Effect of light intensity distribution, due to the absorption of light by the reactant, on the conversion. *Kagaku Kogaku* **35,** 461–466; *Chem. Abstr.* **75,** 13478.

Thiele, R., Weickert, G., Voss, H. and Haeusler, D. (1975). Technical reaction studies of photochemical processes. *Chem. Tech.* (*Leipzig*) **27,** 5–8; *Chem. Abstr.* **82,** 162914.

Thiele, R., Weickert, G., Hertwig, K. and Haeusler, D. (1975). Technical study and mathematical modelling of a photochlorination reaction 2. *Chem. Tech.* (*Leipzig*) **27,** 595–599; *Chem. Abstr.* **84,** 52055.

Tournier, A., Deglise, X., Andre, J. C. and Niclause, M. (1982). Industrial photochemistry. I. Measurement of the absorption profile of the light in a photochemical reactor. *J. Photochem.* **18,** 47–56.

Tournier, A., Deglise, X., Andre, J. C. and Niclause, M. (1982). Experimental determination of the light distribution in a photochemical reactor: influence of the concentration of an absorbing substance on this profile. *AIChE J.* **28,** 156–166.

Weickert, G. and Thiele, R. (1976). Reaction technique of photochemical processes. 3. Modelling of the radiation field of photoreactors with immersion-lamp system. *Chem. Tech.* (*Leipzig*) **28,** 580–584; *Chem. Abstr.* **87,** 76289.

Williams, J. A. (1972). Radiation efficiency of elliptical reflector photoreactors. *AIChE J.* **18,** 643.
Williams, J. A. (1978). The radial light intensity profile in cylindrical photoreactors. *AIChE J.* **24,** 335–338.
Williams, J. A. and Yen, H. C. (1973). Incident wall intensity in elliptical reflector photoreactors. *AIChE J.* **19,** 862–863.
Yokota, T. and Tadaki, T. (1973). New approximate expression of light intensity distribution in tubular photoreactors. *Kagaku Kogaku* **31,** 1138–1142; *Chem. Abstr.* **80,** 151033.
Yokota, T., Iwano, T. and Tadaki, T. (1976). Light intensity in an annular photochemical reactor. *Kagaku Kogaku Rombunshu* **2,** 298–303; *Chem. Abstr.* **90,** 14599.
Yokota, T., Iwano, T., Deguchi, H. and Tadaki, T. (1981). Light absorption rate in a bubble column photochemical reactor. *Kagaku Kogaku Ronbunshu* **7,** 157–163; *Chem. Abstr.* **94,** 183281.
Yokota, T., Iwano, T. and Tadaki, T. (1981). Light absorption rate in a photochemical reactor with lamp placed at the centre. The effect of specular reflection at the mirror wall of reactor. *Kagaku Kogaku Ronbunshu* **7,** 298–303. *Chem. Abstr.* **95,** 88887.
Zolner, W. J., III and Williams, J. A. (1971). Three dimensional light intensity distribution model for an elliptical photoreactor. *AIChE J.* **17,** 502–503.
Zolner, W. J., III and Williams, J. A. (1972). Effect of angular light intensity distribution on the performance of tubular flow photoreactors. *AIChE J.* **18,** 1189–1195.

15 Hazards of Ultraviolet Radiation

15.1 Ionizing and Non-ionizing Radiation

Practical applications of ultraviolet radiation employ sources which radiate almost entirely above 200 nm. This wavelength corresponds to a photon energy of 6 eV, which is below the minimum energy generally required to ionize isolated organic molecules. Therefore UV is classed as non-ionizing radiation.

The principal hazards associated with the radiation itself are brought about by photochemical action, either on components of the atmosphere to produce toxic species (mainly ozone), or on exposed surfaces of the body. If there are dead or dying cells on the surface, there must be penetration to living cells below the surface. Certain substances, known as phototoxins, when present on or in the skin, may produce adverse reactions (phototoxicity) on irradiation.

15.2 Incidental Hazards of UV Sources

There are several hazards involved in the use of UV radiation sources other than those associated with the UV radiation itself:

15.2.1 Electrical hazards

Many discharge lamps operate at voltage well above mains voltage. Lamp voltages of 1500–2000 V are commonplace in medium pressure mercury lamps for large sheet and web applications. Some discharge lamps employ high voltage trigger pulses for ignition. Electrical safety problems can be eliminated by adopting approved standards of construction and use.

15.2.2 Lamp breakages and explosions

Careless handling of lamps can easily result in breakage and sharp slivers of glass or silica. It is therefore advisable to wear gloves and eye protection when lamps are handled. Most lamps contain mercury, which presents a toxic hazard, and in the event of lamp breakage any spilled mercury should be carefully collected and disposed of. Lamp breakage during use is uncommon, except in the case of very high pressure compact arcs. Long medium pressure mercury lamps, if operated horizontally with too high a wall loading or inadequate cooling, tend to sag in the centre. This, combined with the effects of convection currents within the tube, may cause a bubble to be blown in the top surface, which may burst.

15.2.3 Atmospheric pollution from carbon arcs

Michalski (1973) has discussed in some detail the toxic pollutants generated by the operation of the carbon arc. Gaseous pollutants include carbon monoxide and nitrogen dioxide, as well as ozone. There is much dust and ash produced, and most of this has a very small particle size (less than $1{\cdot}5\ \mu m$), so that it remains suspended in air for long periods, and is readily inhaled into the lungs. Most of the ash consists of metal oxides. The carbon arc is a very dirty source of radiation, and although it is still used to some extent, it is being replaced by the various enclosed arc sources.

15.2.4 Thermal burns

High powered lamps which are not water cooled operate at high temperatures. Both the lamps themselves and their housings must be allowed to cool sufficiently after switch off before handling. Care must be taken to eliminate fire hazard if the material being irradiated is inflammable.

15.3 Ozone: Formation

Ozone is formed by the action of short wavelength UV radiation on molecular oxygen:

$$O_2 + h\nu \longrightarrow 2O$$
$$O + O_2 \longrightarrow O_3$$

An upper wavelength limit of 242·2 nm has been quoted (Koller, 1965, p. 118), and this agrees with the absorption spectrum of oxygen shown in Fig. 15.1. If the wavelength limit quoted appears rather too precise, a possible explanation is that it could have been derived from the known dissociation energy of molecular oxygen. With mercury lamps, the 185·0 nm line is strongly absorbed by oxygen and is responsible for at least some of the ozone produced (Washida *et al.*, 1971). The strong short wavelength continuum emitted by medium pressure mercury lamps is also likely to generate ozone.

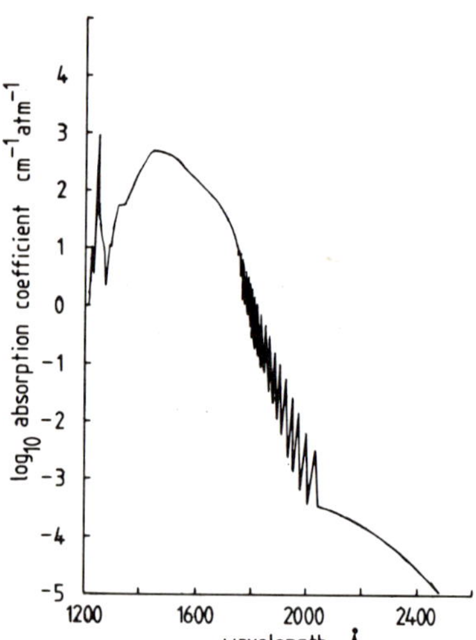

Fig. 15.1. Absorption spectrum of molecular oxygen. (From McNesby and Okabe, 1964. Reproduced with permission of John Wiley & Sons Inc.)

Ozone is moderately stable in the absence of moisture or reducing agents, with a half life of 15 h in the dark at room temperature (Koller, 1965, p. 257). It is rapidly decomposed by ultraviolet light in the region between 200 and 300 nm, by a chain reaction (Norrish and Wayne, 1965). The quantum yield for photolysis is independent of wavelength (Jones and

Wayne, 1970), but the absorption of ozone is greatest in the 250–260 nm region, as shown in Fig. 15.2. The thermal decomposition of ozone is catalysed by metal oxides (Neely *et al.*, 1975).

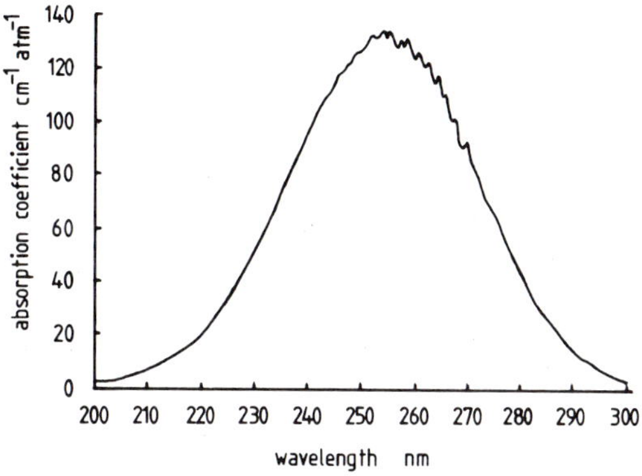

Fig. 15.2. The absorption spectrum of ozone. (From Inn and Tanaka, 1953.) Later measurements by Griggs (1968) gave similar results.

UV lamps may generate equilibrium concentrations of ozone which exceed the recommended threshold levels. This may be avoided either by doping the envelope material to reduce the transmission of ozone-producing wavelengths (see Sections 6.6 and 8.8) or by using an extraction system to remove the ozone as it is produced. Medium pressure mercury lamps generate most ozone during their warm-up period (Diffey and Hughes, 1980), before self-absorption depletes the 185·0 nm line and the increase in temperature causes the transmission cut-off of the envelope to shift to longer wavelengths. As lamps age, the transmission of the envelope for short wavelengths decreases, and less ozone is produced.

15.4 Ozone: Toxicity

Ozone is very toxic because it is one of the strongest oxidizing agents known, and can damage cells at the body's surface. The recommended threshold limit value (TLV) is only 0·1 p.p.m. (Sax, 1979, p. 882; American Conference of Governmental Industrial Hygienists, 1982), and above this level ozone causes smarting of the eyes and discomfort in the respiratory tract. Higher concentrations cause reduction in lung function and ultimately convulsions and coma. Sax (1979, p. 882) has stated that ozone produces no systemic effects. Ozone has a very penetrating smell and can be detected by the average nose at the level of 0·01–0·02 p.p.m., so that it can be detected at levels below the TLV. However, fatigue and tolerance can set in rapidly. Ozone can be measured with a Draeger tube or similar apparatus, or by the absorption of 253·7 nm radiation (Zafonte *et al.*, 1974).

A very detailed review of the toxicity of ozone has been published by Hughes (1979).

15.5 Effects of UV Radiation on the Eye

The eye is the part of the body most at risk from UV, and is in the greatest need of protection. The acute effects on the eye are generally more disabling than the acute effects on the skin, and the eye has no mechanism for acclimatization. There are several distinguishable problems which can occur: (a) photokeratitis (inflammation of the cornea); (b) conjunctivitis (inflammation of the conjunctiva); (c) retinal lesions, caused by exposure to high intensity sources (the lens of the eye focuses radiation to an intense spot on the retina; damage may be a purely thermal effect, or may be photochemical); (d) yellowing of the lens; (e) formation of cataracts. (a), (b) and (c) are acute effects, whereas (d) and (e) result from long term exposure.

The combination of photokeratitis and conjunctivitis, two effects which appear together, is known medically as "kerato-conjunctivitis" and colloquially by such terms as "welder's flash", "arc eye" and "snow blindness". The symptoms are severe irritation, with a sensation of grit in the eye, copious watering (lacrimation), aversion to bright light (photophobia), and spasm of the lid muscles (blepharospasm). There is a reduction in visual acuity, and changes in the epithelial cells which can be observed clinically (Pitts and Tredici, 1971). The symptoms were described vividly by Duke-Elder (1926), whose personal suffering added eloquence to his description.

Many of the characteristics of photokeratitis were defined by Verhoeff *et al.* (1916). Photokeratitis has a latency period which may vary between 0·5 and 24 h, but is most commonly between 6 and 12 h. The latency period varies inversely with exposure. For exposures during a period not exceeding 24 h, the severity of the symptoms has been found to depend on the *dose* of radiation (given in joules per square centimetre), and therefore a reciprocity law applies. Over longer periods the situation will be complicated by repair mechanisms.

The latency period depends also on the wavelength, with wavelengths less than 250 nm giving a shorter latency period than wavelengths greater than 250 nm (Pitts and Gibbons, 1972). The delay in appearance of the symptoms adds to the hazard, as when the symptoms do appear it is too late to avoid their cause. The symptoms usually disappear after 24–36 h, with no permanent effect (Pitts, 1973), except a determination not to repeat the experience. Unlike the skin, the eye does not develop any tolerance to repeated UV exposure (Pitts and Gibbons, 1972).

The cornea absorbs almost all the radiation below 300 nm which enters the eye (Boettner and Wolter, 1962), and it is radiation below this wavelength which is most effective in inducing photokeratitis. Much of the work that has been done to discover which wavelengths are most effective in producing photokeratitis has been done on animals, particularly rabbits and primates. Cogan and Kinsey (1946) determined the threshold dose required to produce photokeratitis in the eyes of albino rabbits for the wavelength region between 240 and 316 nm. Their "action spectrum" shows a peak sensitivity at 288 nm, with a threshold dose at this wavelength of 15 mJ cm^{-2}. More recent work by Dr Pitts and his coworkers at the University of Houston (Pitts and Kay, 1969) produced an action spectrum which differed considerably from that of Cogan and Kinsey (Fig. 15.3). The differences can probably be explained in terms of the methodology used. Both sets of workers used similar criteria for assessing changes in the cornea (e.g. formation of opaque granules in epithelial cells), but whereas Cogan and Kinsey made their first assessment 18 h after exposure, Pitts and Kay examined their subjects at hourly intervals from the time of exposure. It seems likely that, in the experiments of Cogan and Kinsey, the effects of the shorter wavelengths had already begun to subside before the first assessment was made, and this may explain why there is such a discrepancy between the two sets of results. Pitts and Tredici (1971) extended their investigations to rhesus monkeys, and found a lower threshold of 4 mJ cm^{-2}, again at 270 nm (Fig. 15.3). The results of Pitts and Tredici are of considerable significance because it is upon them that the recommended standards for occupational exposure to UV radiation appear to be at least partly based (Sliney and Freasier, 1973). Pitts and Gibbons (1972) and Pitts (1973, 1974) described experiments on the human eye, and found it to be generally more sensitive than the primate eye, but with the same minimum threshold of 4 mJ cm^{-2} at 270 nm (Fig. 15.3).

It seems likely that photokeratitis results from damage to the nucleoprotein in the cornea. Cogan and Kinsey (1946) extracted nucleoprotein from the corneal epithelium and found it to have an absorption peak at 270 nm.

The low pressure mercury lamp with almost all its output at 254 nm is particularly dangerous, as its low wattage and small visible output make it appear deceptively harmless. The irradiance at 20 cm from a 15 W lamp is

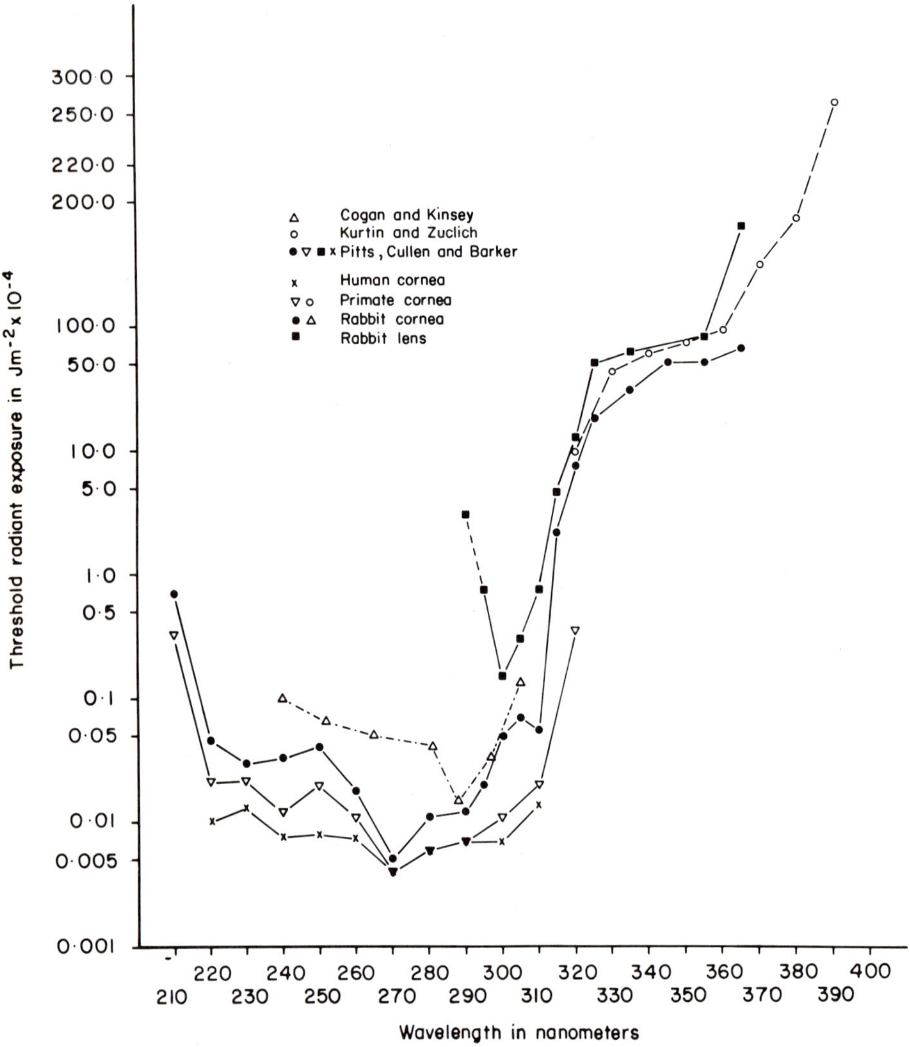

Fig. 15.3. Exposure thresholds for corneal and lens damage. *Author's note*: I originally intended to use for Fig. 15.3 a figure published by Dr D. G. Pitts in *Amer. J. Optometry* **51,** 559–566 (1974), which shows the data given by points ●, ▽ and × in the above figure. When I requested permission to reproduce the original figure, Dr Pitts kindly sent me a bromide of a more recently published figure (D.G. Pitts, *J. Amer. Optometric Assoc.* **52,** 949–957 (1981)), which also shows the data of Cogan and Kinsey, data published by Kurtin and Zuclich (1978) for near ultraviolet-induced corneal damage, and data published by Pitts and Cullen (1977) for lens damage. The figure supplied by Dr Pitts is reproduced above, and I wish to express my gratitude for his generosity.

sufficient for the threshold dose for kerato-conjunctivitis to be reached within a few seconds (Hughes, 1978).

The lens of the eye is protected from UVC and most UVB by the cornea. Long term exposure to UVA is thought to accelerate yellowing of the lens and to cause cataracts (Bachem, 1956). The threshold wavelength for lenticular damage is above 300 nm (Pitts, 1974).

15.6 Effects of UV Radiation on the Skin

Although visible light is known to affect the fundamental biochemical and hormonal rhythms of the body by the daily cycle of light and dark (Wurtman, 1975), and can cause actual photochemical effects in mammalian tissues at considerable depths below the skin, the limited penetration of ultraviolet radiation confines its primary effect to the skin itself.

The short term (acute) effect on the skin is called "erythema", a term derived from the Greek word meaning "being red". Moderate exposure to the active wavelengths causes a reddening of the skin, while grossly excessive exposure may cause blistering (oedema), or even in extreme cases, bleeding.

The reddening of the skin, and the feeling of tenderness and warmth that accompanies it, is caused by the enlargement of the blood capillaries in the inner layer of the skin (the dermis), principally by the action of toxic substances which diffuse from photochemically damaged cells in the outer layer of the skin (the epidermis). Large doses of ultraviolet kill some cells in the epidermis, and the accumulation of serum and white blood cells causes the formation of a blister.

Several other less damaging effects occur when the skin is exposed to ultraviolet radiation. The amount of pigment in the skin increases immediately after exposure due to the photooxidation of a precursor of the pigment melanin, but this is a temporary effect ("immediate tanning").When this initial response has subsided, melanin granules migrate into more superficial cells, and melanocytes in the epidermis begin to divide and increase their synthesis of melanin, which is absorbed into adjacent skin cells. Accelerated cell division causes an increase in the thickness of the epidermis, and eventual scaling of the skin cells. The increase in pigmentation ("delayed tanning") and the thickening of the epidermis last several weeks, and provide protection against erythema. The action spectrum for delayed tanning is similar to that for erythema, except that UVA has more effect (Pathak *et al.*, 1962), so that it is possible to achieve a tan without suffering erythema as well.

The erythemal response of human beings varies greatly. Those with fair skin require a lower threshold dose than those with dark skin. Those of the Celtic origin are particularly susceptible (Giese, 1976, p. 77). Different parts of the body differ in sensitivity. The trunk and head are more sensitive than the extremities, and the concave surfaces of the limbs are more sensitive than the convex surfaces (Olson *et al.*, 1966; Tronnier, 1977). The least sensitive areas are the palms of the hands and the soles of the feet. The natural oils of the skin provide some protection, and if these are removed by cleansing the sensitivity increases (Rottier, 1953).

Work to establish the action spectrum for erythema began in the 1920s, when there was much use of ultraviolet radiation for clinical therapy. Indeed, this was the principal application of the medium pressure mercury lamp prior to 1930 (Tol, 1965). Work by Hausser and Vahle (1922, 1927), Luckiesh *et al.* (1930) and Coblentz *et al.*, (1931) led to the derivation of a "standard erythemal curve" (Coblentz and Stair, 1934) which is shown in Fig. 15.4. This curve was accepted by the Commission Internationale de l'Eclairage (Luckiesh and Taylor, 1939). There are two peaks, one at about 250 nm and another at about 300 nm, with a deep minimum at 280 nm. The standard erythemal curve is not for the threshold dose, but for the dose required to produce moderate erythema. There has been considerable debate as to whether the two peaks in the standard curve represent two

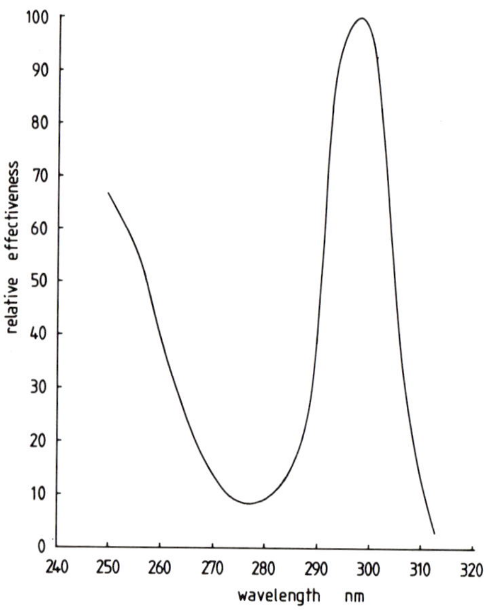

Fig. 15.4. "Standard" erythema action spectrum.

distinct mechanisms for the production of erythema (Parrish *et al.*, 1978, p. 112).

The early workers, from whose data the standard erythemal curve was derived, used mercury arc lamps with filters or monochromators. Later workers (Magnus, 1964; Everett *et al.*, 1965; Freeman *et al.*, 1966; Cripps and Ramsay, 1970; Mackenzie and Frain-Bell, 1973) used continuum radiation from high intensity xenon arcs to illuminate their monchromators, and measured the action spectrum for *minimum perceptible erythema.* Selected results are shown in Fig. 15.5. The results differ considerably from

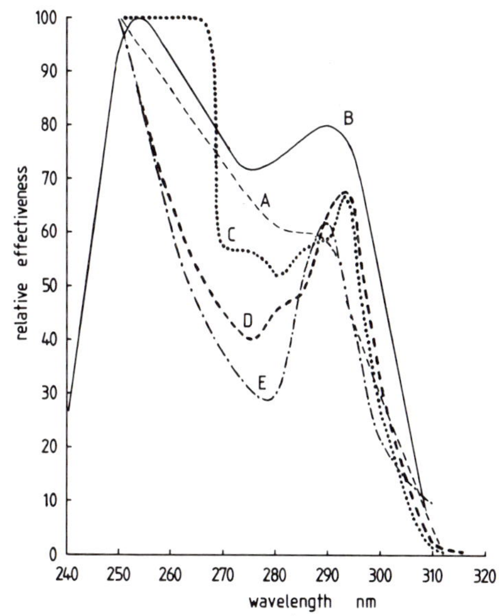

Fig. 15.5. Action spectra for minimum perceptible erythema.

Curve	Authors	Site	Bandwidth (nm)	Time of assessment (h)
A	Everett *et al.* (1965)	Abdomen	10	8
B	Freeman *et al.* (1966)	Abdomen	5	24
C	Cripps and Ramsay (1970)	Back	4	8
D	Cripps and Ramsay (1970)	Back	4	24
E	Mackenzie and Frain-Bell (1973)	Back	6·6	". . . lowest dose producing a reaction regardless of the time of its appearance"

the standard curve. This was not unexpected because earlier work had shown that whereas the dose required at 254 nm to give moderate erythema could be exceeded by a considerable margin without serious effects, even slight overexposure at longer wavelengths could produce painful blisters (Hausser, 1928; Adams *et al.*, 1931; Coblentz *et al.*, 1931). Another complication in the measurement and interpretation of action spectra for erythema is the fact that as with the eye there is a latency period, which is inversely proportional to the exposure (Olson *et al.*, 1966). At short wavelengths the erythema is pinker, the latency period is shorter, the peak effect occurs sooner, and the erythema subsides more quickly than at longer wavelenths (Olson *et al.*, 1966). Therefore the action spectrum depends on the time of assessment (Berger *et al.*, 1968).

UV-induced erythema is dependent on dose rather than dose rate (Hausser and Vahle, 1927; Luckiesh, 1946, p. 80) and therefore follows a reciprocity law with a fixed quantum yield, for the production of toxins which yield a constant erythemal response (Parrish *et al.*, 1978, p. 112). It is of interest that Rottier (1953) concluded that at 300 nm the dose required for minimum perceptible erythema depended on whether the radiation was continuous or intermittent, a phenomenon not observed at 250 nm. This work was repeated by Everett *et al.* (1969), who found no statistically significant difference between the effects of continuous and intermittent radiation, at either wavelength.

The detailed chemistry involved in the production of erythema has been reviewed by Parrish *et al.* (1978, p. 109) and Coyle (1980). As with the eye, damage to nucleoprotein is involved. The differences between the action curves for keratitis and erythema are partly due to the layer of dehydrated cells at the surface of the skin (the stratum corneum, or "horny layer"). The stratum corneum is a "highly concentrated mass of proteinaceous material" about 0·15 mm in thickness (Kligman, 1969). The presence of keratins causes the statum corneum to have an absorption peak at about 270–280 nm (Everett *et al.*, 1969; Pathak and Stratton, 1969), and therefore some protection is provided for the epidermis at these wavelengths.

It is normally assumed that when the skin is irradiated with a polychromatic source, the effects of individual wavelengths are additive. Attempts have been made to prove this experimentally (Sayre *et al.*, 1966). It is difficult to escape the conclusion, after an examination of the literature on photodermatology, that the human skin is so variable that the drawing of generalized *quantitative* conclusions from what has always been limited experimental data may not always be valid. The published data for minimum erythemal dose varies greatly (Magnus, 1964; Hughes, 1977), with the lowest figures quoted only a little above those required for threshold keratitis. It is doubtful whether the concept of a standard erythemal curve has any genuine validity (Blum and Terus, 1946; Magnus, 1964).

Increased sensitivity to erythema, and also the extension of the action spectrum into the UVA and even the visible, can be caused by the presence of certain photosensitizers in the skin. The subject may be divided into *phototoxicity* and *photoallergy* (Parrish *et al.*, 1978, Chapter 7). Certain common drugs, including antiseptics, tranquillizers, antibiotics and antifungal agents, can cause these effects, as can some industrial chemicals such as tars and synthetic dyes. Some components of UV curing formulations have been shown to be phototoxic (Emmett *et al.*, 1977). Some unfortunate individuals suffer from a range of rare metabolic diseases, often of genetic origin, called porphyrias, which cause photosensitizing chemicals, porphyrins or porphyrin precursors, to be released into the blood stream (Ramsay, 1975). Action spectra for light associated diseases have been discussed by Cripps (1981).

Long term exposure to ultraviolet radiation results in premature ageing of the skin, and increases the risk of skin cancer (Giese, 1976; World Health Organization, 1979). However, the incidence of skin cancer can be related to exposure to solar radiation, and there is no documented evidence of cancers being caused by occupational exposure to artificial sources of ultraviolet radiation (Michaelson, 1972). Nevertheless, as Urbach and Wolbarsht (1981) have pointed out: "There is enough UV in the normal environment to pose a hazard. Additional exposure from industrial or home sources is just not needed, and should be eliminated where possible". Solar induced skin cancer is most prevalent among those who have low melanin levels in the skin and live in hot climates. It has reached epidemic proportions among the "Celts of Australia".

15.7 Occupational Exposure to UV Radiation

The widespread use of artificial sources of UV has led to the development of recommended safety standards for occupational exposure. In 1948 the Council of Physical Medicine of the American Medical Association published standards for exposure to the 253·7 nm radiation from germicidal lamps. They recommended that for exposures of less than 7 h, the irradiance should not exceed 0·5 $\mu W\,cm^{-2}$, while for continuous exposure it should not exceed 0·1 $\mu W\,cm^{-2}$. The standard which is now generally accepted for polychromatic sources is that first published in 1971 by the American Conference of Governmental Industrial Hygienists, and later recommended by the National Institute for Occupational Safety and Health (1972) in the USA and the National Radiological Protection Board and the Health and Safety Executive in the UK (National Radiological Protection Board, 1977; McKinlay *et al.*, 1978; McKinlay and Harlen, 1979).

A detailed explanation of the basis of the American Conference of Governmental Industrial Hygienists (ACGIH) standard has been given by the National Institute for Occupational Safety and Health (1972), Sliney (1972) and Sliney and Freasier (1973). The purpose of the standard is to maintain irradiance levels below those required to produce a perceptible response in either the unprotected eye or the most sensitive untanned region of the skin of the normal Caucasian adult human. For the purpose of the standard, the UV region is divided into the UVA (315–400 nm) and UVC + UVB (200–315 nm):

(1) *Wavelength region 315–400 nm* (a) Total irradiance of unprotected eyes and skin for periods greater than 1000 s should not exceed 1 mW cm^{-2}. (b) Total radiant exposure of unprotected eyes and skin for periods of less than 1000 s should not exceed 1 J cm^{-2}. There is so little reliable data on the effects of UVA that the standard in this region is recommended mainly to prevent occupational exposure exceeding levels normally encountered out of doors (National Institute of Occupational Safety and Health, 1972).

(2) *Wavelength region 200–315 nm* The eye tends to be more sensitive than the skin and lacks a tolerance-developing mechanism. The effects of overexposure are more painful and debilitating. For this reason the standard in the region below 300 nm is based on the action curve for photokeratitis developed by Pitts and his coworkers. The Pitts action spectrum, when weighted against the ultraviolet spectrum of indirect sunlight, would suggest that almost anyone spending a few hours out of doors would develop kerato-conjunctivitis (Sliney and Freasier, 1973), and therefore in the 300–315 nm region more weight is given to the action curves for minimum perceptible erythema of Everett *et al.* (1965) and Freeman *et al.* (1966). The reciprocity law is assumed and an envelope action spectrum of dose against wavelength is drawn. The curve is drawn somewhat below the minimum energy dose required to produce any observable reaction. The envelope action curve is shown in Fig. 15.6. The data of the envelope action curve can be expressed in tabular form. Table 15.1 lists threshold limit values (TLVs) for wavelengths in the 200–315 nm region, and also the relative spectral effectiveness (S_λ), normalized so that S_λ for 270 nm is unity. For exposure to monochromatic radiation sources, the TLV can be read directly from the envelope curve. It is then possible either to

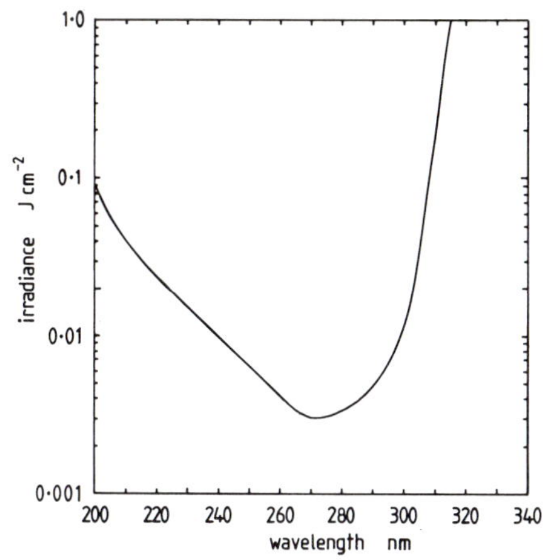

Fig. 15.6. ACGIH threshold limit values for ultraviolet radiation. (© ACGIH 1982 Reproduced with permission.)

calculate the maximum permissible irradiance for a fixed exposure time or, alternatively, if the irradiance is fixed, to calculate the maximum permissible exposure time. If the source is polychromatic, the assumption is made that the effects of individual wavelengths are additive (Sayre *et al.*, 1966), and the following formula is used:

$$E_{\text{eff}} = \Sigma E_\lambda S_\lambda \Delta\lambda,$$

Table 15.1. Threshold limit values and relative spectral efiectiveness

Wavelength (nm)	TLV (mJ cm^{-2})	Relative spectral effectiveness, S_λ
200	100	0·03
210	40	0·075
220	25	0·12
230	16	0·19
240	10	0·30
250	7·0	0·43
254	6·0	0·5
260	4·6	0·65
270	3·0	1·0
280	3·4	0·88
290	4·7	0·64
300	10	0·30
305	50	0·06
310	200	0·015
315	1000	0·003

where E_{eff} is the effective irradiance relative to a monochromatic source at 270 nm; E_λ is the spectral irradiance (in watts per square centimetre per nanometre); S_λ is the relative spectral effectiveness; $\Delta\lambda$ is the band width (in nanometres). The TLV for 270 nm is 3 mJ cm^{-2}, and therefore the relationship between E_{eff} and exposure times is E_{eff} (in watts per square centimetre) × exposure time (in seconds) should not exceed 3 mJ cm^{-2}. For an 8 h working day the maximum allowed value of E_{eff} is 0·1 μW cm^{-2}.

There is little doubt that the ACGIH standard, which has the status of a voluntary standard in the UK, is based on rather limited knowledge of the effects of occupational exposure to UV radiation, and for this reason errs, quite correctly, on the side of conservatism (Parr, 1975). Also, the assumption of reciprocity and the additivity of the effects of individual wavelengths are difficult to vindicate with complete confidence.

The standard has not been without its critics, particularly those feeling that its conservatism may have unfortunate implications for the illumination industry (Bickford *et al.*, 1974). It is valid to ask how the standard should be applied in situations where eye protection is worn. It is also significant that in almost all situations of occupational exposure only the hands, forearms and head are likely to be exposed, and these are areas of the body acclimatized to solar radiation.

The ACGIH standard was devised for normal healthy people, and is not applicable to individuals who suffer from phototoxic or photoallergic reactions, which may affect not only the skin but also the eye (Parrish *et al.*, 1978, p. 252). The ACGIH standard as described is not applicable to radiation from UV lasers, for which separate recommendations are provided.

15.8 Applying the ACGIH Standard

There are severe problems involved in actually applying the standard, and ensuring that the product of E_{eff} and the time of exposure does not exceed 3 mJ cm^{-2}. It is difficult enough to measure in absolute terms the spectral irradiance of a fixed surface. The human body in the working environment is often in motion and the irradiance of the body due to the UV source will constantly change. It is inevitable that certain value judgments are required. At Metal Box great care is taken to ensure that stray radiation from UV driers does not present a hazard to factory personnel. In order to satisfy ourselves on this point it has been our aim to provide a level of shielding which prevents the value of E_{eff} exceeding the 8h limit of $0{\cdot}1\ \mu W\ cm^{-2}$ in any position in which an operator will be situated for any appreciable length of time. This is normally achieved, but if it proves difficult account is taken of the time of exposure and sufficient shielding provided to maintain the product of E_{eff} and the exposure time safely within the limit of $3\ mJ\ cm^{-2}$. Other radiation sources with a potentially hazardous UV content in their spectral output are pulsed xenon camera lamps and metal halide platemaking lamps. Whereas a UV drier operates almost continuously, these sources are used only for brief periods each day, and therefore a value of E_{eff} very considerably higher than the 8h limit may be tolerated, again with the proviso that the product of E_{eff} and the exposure time does not exceed the limit of $3\ mJ\ cm^{-2}$.

It is clearly impractical routinely to make absolute measurements of spectral irradiance in many positions around a UV radiation source, particularly in a factory situation. What is required is a portable meter with a photodetector giving a spectral response corresponding to the variation of S_λ with wavelength. With proper calibration this will give a direct reading of E_{eff} (National Institute for Occupational Safety and Health, 1975).

The National Radiological Protection Board use a commercial instrument, the International Light IL730 Radiometer (International Light Corporation, Newburyport, Mass., USA). This is fitted with a PT171 vacuum photodiode, an ACT270 interference filter and a cosine correction diffuser. The combination gives a spectral sensitivity which is a reasonable approximation to the ACGIH curve (McKinlay *et al.*, 1978). The small discrepancies between the two spectral sensitivity curves lead to errors the magnitudes of which depend on the spectral distribution of the radiation. McKinlay *et al.* found that with their detector, for sources rich in radiation below 250 nm, the instrument would underestimate the value of E_{eff}, while for sources rich in radiation above 280 nm, E_{eff} would be overestimated.

At Metal Box we use a similar instrument, the IL745 Radiometer, fitted with the same combination of photodiode, filter and diffuser. The instrument is calibrated for monochromatic radiation at 270 nm. The spectral sensitivity curve (supplied by International Light Corp.) is shown in Fig. 15.7. Also shown is the curve for S_λ, and a 1 nm histogram of the output of a typical medium pressure mercury lamp (Lamp V, Table 8.1). Calculations show that because of the discrepancies between the two sensitivity curves, the radiometer overestimates the value of E_{eff} for direct radiation from the lamp by about 25%. It is easy to see that, with this particular detector, E_{eff} for monochromatic radiation from a germicidal lamp, or for solar radiation, will be overestimated by a greater margin, while E_{eff} for radiation from a deuterium lamp, with a large output below 240 nm, could

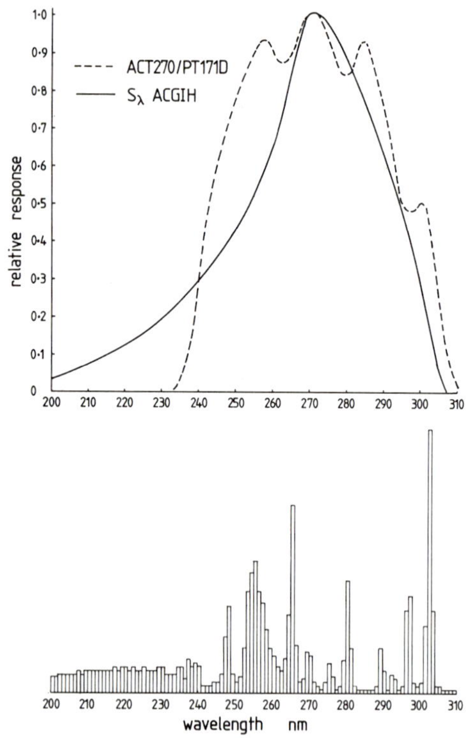

Fig. 15.7. Comparison of spectral response of ACT270/PT171 detector with ACGIH action curve. 1 nm histogram of spectral output of medium pressure mercury lamp.

be underestimated. It is worth bearing in mind that stray radiation has often suffered at least one reflection, and therefore may have a spectral distribution which differs from that of radiation direct from the lamp.

References

Adams, E. Q., Barnes, B. T. and Forsythe, W. E. (1931). *J. Opt. Soc. Amer.* **21,** 207–222. German translation in *Strahlentherapie* **48,** 235–249 (1933).

American Conference of Governmental Industrial Hygienists (1971–82). "Threshold Limit Values for Chemical Substances and Physical Agents in the Workroom Environment". ACGIH, Cincinatti, Ohio (published annually).

Bachem, A. (1956). *Amer. J. Ophthalmol.* **41,** 969–975.

Berger, D., Urbach, F. and Davies, R. E. (1968). *In* "Proceedings of the 13th International Congress of Dermatology, Munich (1967)", pp. 1112–1117. Springer Verlag, Berlin.

Bickford, E. D., Clark, G. W. and Spears, G. R. (1974). *J. Illum. Eng. Soc.* **4,** 43–48.

Blum, H. F. and Terus, W. S. (1946). *Amer. J. Physiol.* **146,** 107–117.

Boettner, E. A. and Wolter, J. R. (1962). *Invest. Ophthal.* **1,** 776–783.

Coblentz, W. W., Stair, R. and Hogue, J. M. (1931). *Proc. Nat. Acad. Sci. U.S.* **17,** 401–405.

Coblentz, W. W. and Stair, R. (1934). *J. Res. Nat. Bur. Stand.* **12,** 13–14.

Cogan, D. G. and Kinsey, V. E. (1946). *Arch. Ophthal.* **35,** 670–677.

Coyle, J. (1980). *Chem. Brit.* **16,** 460–464.

Cripps, D. J. (1981). *J. Invest. Dermatol.* **77,** 20–31.

Cripps, D. J. and Ramsay, C. A. (1970). *Brit. J. Dermatol.* **82,** 584–592.

Diffey, B. L. and Hughes, D. (1980). *Phys. Med. Biol.* **25,** 559–561.

Duke-Elder, W. S. (1926). *Lancet i*, 1137–1141.

Emmett, E. A., Taphorn, B. R. and Kominsky, J. R. (1977). *Arch. Dermatol.* **113,** 770–775.

Everett, M. A., Olson, R. L. and Sayre, R. M. (1965). *Arch. Dermatol.* **92,** 713–719.

Everett, M. A., Sayre, R. M. and Olson, R. L. (1969). *In* "The Biologic Effects of Ultraviolet Radiation (with Emphasis on the Skin)" (Ed. F. Urbach), pp. 181–186. Pergamon, Oxford.

Freeman, R. G., Owens, D. W., Knox, J. M. and Hudson, H. T. (1966). *J. Invest. Dermatol.* **47,** 586–592.

Giese, A. C. (1976). "Living with Our Sun's Ultraviolet Rays". Plenum, New York.

Griggs, M. (1968). *J. Chem. Phys.* **49,** 857–859.

Hausser, K. W. (1928). *Strahlentherapie* **28,** 25–44.

Hausser, K. W. and Vahle, W. (1922). *Strahlentherapie* **13,** 41–71.

Hausser, K. W. and Vahle, W. (1927). *Wissenschaft. Veroffn. Siemens Konzern* **6,** 101–120. English translation in "The Biologic Effects of Ultraviolet Radiation (with Emphasis on the Skin)" (Ed. F. Urbach) pp. 3–21. Pergamon, Oxford, 1969.

Hughes, D. (1977). *Chem. Brit.* **13,** 134–137.

Hughes, D. (1978). "Hazards of Occupational Exposure to Ultraviolet Radiation", Occupational Hygiene Monograph No. 1. University of Leeds, Leeds.

Hughes, D. (1979). "The Toxicity of Ozone", ULIS Occupational Hygiene Monograph No. 3. Science Reviews Ltd, London.

Inn, E. C. Y. and Tanaka, Y. (1953). *J. Opt. Soc. Amer.* **43,** 870–873.

Jones, I. T. N. and Wayne, R. P. (1970). *Proc. Roy. Soc. A*, **319,** 273–287.

Kligman, A. M. (1969). *In* "The Biologic Effects of Ultraviolet Radiation (with Emphasis on the Skin)" (Ed. F. Urbach), pp. 165–167. Pergamon, Oxford.

Koller, L. R. (1965). "Ultraviolet Radiation", 2nd ed. John Wiley, New York.

Kurtin, W. E. and Zuclich, J. A. (1978). *Photochem. Biochem.* **27,** 329–333.

Luckiesh, M. (1946). "Applications of Germicidal, Erythemal and Infra-Red Energy". Van Nostrand, New York.

Luckiesh, M. and Taylor, A. H. (1939). *General Electric Rev.* **42,** 274–278.

Luckiesh, M., Holladay, L. L. and Taylor, A. H. (1930). *J. Opt. Soc. Amer.* **20,** 423–432.

Mackenzie, L. A. and Frain-Bell, W. (1973). *Brit. J. Dermatol.* **89,** 251–264.
McKinlay, A. F., Harlen, F. and Clark, I. M. (1978). "A Limited Survey and Evaluation of Ultraviolet Radiation Hazards in University Laboratories", NRPB-R72. National Radiological Protection Board, Harwell.
McKinlay, A. F. and Harlen, F. (1979). *Occupational Health* **31,** 454–461.
McNesby, J. R. and Okabe, H. (1964). *Advanc. Photochem.* **3,** 157–240.
Magnus, I. A. (1964). *Brit. J. Dermatol.* **76,** 245–264.
Michaelson, S. M. (1972). *Proc. IEEE* **60,** 389–421.
Michalski, M. (1973). *Proc. Tech. Assoc. Graphic Arts* **1973,** 122–139.
National Institute for Occupational Safety and Health (1972). "Criteria for a Recommended Standard: Occupational Exposure to Ultraviolet Radiation". US Department of Health, Education and Welfare, Washington, D.C.
National Institute for Occupational Safety and Health (1975). "Ultraviolet Transfer Standard Detectors and Evaluation and Calibration of NIOSH UV Hazard Monitor", NIOSH-IA-73-20. US Department of Health, Education and Welfare, Washington, D.C.
National Radiological Protection Board (1977). "Protection against Ultraviolet Radiation in the Workplace". National Radiological Protection Board, Harwell.
Neely, W. C., West, A. D. and Hall, T. D. (1975). *J. Phys. E* **8,** 543.
Norrish, R. G. W. and Wayne, R. P. (1965). *Proc. Roy. Soc. A* **288,** 200–211.
Olson, R. L., Sayre, R. M. and Everett, M. A. (1966). *Arch. Dermatol.* **93,** 211–215.
Parr, W. (1975). *J. Illum. Eng. Soc.* **4,** 234.
Parrish, J. A., Anderson, R. R., Urbach, F. and Pitts D. (1978). "UV-A: Biological Effects of Ultraviolet Radiation with Emphasis on Human Responses to Longwave Ultraviolet". John Wiley and Sons, Chichester.
Pathak, M. A. and Stratton, K. (1969). *In* "The Biologic Effects of Ultraviolet Radiation (with Emphasis on the Skin)" (Ed. F. Urbach), pp. 207–222. Pergamon, Oxford.
Pathak, M. A., Riley, F. J., Fitzpatrick, T. B. and Curwen, W. L. (1962). *Nature* **193,** 148–150.
Pitts, D. G. (1973). *Health Phys.* **25,** 559–566.
Pitts, D. G. (1974). *Amer. J. Optometry* **51,** 946–960.
Pitts, D. G. and Cullen, A. P. (1977). Ocular ultraviolet effects from 295 nm to 400 nm in the rabbit eye. DHEW (NIOSH) Publication No. 77–175.
Pitts, D. G. and Gibbons, W. D. (1972). The human, primate and rabbit ultraviolet action spectra. University of Houston College of Optometry NASA Contract No. NA39-10836, 31 March 1972.
Pitts, D. G. and Kay, K. R. (1969). *Amer. J. Optometry,* **46,** 561–572.
Pitts, D. G. and Tredici, T. J. (1971). *J. Amer. Industrial Hygiene Assoc.* **32,** 235–246.
Ramsay, C. A. (1975). *Brit. J. Hosp. Med.* **13,** 536–538, 541, 543–544.
Rottier, P. B. (1953). *J. Clin. Invest.* **32,** 681–689.
Sax, N. I. (1979). "Dangerous Properties of Industrial Materials", 5th edn. Van Nostrand–Reinhold, New York.
Sayre, R. M., Olson, R. L. and Everett, M. A. (1966). *J. Invest. Dermatol.* **46,** 240–244.
Sliney, D. H. (1972). *Proc. Amer. Ind. Hyg. Assoc. Conf. San Francisco 1972* 28.
Sliney, D. H. and Freasier, B. C. (1973). *Appl. Opt.* **12,** 1–24.
Tol, T. (1965). *In* "High Pressure Mercury Vapour Lamps and their Applications"

(Ed. W. Elenbaas), pp. 203–239, Philips Technical Library. Centrex Publishing Co., Eindhoven.
Tronnier, H. (1977). *In* "Ultraviolette Strahlen" (Ed. J. Kiefer), pp. 567–598. Walter de Gruyter, Berlin.
Urbach, F. and Wolbarsht, M. L. (1981). *Proc. Soc. Photo-opt. Instr. Eng.* **279,** 201–215.
Verhoeff, F. H., Bell, L. and Walker, C. B. (1916). *Proc. Amer. Acad. Arts Sci.* **51,** 630–818.
Washida, N., Mori, Y. and Tanaka, I. (1971) *J. Chem. Phys.* **54,** 1119–1122.
World Health Organization (1979). Environmental Health Criteria, Vol. 14: Ultraviolet Radiation. World Health Organization, Geneva.
Wurtman, R. J. (1975). *Sci. Amer.* **233** (1), 68–77.
Zafonte, L., Long, W. and Pitts, J. N., Jr (1974). *Anal. Chem.* **46,** 1872–1874.

Index